Superconducting Radiofrequency Technology for Accelerators

Superconducting Radiofrequency Technology for Accelerators

State of the Art and Emerging Trends

Hasan Padamsee

WILEY-VCH

Author

Dr. Hasan Padamsee
Cornell University
Newman Laboratory
Cornell University
NY
United States

Cover Image: Courtesy of Fermi National Accelerator Laboratory

Library of Congress Card No.: applied for

British Library Cataloguing-in-Publication Data
A catalogue record for this book is available from the British Library.

Bibliographic information published by the Deutsche Nationalbibliothek
The Deutsche Nationalbibliothek lists this publication in the Deutsche Nationalbibliografie; detailed bibliographic data are available on the Internet at <http://dnb.d-nb.de>.

Print ISBN: 978-3-527-41409-3
ePDF ISBN: 978-3-527-83629-1
ePub ISBN: 978-3-527-83630-7
oBook ISBN: 978-3-527-83631-4

Typesetting Straive, Chennai, India
Printing and Binding CPI Group (UK) Ltd, Croydon, CR0 4YY

C117725_190223

Contents

Preface

It has been more than 20 years since Wiley's 1998 publication and enthusiastic reception of *RF Superconductivity for Accelerators* [1] and now more than 10 years since the sequel in 2009: *RF Superconductivity – Science, Technology, and Applications* [2]. Many aspects of superconducting RF (SRF) development are thoroughly covered in these two books, plus many review papers [3–6], and most completely in the proceedings of international SRF conferences (1980–2021) published on JACoW.org [7].

Over the period 2010–2022 there has been spectacular progress in terms of the performance of SRF structures, scientific understanding of the improvements, innovative cavity designs for new applications, and wide exploration of new material avenues to take us beyond the capabilities of the popular standard of niobium, as well as the large scale, worldwide implementation of the mature technology to many new accelerators. Exciting new prospects are on the horizon.

It is time for a new volume on RF Superconductivity to provide a comprehensive update for more than a decade of advances carried out by enthusiastic researchers all over the world. A large fraction of the progress in SRF performance reported here is a testament to the creativity and success of imaginative researchers who have worked on innovative treatments, pursued efforts to gain understanding, and opened the door to new applications. Our review of the field covers progress till January 2022. No doubt there will be much additional progress reported in upcoming coming meetings such as Tesla Technology Collaboration (TTC) Meetings, as well as Thin Film SRF Conferences. We look forward to many new results by the time of the next SRF Conference in 2023.

Experts as well as newcomers to the field, including students, will benefit from the discussion of progress, as well as recent and forthcoming applications. Researchers in accelerator physics may also find much that is relevant to their discipline. There are now more than a thousand practitioners of the SRF field at more than 150 institutions and industries worldwide.

The book has four parts. Part I is the introduction and update of SRF fundamentals. Many of the SRF basics covered in the first two books will only be briefly touched, although essentials will be summarized for the sake of completeness. Part II covers performance advances and understanding at the *high Q frontier*. Part III covers performance advances and understanding at the *high gradient* frontier.

Part IV covers new cavity and new treatment developments, as well as ongoing applications and future prospects.

An exciting new development discussed briefly in Part IV is the use of SRF cavities for quantum computing. Nb cavities offer a transformative vehicle for increasing the coherence times of qubits from sub-milliseconds to seconds, promising to bring the quantum computing field to quantum advantage over classical computers.

October 26, 2022 *Hasan Padamsee*

References

1 Padamsee, H., Knobloch, J., and Hays, T. (1998). *RF Superconductivity for Accelerators*, 2e, 2008 Wiley.

2 Padamsee, H. (2009). RF Superconductivity: Science, Technology, and Applications, 2e 2008. Wiley.

3 Kelly, M. (2012). Superconducting radio-frequency cavities for low-beta particle accelerators. *Rev. Accel. Sci. Technol.* 5: 185–203. https://doi.org/10.1142/9789814449953_0007.

4 Belomestnykh, S. (2012). Superconducting radio-frequency systems for high-ß particle accelerators. In: *Reviews of Accelerator Science and Technology: Applications of Superconducting Technology to Accelerators* (eds. A. Choa and W. Chou) vol. 5, 147–184. World Scientific. https://doi.org/10.1142/S179362681230006X.

5 Reece, C.E. and Ciovati, G. (2012). Superconducting radio-frequency technology R&D for future accelerator applications. In: *Reviews of Accelerator Science and Technology: Applications of Superconducting Technology to Accelerators*, vol. 5 (ed. A. Chao and W. Chou), 285–312. World Scientific. https://doi.org/10.1142/S1793626812300113.

6 Gurevich, A. (2012). Superconducting radio-frequency fundamentals for particle accelerators. *Rev. Accel. Sci. Technol.* 5: 119–146. https://doi.org/10.1142/S1793626812300058.

7 SRF Conference Proceedings.

Part I
Update of SRF Fundamentals

1
Introduction

Discovered in 1911, superconductivity is a fascinating phenomena of modern physics with marvelous scientific and technological applications, such as powerful magnets for medical imaging (magnetic resonance imaging [MRI]), for high energy physics, in particular, the large hadron collider (LHC), for nuclear fusion, and a wide range of modern applications.

The first major milestone in the history of superconductivity was the discovery by Kamerlingh Onnes [1, 2] that the electrical resistance of various metals, such as mercury, lead, and tin disappears when the temperature is lowered below some critical temperature value, T_c. Zero electrical resistance allows persistent currents in superconducting rings. These currents flow without any measurable decrease up to one year, allowing a lower bound of 10^5 years on their decay time. Compared to good conductors, such as copper, which have a residual resistivity at low temperature of the order of 10^{-6} Ω-cm, the resistivity of a superconductor is lower than 10^{-23} Ω-cm.

Subsequently, Meissner and Ochsenfeld [3] discovered perfect diamagnetism in superconductors. Magnetic fields are excluded from superconductors. Any field originally present in the metal is expelled from the metal when lowering the temperature below its critical value. Expulsion of magnetic field from walls of superconducting cavities via the Meissner effect will be an important topic in Chapter 4.

Starting with pioneering efforts in the 1960's, RF superconductivity (SRF) finally catapulted to an enabling technology since the 1980's. SRF has since equipped frontier accelerators in high energy physics, nuclear astrophysics, nuclear physics, as well as light sources and neutron sources for materials and life sciences. New applications are coming on line to intense proton sources for neutrino beams, and transmutation of nuclear waste, as well as for deflecting cavities for beam tilts for higher luminosity at LHC.

The primary advantages of the SRF technology have been discussed in the two previous books [4, 5]. The most attractive features of applying SRF to particle accelerators lie in the high accelerating gradient, E_{acc}, possible in continuous wave (cw) and long-pulse operating modes, along with extremely low RF losses in the cavity walls at cryogenic temperatures. There is another important advantage. The presence of accelerating structures has a disruptive effect on the beam, limiting the quality of the beam in aspects such as energy spread, beam halo, or the maximum current. SRF systems can be shorter, and thereby impose less disruption to the beam. By virtue of

Superconducting Radiofrequency Technology for Accelerators: State of the Art and Emerging Trends, First Edition. Hasan Padamsee.

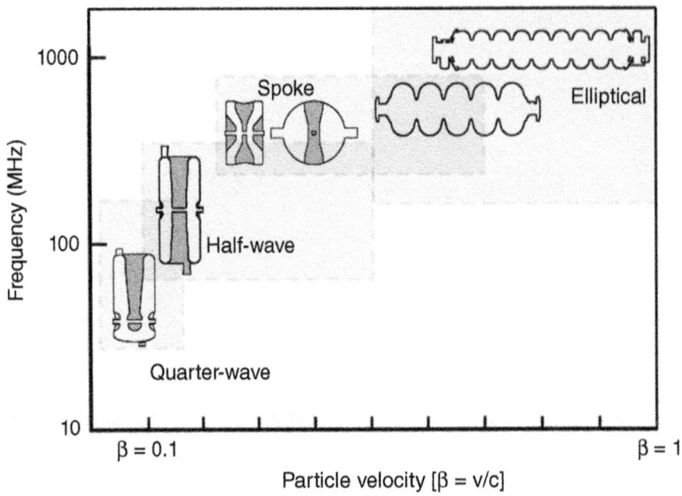

Figure 1.1 Superconducting cavities spanning the full range of particle velocities. Source: [6]/M. Kelly, Argonne National Lab/with permission from World Scientific Publishing.

low wall losses, SRF cavities can be designed with large beam holes (apertures) to further reduce beam disruption and allow higher beam currents desirable.

There are two distinct types of superconducting cavities. The first type, TM-mode cavities, is for accelerating charged particles that move at nearly the speed of light, such as electrons in a high-energy linear accelerator (linac) or a storage ring. The second type, TEM-mode cavities, is for particles that move at a small fraction (e.g. 0.01–0.5) of the speed of light, such as the heavy ions. Structures for these applications are the quarter wave resonator (QWR), the half wave resonator (HWR) and the single spoke resonator (SSR), or one with multiple spokes. At intermediate velocities, both TM and TEM types could be used, depending on the application. Figure 1.1 [6] shows practical geometry sketches, and typical RF frequencies for each cavity type, depending on the velocity of the particles spanning the full velocity range of particles.

The QWR is the compact choice for low-β applications ($\beta < 0.15$) requiring ~50% less structure with less overall RF dissipation compared to the HWR for the same frequency and β. (Here $\beta = v/c$, where v is the speed of the particle under acceleration, and c is the speed of light.) But the asymmetric field pattern in the accelerating gaps produces vertical steering that increases with velocity. The QWR is less mechanically stable than the HWR due to the unsupported end at the bottom in Figure 1.1. Hence the HWR is more suitable in the mid-velocity range ($\beta > 0.15$) or where steering must be eliminated (i.e. for high intensity). It has a symmetric field pattern and provides higher mechanical rigidity. But the HWR is larger, requires a larger cryomodule (CM), and has roughly twice the dissipation for the same β and frequency. The SSR is a more compact variant of the HWR. It opens a path to extension to several accelerating gaps along the beam in a single resonator, using multiple spokes. It provides a higher effective voltage, but with a narrower transit time acceptance.

This book will mostly focus on a review for the near velocity-of-light, or high-β accelerating cavities, and to particle accelerators that use these structures. We only briefly cover some of the latest applications of low-β structures to major facilities. For in-depth coverage of low-β cavities, we refer the reader to excellent articles [6], and tutorials at International SRF conferences [7, 8].

This book will not cover many important topics in SRF, such as input couplers, higher-order-mode couplers, tuners, and cryomodules. For latest developments in these areas, we refer the reader to many papers published in the Proceedings of the International SRF Conferences. The proceedings are available on the JACoW website [9].

2

SRF Fundamentals Review

2.1 SRF Basics

We briefly review the key figures of merit that characterize the performance of an SRF cavity or structure, referring the reader to [4, 5] for in-depth coverage. The first important parameter – the *accelerating voltage* V_c – is the ratio of the maximum energy gain that a particle moving along the cavity axis can achieve, to the charge of that particle. As all existing high-β multicell SRF structures operate in a π standing-wave mode, the optimal length (active length) of the cavity cells is $\beta\lambda/2$. Here λ is the rf wavelength. Next, the *accelerating gradient* is the ratio of the accelerating voltage per cell to the cell length, or $E_{acc} = V_c/(\beta\lambda/2)$. The cavity *quality factor* Q_0 determines the number of rf cycles (multiplied by 2π) required to dissipate the energy stored in the cavity. The key performance factor of an SRF cavity is typically given by the Q_0 versus E curve, showing how rf losses change as the gradient (E_{acc}) rises. The quality factor (Q_0) is derived as a ratio of two values via $R_s = G/Q_0$, where G is the geometry factor, and R_s is the surface resistivity. As the name suggests, the geometry factor is determined only by the shape of the cavity. Surface resistivity (often referred to as surface resistance, R_s) depends only on material properties and the rf frequency. The physics of surface resistance is dominated by the physics of superconductors, and so will be a major topic of the book. The cavity's shunt impedance, R_{sh}, determines how much acceleration a particle can derive from a cavity for a given power dissipation, P_c in the cavity walls. Hence $R_{sh} = V_c^2/P_c$. A related quantity is the geometric shunt impedance R_{sh}/Q_0, or simply R/Q, which depends only on the cavity shape. Two other important figures of merit are the ratios E_{pk}/E_{acc} and B_{pk}/E_{acc} of the peak surface electric field E_{pk} and magnetic field B_{pk} to the accelerating gradient E_{acc}. The typical distributions of the electric and magnetic field in a single cell $\beta = 1$ cavity are shown in Figure 2.1a,b, as well as for a low-β QWR in Figure 2.1c. Note that for the single cell $\beta = 1$ cavity, the magnetic field is maximum near the equator, whereas the electric field is at a peak near the iris. Maximum electric field locations for the QWR are shown in red.

For a given accelerating field, both E_{pk} and H_{pk} need to be minimized for a good design. A high surface electric field can cause field emission of electrons, which

Note: Q_0 and Q will be used interchangeably throughout the book.

Superconducting Radiofrequency Technology for Accelerators: State of the Art and Emerging Trends, First Edition. Hasan Padamsee.

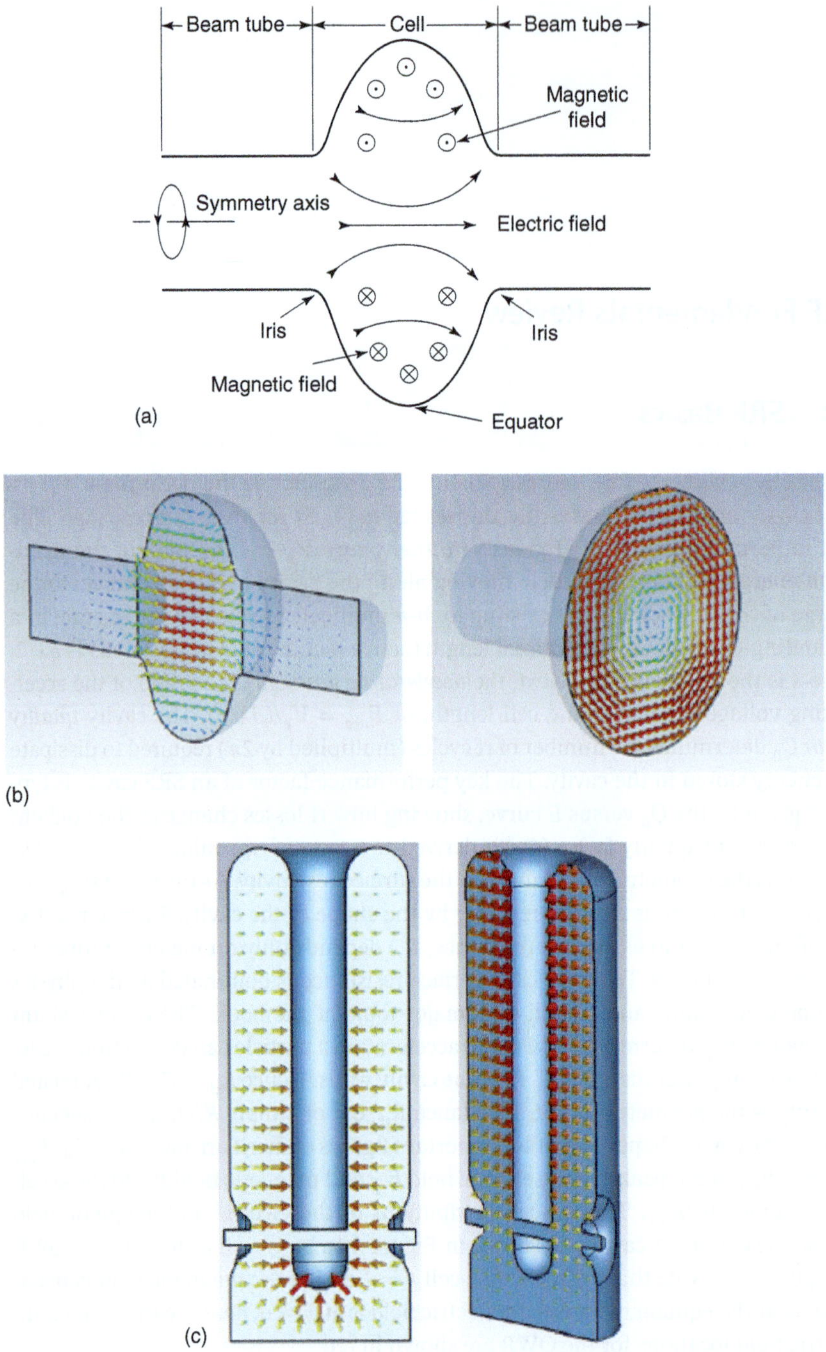

Figure 2.1 (a) Electric and magnetic field distributions for a single-cell TM_{010} cavity. Source: [10] Courtesy of J. Knobloch, Cornell University. (b) Microwave Studio® [11] simulations of the electric field (left) and magnetic field (right) in a TM_{010} mode [12] Courtesy of D. Bafia, Illinois Institute of Technology. The phase of the magnetic field is 90° shifted relative to the phase of the electric field. (c) Electric field (left) and magnetic field (right) simulation for the QWR [13]. Zhang and Venturini Delsolaro/JACoW/CC BY 3.0.

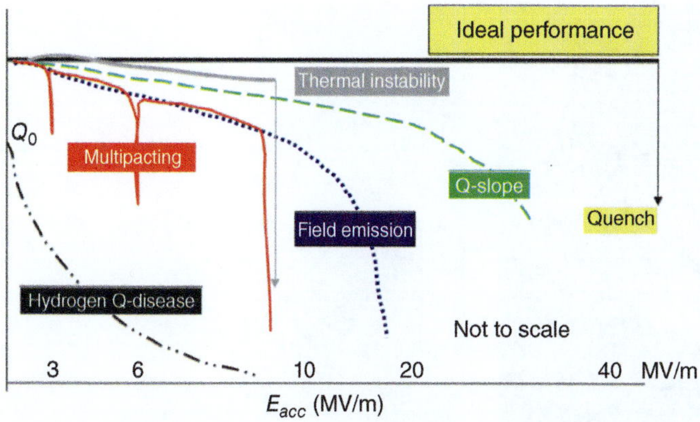

Figure 2.2 Typical Q versus E curves obtained for cavities exhibiting various performance limiting phenomena such as: hydrogen-Q-disease, multipacting, thermal instability (or quench), field emission, or high field Q-slope (HFQS). The flat curve depicting ideal performance is rarely (or never) achieved. The X-axis for gradient is not to scale [14].

impact and heat the cavity wall, often leading to a premature breakdown of superconductivity (called "quench"). Field emission electrons also generate undesirable "dark current" in the accelerator. A high surface magnetic field may limit the cavity's performance at high gradients if rf heating from a high resistance region (such as a defect) triggers a quench of superconductivity, or if the local field approaches the critical rf magnetic field, discussed in more detail in later chapters.

The key performance of an SRF cavity is expressed by measuring the Q_0 versus E_{acc} curve. As shown in Figure 2.2, the Q_0 departs from the ideal flat curve due to limitations arising from various phenomena such as the hydrogen-related Q-disease, multipacting, breakdown from a defect, field emission, high field Q-slope (HFQS), and medium field Q-slope (MFQS). Each of these phenomena has been extensively studied with great progress in understanding the fundamental causes. Remedies have been developed to overcome the limitations and to return cavity behavior toward the ideal, flat Q_0 versus E_{acc} curve.

Temperature mapping of the outer wall of the cavity has played a crucial role in understanding and curing many of these limitations. Figures 2.3 and 2.4 show the earliest system [15] for *rapid* mapping the outer-wall temperature below the lambda point of liquid He (2.2 K). Figure 2.4 also shows a temperature map when there is heating at a defect that eventually leads to a quench at a higher field. The thermometry system shown here has been improved [16] and adopted by many labs [17–19].

The performance of an SRF cavity depends on the maximum values of the peak surface fields that can be tolerated without increasing the microwave surface resistance substantially, or without causing a breakdown of superconductivity. A high surface electric field can cause field emission of electrons, degrading the Q_0. A high surface magnetic field may limit the gradient of the cavity through heating at a defect followed by thermal runaway (Figure 2.4), or through a magnetic transition to the normal state at the local critical magnetic field. The ultimate accelerating field achievable for an ideal Nb cavity is set by the rf critical magnetic field, theoretically

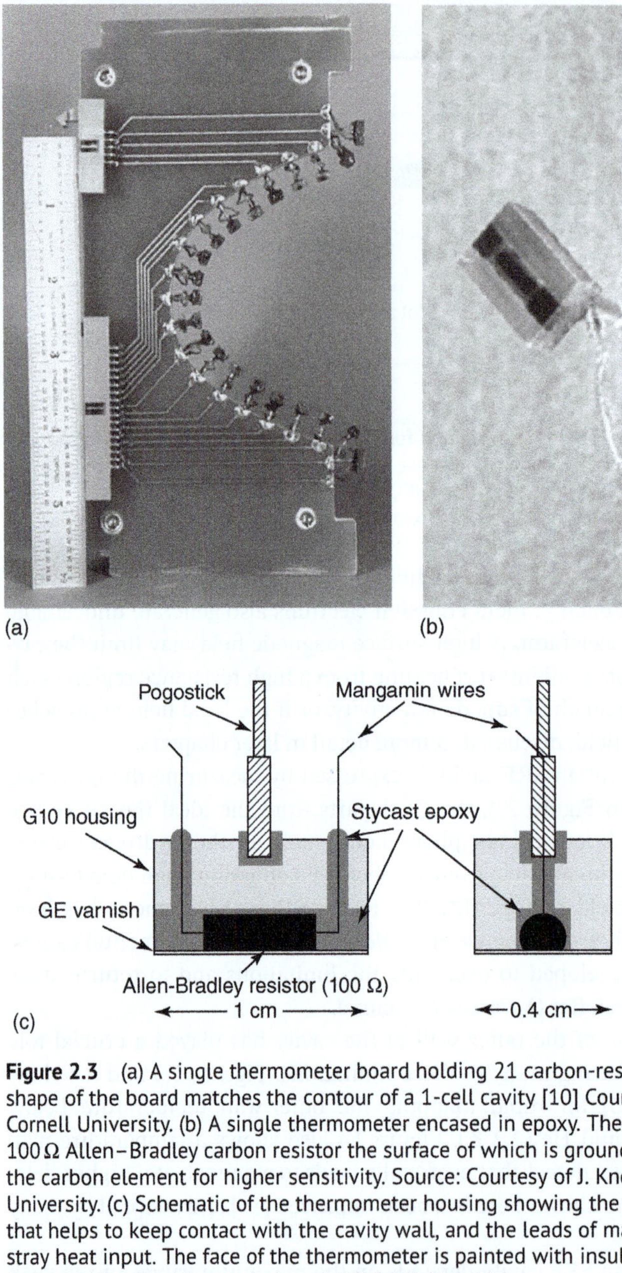

Figure 2.3 (a) A single thermometer board holding 21 carbon-resistor thermometers. The shape of the board matches the contour of a 1-cell cavity [10] Courtesy of J. Knobloch, Cornell University. (b) A single thermometer encased in epoxy. The sensing element is a 100 Ω Allen–Bradley carbon resistor the surface of which is ground down to just expose the carbon element for higher sensitivity. Source: Courtesy of J. Knobloch, Cornell University. (c) Schematic of the thermometer housing showing the spring-loaded pogo stick that helps to keep contact with the cavity wall, and the leads of manganin wire to limit the stray heat input. The face of the thermometer is painted with insulation. Source: [10, 16]/with permission of AIP Publishing LLC.

equal to the superheating critical magnetic field [21], H_{sh}. For ideal niobium, H_{sh} at 2 K is about 0.22 T, which translates to a maximum accelerating field of about 52 MV/m for a typical shape $\beta = 1$ niobium structure, and roughly 30 MV/m for a typical $\beta < 1$ Nb structure.

Other important design features for an SRF structure discussed further in [22] are cell-to-cell coupling for multicell structures, Lorentz-force (LF) detuning coefficient,

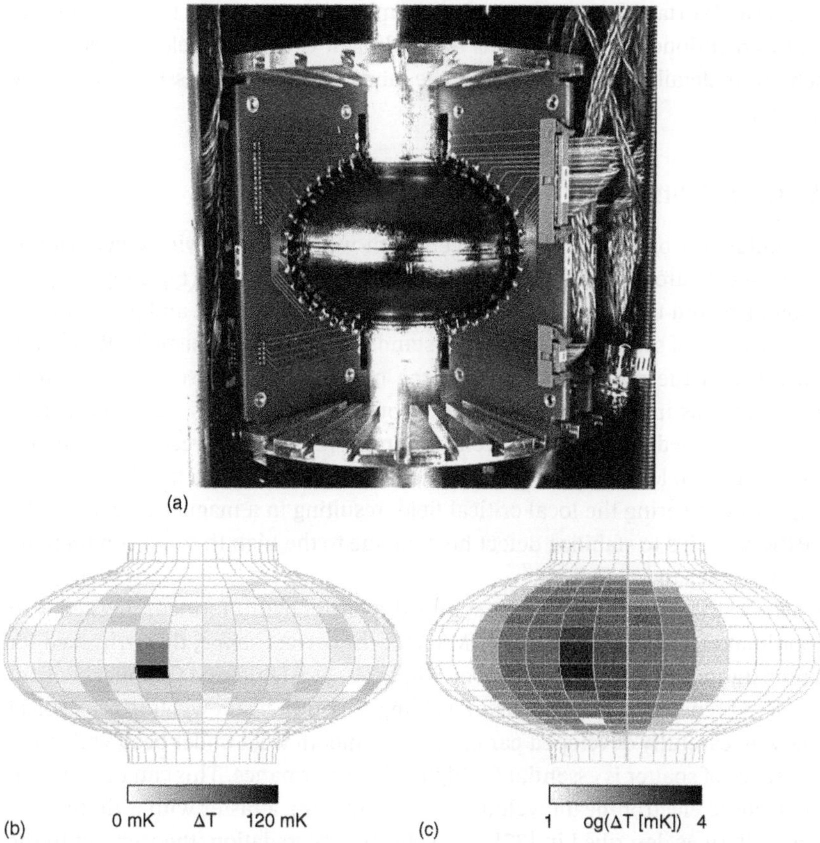

Figure 2.4 (a) Thermometers positioned on a cavity wall. Apiezon-N grease promotes thermal contact between the thermometer and the cavity wall. Some boards are removed to expose the cavity. [10, 16]. Courtesy of J. Knobloch, Çornell University & with permission of AIP through CCC. (b) Sample temperature map showing heating at a sub-mm defect site that leads to quench at higher fields. (c) At higher RF field, the defect heating grows to cause a quench of superconductivity, and a large region of the cavity surface around the defect shows high temperatures. Source: [20]/H. Padamsee, Cornell University.

input power required for beam power (P_b), coupling strength of input coupler (Q_{ext}), higher order mode (HOM) frequencies, HOM shunt impedances and HOM Q values. Mechanical properties also play a role in ensuring stability under atmospheric loading and temperature differentials, to minimize Lorentz-force detuning, and to keep microphonics detuning under control.

2.2 Fabrication and Processing on Nb-Based SRF Structures

To appreciate the latest progress in the performance and applications of SRF cavities it is helpful to briefly review the main features of customary fabrication and processing methods. The short review will help understand how the evolution of fabrication and surface treatment practices couple to the solution of the performance difficulties mentioned above, such as the hydrogen Q-disease, field emission and quench. More detail information about the fabrication and processing is available in [4, 5, 22].

2.2.1 Cavity Fabrication

Several industries provide niobium sheets with well-defined cavity specifications [23]. The sheets are inspected for flatness, uniform grain size (typically 50 μm), near-complete and uniform recrystallization, RRR value (>300), and good surface quality (absence of scratches). Here RRR stands for Residual Resistivity Ratio, and is a measure of the purity of Nb. Since the many fabrication stages can embed "defects," such as impurity inclusions, pits, bumps, or scratches, each sheet from industry is scanned with eddy-current scanning [24, 25] to weed out defective sheets. Defects can lead to breakdown of superconductivity (quench) either by overheating, or by lowering the local critical field, resulting in a magnetic quench. The high RRR Nb helps to stabilize defect heating due to the high thermal conductivity that accompanies the high RRR.

For a $\beta \sim 1$ structures, half-cells are stamped, spun, or hydro-formed, checked with the coordinate measuring machine (CMM) for the correct shape, then trimmed for weld preparations. Cavity parts are given a light (20 μm) (Buffered Chemical Polish) BCP etch to prepare for electron beam welding. Electron beam welding is a critical process with carefully developed parameters. A smooth weld under bead with complete absence of spatter is essential for high field performance. This can be achieved with defocused electron-beam welding [26], or by using a raster with a rhombic or circular pattern as described in [27]. To avoid RRR degradation, the vacuum in the electron-beam welder should be better than 2×10^{-5} Torr. All welds are inspected for complete, smooth under bead, flat on the inside, and no weld spatter. After completing a single-cell or multicell structure, the inside surface is inspected optically. A special optical inspection apparatus has been developed and widely adopted [28]. Mechanical measurements ensure straightness and correct dimensions. The electric field profile along the beam axis is checked and adjusted. The goal is usually 98% field flatness. A "flat" field profile is achieved by tuning the cells relative to each other by squeezing or stretching the cells mechanically to adjust and properly match the frequency of each cell.

Most low-β resonators are made from bulk niobium with high RRR (150–300). Fabrication of parts include machining, forming, rolling, and welding. Recently, wire electric discharge machining (EDM) has been developed together with industry [6] which has little possibility for foreign material inclusions as compared to traditional machining. Parts are joined together by electron beam welding in high vacuum.

2.2.2 Preparation

Niobium cavities undergo a first stage etching (100–150 μm) to remove the "surface damage" layer. Methods used for material removal are standard buffered chemical polishing (BCP), electropolishing (EP) [29], and centrifugal barrel polishing (CBP) [30]. By far the best method proven is EP [31], giving the smoothest surface (roughness <0.3 μm) and leading eventually to highest gradients.

BCP is a technically simpler process and used for pre-etching parts for cleaning and welding. BCP is chemical etching with a mixture of HF (40% concentration), HNO_3 (65%), and H_3PO_4 (85%) acids in a volumetric ratio of 1 : 1 : 2. The process is exothermic so that good heat exchange and stirring are necessary for uniform material removal, and to keep the acid temperature below 15 °C to prevent excess hydrogen take-up (Section 6.11). BCP yields sharp grain boundary steps of 1–2 μm, and several μm at the weld because the etch rate depends on the crystal grain orientation. Such sharp steps are undesirable due to local field enhancement and lower quench fields.

EP is carried out with an acid mix of HF (40%) and H_2SO_4 (98%) in a ratio of 1 : 9 as the electrolyte. The niobium cavity serves as the anode and a high-purity aluminum cathode is inserted into the cavity. A typical arrangement has the cavity and the cathode in horizontal orientation with electrolyte filling about 60% of the cavity. The assembly is slowly rotated to allow uniform etching and polishing of the surface. The acid is circulated to an external reservoir where it is also cooled. Nitrogen gas is circulated via the cathode structure to expel hydrogen produced. Steps at the grain boundaries are reduced to below 0.2 μm (Figure 2.5). After EP, removal of sulfur

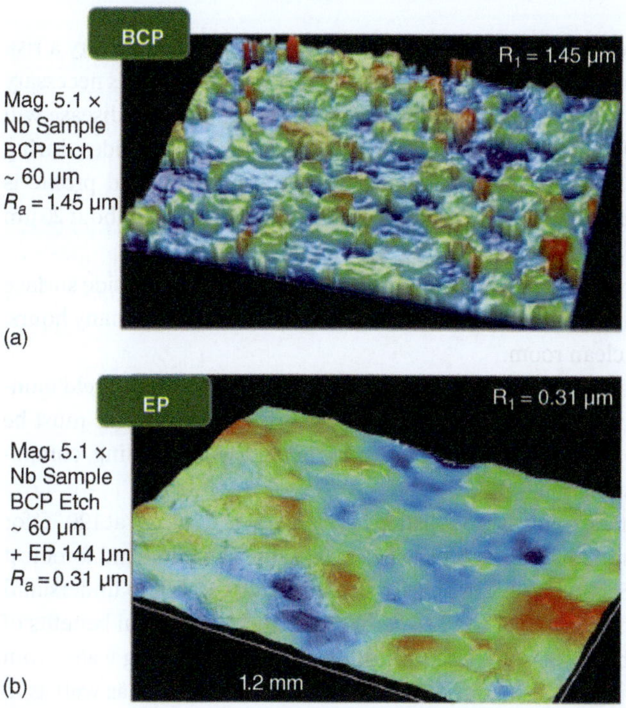

Figure 2.5 Comparison of surface roughness Nb surface treated with (a) BCP and (b) EP. Source: [35]/Courtesy of L. Popielarski, FRIB, MSU/U.S. Department of Energy, Office of Science.

residues [32] must be carried out with ultrasonic degreasing for a couple of hours in ethanol [33] or in detergent and water [34].

Tumbling or CBP is used to remove irregularities at welds, as well as pits, and scratches. The slow rate of material removal is highly dependent on the tumbling medium and rotation speeds. The final finish can be mirror-like, with 10 nm roughness, but the mirror quality by itself has not been shown to yield higher performance than EP. A final stage of light (5–20 μm) chemical etching via EP is still necessary to remove the tumbling abrasive embedded in the surface [36]. CBP has successfully been used to repair cavities with mechanical defects, such as pits, and bring them to high field performance [36, 37]. All chemical treatment or mechanical abrasion admits hydrogen into the bulk, leading to a Q-disease (Section 6.11).

After chemical etching and rinsing, high pressure water rinsing (HPR) is carried out at 100 atm water pressure for several hours to remove chemical residues and attached particles to avoid field emission or thermal breakdown. HPR is effective in scrubbing the surface free of impurities from chemical processing. The water must be particulate-free (using 0.1 μm or better filters) deionized water at a nozzle pressure of ~100 bar to remove chemical residues [38, 39]. HPR is performed in an ISO5 (Class 100) or better clean room to prevent dust contamination during the process. Dust particles and chemical surface contamination results is heavy field emission, so both HPR and clean room environments are essential for good performance.

All preparation procedures for bulk material removal (BCP, EP, CBP) carry a risk of H evolution and absorption so that a furnace treatment at 600–800 °C is necessary to remove the H and to avoid the Q-disease. As shown in Figure 2.2, Q-disease will cause the Q to fall at low fields due to the formation of niobium hydrides during cool-down (Section 6.11). After furnace treatment, the cell-to-cell field profile is remeasured and readjusted. The final chemical treatment is a light etch (about 20 μm material removal) by EP to reach the highest field levels.

After final EP, the cavity is transported to the clean room where the inside surface is once again given a high-pressure rinse with high purity water jets for many hours. The cavity dries in the clean room.

A final HPR takes place after assembly of the necessary flanges, and the field monitor probe. The cavity is then ready for evacuation for rf tests. Great care must be exercised to avoid recontamination during the subsequent cavity handling, component assembly, and installation [40].

To reach the highest fields, an electropolished cavity needs to be baked at 120 °C for 48 hours [17, 41]. The mild baking step provides several benefits: first and foremost to remove the HFQS (Section 6.4). Chapter 6 discusses new research to understand the origins of HFQS and physics of the 120 °C bake benefits. Additional benefits of baking are to reduce the processing time for multipacting by degassing water from the surface, and lowering the secondary electron emission coefficient, as well as to reduce the BCS surface resistance for higher Q values by lowering the electron mean free path (Section 2.3.5). The bake is normally carried out with the inside of the cavity in a good vacuum (~10^{-8} Torr).

2.2.3 A Decade of Progress

The decade 2010–2020 has brought enormous progress to the physics, technology, and applications of SRF cavities, the major rationale for this book. Here we summarize some of the highlights. Typical Q versus E curves for niobium cavities shows three distinct regions of Q-slope changes (Figure 2.6 [a]): low, medium, and high field Q-slopes [5], abbreviated as low-field Q-slope (LFQS), MFQS, and HFQS. The book reviews progress in understanding these Q-slope regimes.

Unprecedented Q values ($>2\times10^{11}$ at 1.3 GHz and 1.5 K) have been attained up to $E_{acc}=20$–30 MV/m. These advances were achieved by novel surface preparation techniques, such as nitrogen doping (Chapter 3), and 300 °C (mid-T) baking (Chapter 4), along with special cavity cool-down procedures to eliminate the residual resistance contribution from trapped dc magnetic flux (Chapter 5). These recent accomplishments have translated into significant increases (factors >2–3) in the efficiency of cw particle accelerators (e.g. LCLS-II at SLAC) operated at medium accelerating fields of about 20 MV/m.

On the high gradient frontier, new treatments of nitrogen infusion (Section 7.3) and two-step baking (75/120C) (Section 7.1) have paved the way for gradients near 50 MV/m. The benefits of these new discoveries are collected in a Q versus E collage shown in Figure 2.6b. The proximity effect, nano-hydride model (Section 6.13) for the HFQS, and the 120 °C mild baking cure, which inhibits the formation of the harmful hydrides, have gained much experimental support. The model provides a platform to account for the gradient improvements with new techniques.

Another mystery from the past, the LFQS, is now fully understood as originating from two-level states (TLS) in the niobium pentoxide (Section 4.6). A technique (340 °C baking) has been found to substantially remove the oxide and reduce the TLS losses to achieve record low residual resistance values under one nano-Ohm. A very important benefit of the advances in the low field arena is to improve the lifetime of qubits (Section 13.9) opening the door to higher coherence times for quantum computing (Chapter 13).

2.3 SRF Physics

2.3.1 Zero DC Resistance

The two-hallmark features of superconductivity are zero resistance for dc currents and the Meissner effect. An early treatment for superconductivity comes from the London equations [43, 44] that account for these two salient properties. The London brothers provided a phenomenological description of superconductivity based on a two-fluid type concept proposed by Gorter and Casimir [45]. Here the superfluid and normal fluid densities n_s and n_n are associated with velocities v_s and v_n. The densities satisfy

$$n_s + n_n = n \tag{2.1}$$

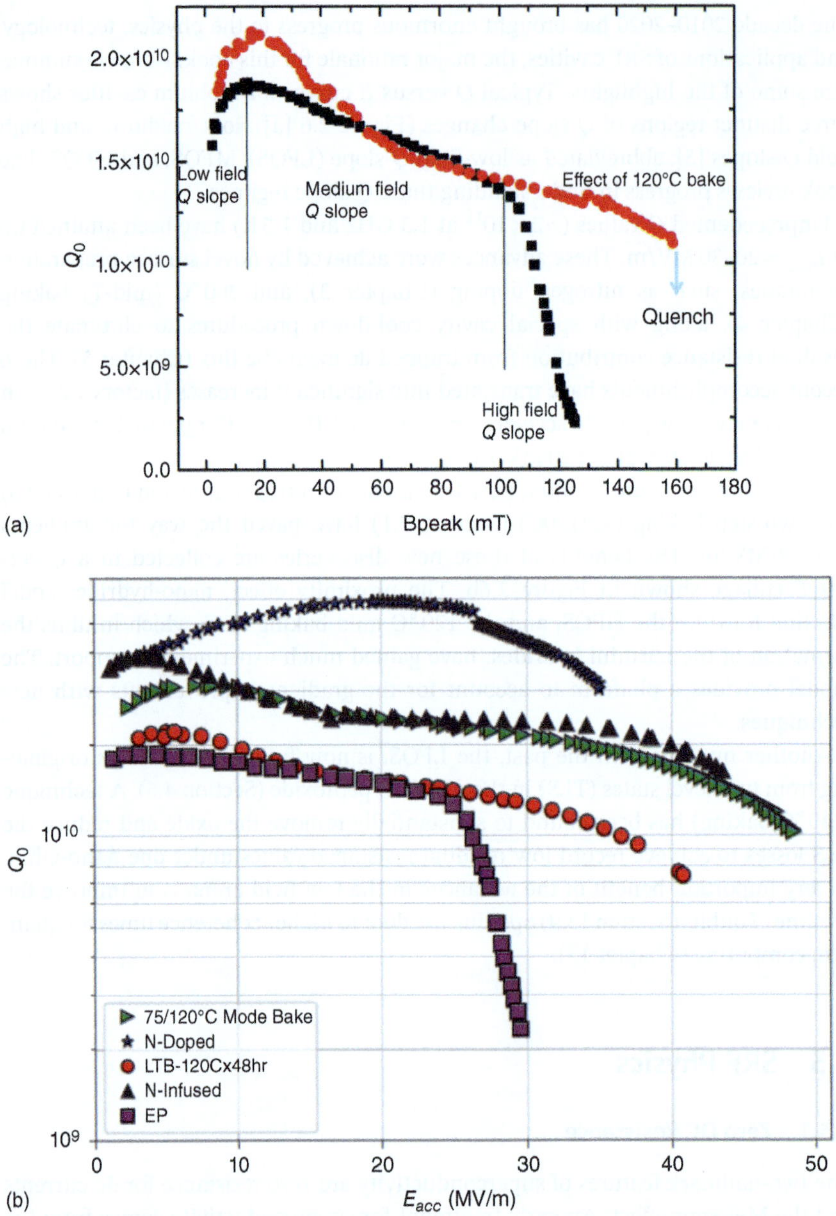

Figure 2.6 (a) Low field, medium field, and high field Q-slopes observed after standard treatment of EP. 120 C bake removes the HFQS, leaving an extended region of MFQS [42] Courtesy of A. Grassellino, Fermilab. (b) Q_0 versus E_{acc} at 2 K of 1.3 GHz Nb SRF cavities treated at FNAL with state-of-the-art surface treatments, such as nitrogen-doping, two-step baking, and nitrogen infusion, compared to standard treatments of EP or EP followed by 120 C baking [12] Courtesy of D. Bafia, Fermilab.

where n is the average electron number per unit volume. The two current densities satisfy

$$J_s = -en_s v_s, \quad J_n = -en_n v_n \tag{2.2}$$

We can think of the first and second London equation pair below (2.3) as describing the electric and magnetic fields inside a superconductor

$$\vec{E} = \frac{\partial}{\partial t}\left(\frac{m_e}{n_s e^2}\vec{J}_s\right)$$
$$\vec{B} = -\frac{m_e}{n_s e^2}\nabla \times \vec{J}_s \tag{2.3}$$

The electric field equation of the pair accounts for zero resistivity in superconductors that carry a direct current (dc) because using the supercurrent component of Eq. (2.2), it can be re-expressed as:

$$m\frac{\partial v}{\partial t} = -eE \tag{2.4}$$

Equation (2.4) implies that in the presence of an electric field, E, the electrons in an ideally perfect conductor are accelerated freely, with zero resistance. The first London equation also implies that when the current is constant in time, as in the case of dc, there is no electric field inside the superconductor, and conduction takes place without losses.

$$\frac{\partial j_s}{\partial t} = 0, E = 0 \tag{2.5}$$

Equation (2.5) should be contrasted with the conductivity equation for a normal metal or Ohm's Law. In the Drude formulation [46, 47]

$$j_n = \frac{n_n e^2 \tau}{m}E = \sigma_n E \tag{2.6}$$

$$\sigma_n = \frac{n_n e^2 \tau}{m} = \frac{n_n e^2 \ell}{m v_F} \tag{2.7}$$

where τ is the average time between collisions or the electron relaxation time, m the electron mass, ℓ is the electron mean free path, and v_F is the Fermi velocity.

2.3.2 Meissner Effect

Using Maxwell's equations with the London equations, we derive the Meissner effect. Starting with Maxwell's equation

$$\nabla \vec{\times} B = \mu_0 \vec{J}_s \tag{2.8}$$

Apply the curl

$$\nabla \times \vec{\nabla} \times B = \mu_0 \nabla \times \vec{J}_s = -\mu_0 \frac{n_s e^2}{m_e} B \tag{2.9}$$

Using the second London equation for B,

$$\nabla(\nabla \cdot B) - \nabla^2 \vec{B} = -\mu_0 \frac{n_s e^2}{m_e} B \tag{2.10}$$

Since $\nabla \cdot B = 0$, this leads to a wave – type differential equation

$$\nabla^2 \vec{B} = \mu_0 \frac{n_s e^2}{m_e} B = \frac{1}{\lambda_L^2} \vec{B} \tag{2.11}$$

where

$$\lambda_L = (m/n_s e^2 \mu_0)^{1/2} \tag{2.12}$$

λ_L is defined as the London penetration depth. (The value is 39 nm for high-purity niobium.) In the case of a semi-infinite superconducting slab that occupies the space $z > 0$, the solution to Eq. (2.11) is

$$\vec{B}(z) = B_0 e^{-z/\lambda_L} \hat{x} \tag{2.13}$$

which states that the parallel component of the B field decays exponentially with distance from the surface, to satisfy $B = 0$ inside the superconductor. This is the Meissner effect.

There are several consequences of the Meissner effect – important for Chapter 4. When a superconducting material is cooled through its transition temperature, T_c, in the presence of an external dc magnetic field, the magnetic flux is abruptly expelled from the volume of the superconductor when the ideal conditions are fulfilled. (As we will see for Nb cavities, this condition is met only in certain circumstances.) The second aspect of the Meissner effect: An externally applied magnetic field (less than the critical field) cannot penetrate into the interior of a superconductor when it is below T_c. The field is excluded from the volume of the superconductor. A metal in the superconducting state therefore never allows magnetic flux density to exist in the interior. In a superconductor,

$$\partial \mathbf{B}/\partial t = 0 \text{ inside and } \mathbf{B} = 0 \text{ inside} \tag{2.14}$$

The external magnetic field is confined to a thin layer characterized by the penetration depth, λ_L, near the surface of the superconductor. The London penetration depth sets the characteristic decay length for the magnetic field. Any supercurrents in the superconductor are also restricted to flow in this near-surface region.

Physically, the exponentially decaying magnetic field can be explained by "screening currents" that are excited near the surface and flow with zero resistance, generating their own magnetic fields that perfectly cancel out the applied field far from the surface.

The London equations can also be motivated from a quantum-mechanical argument. In the absence of a magnetic field, the ground state has zero net momentum, with the canonical momentum:

$$\boldsymbol{p} = m\boldsymbol{v} + \frac{-e\boldsymbol{A}}{mc} = 0 \tag{2.15}$$

with m the electron mass and $\boldsymbol{A}$ is the vector potential. Thus the local average velocity becomes

$$\langle \boldsymbol{v}_s \rangle = \frac{-e\boldsymbol{A}}{mc} \tag{2.16}$$

As before (Eq. [2.2]), the current density J_s can be expressed as

$$\boldsymbol{J}_s = -n_s e\langle \boldsymbol{v}_s\rangle = \frac{-n_s e^2 \boldsymbol{A}}{mc} = \frac{-\boldsymbol{A}}{\Lambda c} \tag{2.17}$$

$$\Lambda = \frac{m}{n_s e^2} \tag{2.18}$$

Note that in this case, the relationship between the supercurrent and the vector potential is "local," i.e. the supercurrent depends on the local value of the vector potential only. Taking the time derivative of both sides, leads to the first London equation, as before:

$$\boldsymbol{E} = \frac{\partial}{\partial t}(\Lambda \boldsymbol{J}_s) \tag{2.19}$$

By taking the curl of Eq. (2.19), one obtains the second London equation, as follows:

$$\nabla \times \boldsymbol{E} = -\frac{\partial}{\partial t}(\boldsymbol{B}) = \nabla \times \frac{\partial}{\partial t}(\Lambda \boldsymbol{J}_s) \tag{2.20}$$

$$\boldsymbol{B} = -\nabla \times (\Lambda \boldsymbol{J}_s) \tag{2.21}$$

2.3.3 Surface Resistance and Surface Impedance in RF Fields

In the earliest and simplest picture of the two-fluid model [45], the current flowing through a superconductor under the influence of an rf field can be separated into two noninteracting fluids: a "superfluid" that moves with zero resistance, and a "normal fluid" that experiences dissipation. As the temperature of the superconductor decreases below the transition temperature, T_c, the density of the normal fluid carriers (n_n) decreases, and the density of the superfluid (n_s) carriers increases. The total density remains constant so that $n_s + n_n = n_0$. In the two-fluid model, the temperature dependence of the superfluid density is postulated to be

$$n_s = n_0\left(1 - \left(\frac{T}{T_c}\right)^4\right) \tag{2.22}$$

The penetration depth has been empirically shown [48] to have the same temperature dependence observed in the density of superfluid electrons in the two-fluid model:

$$\lambda = \frac{\lambda_L}{\sqrt{1 - \left(\frac{T}{T_c}\right)^4}} \tag{2.23}$$

As seen in Figure 2.7a, the penetration depth is approximately independent of temperature until $T/T_c \approx 0.6$. Above this temperature, it quickly diverges toward infinity at T_c.

In Section 2.3.1 we discussed dc resistance for dc currents. For the case of rf currents, let

$$j_s = j_{s0} e^{i\omega t} \tag{2.24}$$

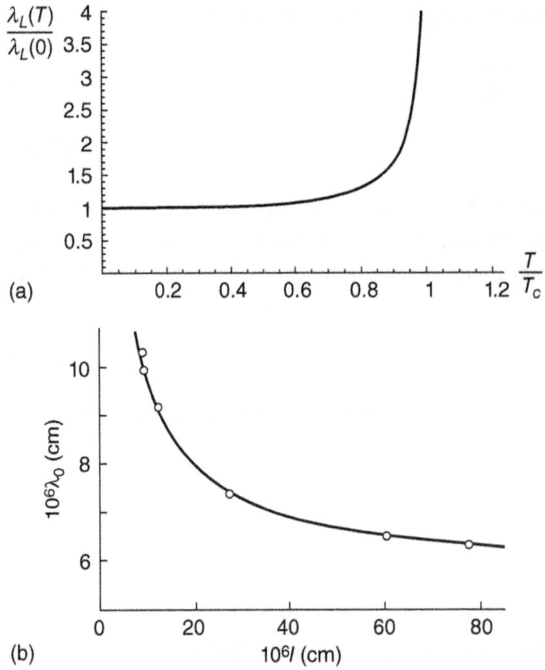

Figure 2.7 (a) Variation of penetration depth versus reduced temperature T/T_c [12] Courtesy of D. Bafia, Illinois Institute of Technology. (b) Variation of the penetration depth with the mean free path. Source: [48, 49]/with permission of the Royal Society.

From the first London equation, we obtain

$$\frac{\partial j_s}{\partial t} = \frac{1}{\mu_0 \lambda_L^2} E, \quad i\omega j_s = \frac{1}{\mu_0 \lambda_L^2} E \tag{2.25}$$

giving an expression for the supercurrent component:

$$j_s = \frac{-i}{\omega \mu_0 \lambda_L^2} E = -i\sigma_s E \tag{2.26}$$

Using the definition of λ_L^2 from Eq. (2.12), we obtain:

$$\sigma_s = \frac{1}{\omega \mu_0 \lambda_L^2} = \frac{n_s e^2}{m\omega} \tag{2.27}$$

to give the total two-fluid current as:

$$J = (\sigma_n - i\sigma_s)E \tag{2.28}$$

For alternating fields,

$$E = E_0 e^{i\omega t} \tag{2.29}$$

$$\nabla^2 E = \mu_0 \omega i(\sigma_n - i\sigma_s)E \tag{2.30}$$

Consequently, the surface impedance of a superconductor is given by

$$Z_s = \sqrt{\frac{i\omega\mu_0}{\sigma_n - i\sigma_s}} \tag{2.31}$$

To derive a simpler expression for Z_s, a useful approximation is:

$$\sigma_n \ll \sigma_s \tag{2.32}$$

i.e. the conductivity of the normal fluid is very much smaller than the conductivity of the superfluid, for the two reasons discussed below. From

$$\sigma_n = \frac{n_n e^2 \tau}{m} \text{ and } \sigma_s = \frac{n_s e^2}{m\omega} \tag{2.33}$$

We note first that at $T \ll T_c$ the number of unpaired electrons, n_n, is very much smaller than the number of paired electrons, n_s. Secondly, for the normal conducting electrons, the relaxation time (10^{-14} s) between collisions is very much smaller than the rf period (10^{-9} s), i.e.

$$\tau \ll 1/\omega \tag{2.34}$$

In analogy with normal conductors,

$$\nabla^2 E = \tau^2 E \tag{2.35}$$

where

$$\tau = \sqrt{\mu_0 \omega i(\sigma_n - i\sigma_s)} \tag{2.36}$$

We define the real and imaginary parts of the impedance

$$Z_s = R_s + iX_s \tag{2.37}$$

Using the above approximations, and with some complex number algebra manipulations (given in a separate Section 2.3.11 to avoid interrupting the flow of concepts), the real and imaginary parts of the impedance become

$$R_s = \frac{1}{2}\sigma_n \omega^2 \mu_0^2 \lambda_L^3 \text{ and}$$
$$X_s = \omega \mu_0 \lambda_L \tag{2.38}$$

We can rewrite the surface resistance as:

$$R_s \approx \frac{\sigma_n}{\lambda_L \sigma_s^2} \tag{2.39}$$

Note the important consequences of Eq. (2.38) are the two salient features of the surface resistance of a superconductor:

$$R_s \propto \omega^2 \text{ and } R_s \propto \sigma_n \tag{2.40}$$

Thus the superconducting state surface resistance (i) increases as the square of the rf frequency and (ii) is proportional to the dc normal state conductivity, an unexpected trend.

According to Eq. (2.7), σ_n is directly proportional to the electronic mean free path, ℓ, so R_s increases with ℓ; *hence dirty superconductors (small ℓ) show lower SRF dissipation*. For example, a niobium cavity made with high purity material and longer mfp will have a *lower* Q than a cavity made from niobium with impurities that lower the mfp. A simple way to interpret this unexpected trend is as follows: when the normal electrons have a higher conductivity, they draw a relatively higher fraction of the current, contributing more to surface resistance, and thereby increasing the SRF surface resistance.

The electron mean free path ℓ (mfp) represents the average distance traveled by an electron moving through the material at the Fermi velocity v_F between subsequent scattering events. The mfp is dependent on material properties, such as the interstitial impurity content. The mfp will feature prominently in Chapter 3 where N-doping introduces N as an interstitial impurity to have a substantial effect on cavity performance.

2.3.4 Nonlocal Response of Supercurrent

From microwave wave absorption measurements of the penetration depth in samples of Sn doped with In, Pippard and Bragg [48] observed that impurities increase the penetration depth of superconductors, as shown in Figure 2.7b above. But the London penetration depth picture contains no mfp dependence. In the London model, Eq. (2.17) shows that the current density depends on the local value of vector potential. Therefore, the London model is called “local.”

To develop an explanation for the mfp dependence of the penetration depth, Pippard proposed a new “nonlocal” description to relate the supercurrents to an average of the vector potential calculated over a region. He defined the size of this region by the “coherence length,” ξ, which also depends on the mean free path ℓ. Pippard generalized the London theory to include these nonlocal effects in analogy with the anomalous skin effect in normal metals [50, 51]. In this effect, the microwave surface resistance of clean normal metals reaches an anomalous limit instead of continuing to fall steadily with temperature, as does the dc resistivity. Pippard realized that the saturation occurs when the electron mean free path becomes larger than the skin depth so that only those electrons moving parallel to the sample surface remain effective in responding to the electric field. The local model is applicable when the mfp is short.

Pippard introduced a similar nonlocal effect for superconductors. The current in a superconductor at a given point depends on the electric field at neighboring points through which the electrons have traveled since last being scattered. Thus the superconducting current depends on the electric field at distant points – the range being limited by an electromagnetic coherence length, ξ. The response of the supercurrent to the vector potential is smeared out over a volume.

With such a description, the local form of the proportionality between $\boldsymbol{j}(\boldsymbol{r})$ and $\boldsymbol{A}(\boldsymbol{r})$ in Eq. (2.17), repeated below

$$\boldsymbol{J}_s(\boldsymbol{r}) = -\frac{1}{c\Lambda(T)}\boldsymbol{A}(\boldsymbol{r}), \quad \Lambda(T) = \frac{e^2 n_s(T)}{m} \tag{2.41}$$

is changed to the nonlocal relation based on a modification to Ohm's law similar to that adopted for the anomalous skin effect. In the nonlocal description,

$$\overline{\boldsymbol{J}}(\overline{\boldsymbol{r}}) = -\frac{3}{4\pi\mu_0\lambda^2\xi_0}\int \frac{\overline{R}[\overline{R}\cdot\overline{\boldsymbol{A}}(\overline{\boldsymbol{r}}')]e^{-R/\xi}}{R^4}d\overline{r}' \tag{2.42}$$

where $\overline{R} = \overline{r} - \overline{r}'$, takes into account that the current at a point $\overline{r}$ depends on the vector potential at a point $\overline{r}'$ within a distance ξ_0, called the coherence length. Thus $\vec{J}(\vec{r})$ depends on a weighted average of $\vec{A}$ over a range ξ. Figure 2.8 contrasts the nonlocal and local behaviors for one dimension.

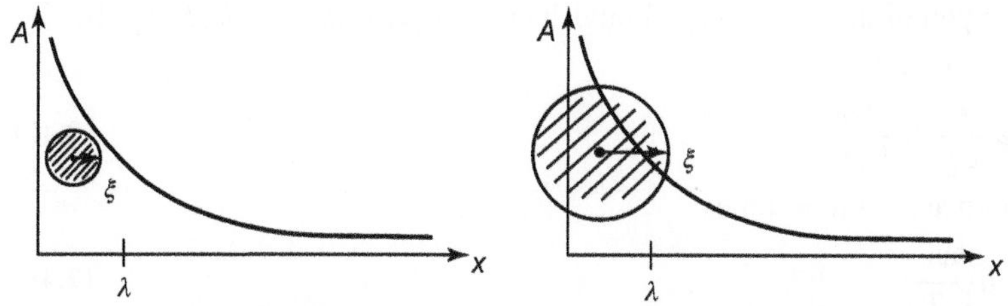

Figure 2.8 (a) In the London (local) model, the supercurrent density depends on the local value of the vector potential, A (b) In Pippard's nonlocal model, the supercurrent density depends on the vector potential over a region with size defined by the larger coherence length, ξ. Source: [52]/with permission of Dale J. Van Harlingen.

Because of the nonlocality, the new length scale, ξ can change due to scattering with impurities. Assuming that the average distance it takes to scatter electrons is the electronic mean free path (ℓ), the coherence length is modified by

$$\frac{1}{\xi} = \frac{1}{\xi_0} + \frac{1}{l} \tag{2.43}$$

Using the Maxwell's equation $\nabla^2 \overline{A} = -\mu_0 J$, the nonlocal equation between J and A can be solved for the penetration depth λ, which is defined as:

$$\lambda = \frac{1}{H_y(0)} \int_0^\infty H_y dz \tag{2.44}$$

The nonlocal result is simplified in the two limits:

$$\text{Short mfp: } \xi \ll \lambda : \lambda = \lambda_L \sqrt{\frac{\xi_0}{\xi}} = \lambda_L \sqrt{1 + \frac{\xi_0}{\ell}} \tag{2.45}$$

$$\text{Long mfp: } \xi \gg \lambda : \lambda = \left(0.62 \lambda_L^2 \xi_\circ\right)^{1/3} \tag{2.46}$$

Nonlocality over the coherence length modifies the penetration depth depending on the impurity concentration, and the mfp. Thus, material purity varies the effective size of superconducting carriers and the effective distance into which the penetration depth extends. For $\xi \ll \lambda$, the penetration depth increases as the mean free path decreases, while for $\xi \gg \lambda$, λ becomes independent of mfp, in agreement with the experimental result shown in Figure 2.7b. Pippard's λ is sometimes called the effective penetration depth to distinguish it from the London's definition. The London local theory can be considered as the limit of the Pippard theory for dirty superconductors ($\xi_0 \gg \ell$).

After the BCS theory (Section 2.3.5) it was realized that the nonlocality originates from the fact that superconductivity is carried by Cooper pairs formed by two electrons that can be hundreds of nanometers apart. If they are far apart, two electrons of the same pair "feel" very different values of the vector potential.

Pippard estimated the scale of ξ from the uncertainty principle. By assuming that only electrons within a very thin shell of thickness $k_b T_c$ around the Fermi surface partake in superconductivity, they should have momentum $\Delta p \approx k_B T_c / v_F$. By the uncertainty principle

$$\Delta x \geqslant \frac{\hbar}{\Delta p} \approx \frac{\hbar v_F}{k_B T_c} \tag{2.47}$$

The coherence length becomes

$$\xi_0 \approx a \frac{\hbar v_F}{k_b T_c} \tag{2.48}$$

Pippard found that he could fit the experimental data for tin and aluminum by using the same value of $a = 0.15$. Later BCS confirmed this result by deriving the value $a = 0.18$.

2.3.5 BCS

Many brilliant physicists – Bohr, Einstein, Feynman, Born, and Heisenberg – tried to develop a microscopic theory of superconductivity; but only Bardeen, Cooper, and Schrieffer (BCS) [53, 54] were successful in 1957 with their Nobel prize-winning theory. This was almost 50 years after the experimental discovery of superconductivity by Kamerlingh–Onnes!

One of the basis of the BCS theory was the fundamental theorem [55] that for a system of many electrons at small T where thermal excitations are feeble, any weak attraction, no matter how small, can bind two electrons together, forming the so-called "Cooper pair".

There were two key experimental discoveries that made some of the main properties of superconductivity transparent before the BCS theory came along. The exponential decay of the specific heat at low temperatures showed that the energy spectrum of a superconductor must have a gap. The first experimental evidence for a gap is due to Corak et al. [56] who measured the specific heat of a superconductor. Below T_c the specific heat has an exponential behavior

$$c_s \approx a\gamma T_c e^{-bT_c/T} \tag{2.49}$$

with $b \approx 1.5$, whereas in the normal state $c_n \approx \gamma T$ where $\gamma = \left((\pi^2/3)k_B^2 g(\varepsilon_F)\right)$ is the coefficient of the normal state electronic specific heat.

The second key was the isotope effect. The transition temperature of materials containing a different element isotope depends on the mass M of the isotope, as $M^{-1/2}$. This makes the critical temperature and critical field larger for lighter isotopes. Since M is related only to the ions in the lattice, the isotope effect indicates that the lattice must play a key role in the formation of the superconducting state via lattice vibrations (phonons). Frölich suggested that the mechanism behind the weak attractive force binding the Cooper pairs is the same mechanism responsible for electrical resistivity in metals, i.e. the interaction of conduction electrons with lattice vibrations. These observations led to the role of the phonons in superconductivity [57].

An illustrative picture for the BCS theory follows. An electron interacts with the lattice by virtue of the Coulomb attraction with the metallic ions. The lattice deforms, generating phonons. A second electron in the vicinity of the deformed lattice lowers its energy, resulting in an overall electron–electron attraction via a phonon. The attractive interaction results in a negative potential energy contribution, lowering the total energy of the electron system. The negative potential energy associated with a Cooper pair is the *binding energy* of that pair. In the quantum-mechanical picture, the attractive interaction is due to exchange of lattice phonons between electrons, in analogy with exchange of photons that mediate the electromagnetic force.

The main element of the BCS theory is that the Cooper pairs are bound states formed by two electrons of opposite spins and opposite momenta. The Cooper pairs form a coherent macroscopic ground state with a gapped spectrum. The energy gap between the BCS ground state and the first excited state is one of the most remarkable features of the BCS theory. The gap energy is the minimum required to create an excitation from the superconducting ground state. The binding energy of a Cooper pair is two times the energy gap.

The size of the gap depends on the strength of the electron–phonon interaction; hence it is different for each material. The energy gap also depends on the temperature: as the temperature increases from 0 to T_c, the gap slowly decreases, dropping rapidly to zero near T_c. A useful approximate formula [58] describing the temperature dependence of the gap is:

$$\frac{\Delta(T)}{\Delta_0} = \sqrt{\cos\left(\frac{\pi t^2}{2}\right)} \tag{2.50}$$

Cooper pairs are Bosons and therefore condense according to Bose–Einstein statistics when the temperature decreases below the critical temperature T_c. With charge $q = 2e$, Cooper pairs propagate through the lattice carrying current, but with zero resistance. They are analogous to the superfluid carriers in the two-fluid model. When current flows in the superconductor, each Cooper pair acquires the same momentum. The unpaired electrons behave as normal electrons in the two-fluid model, and are called quasiparticles. The number density of quasiparticles follows:

$$n_n \propto \exp\left(-\frac{\Delta(T)}{k_B T}\right) \tag{2.51}$$

Cooper pairs form when electrons interact with each other within a certain dimension, the coherence length ξ_0. Therefore, the coherence length may be thought as the dimension of a single pair. BCS estimated its magnitude as:

$$\xi_0 = \frac{\hbar v_F}{\pi \Delta_0} \approx \frac{\hbar v_F}{\kappa_B T_c} \tag{2.52}$$

in agreement with Pippard (Eq. [2.47]). The separation between electrons in a Cooper pair is large enough that millions of other pairs have their centers of mass positioned between them so that Cooper pairs substantially overlap with each other.

In the normal ground state of a metal, the kinetic energy (and hence the total energy) of the system is minimum when the momenta of the plane wave states of the electrons fill up a sphere of radius $p_F = \hbar k_F$ (i.e. a Fermi sphere) in three-dimensional momentum space. Figure 2.9a illustrates a Fermi sphere of radius k_F in three-dimensional k-space for a free electron gas at $T = 0\,°\mathrm{K}$. Electrons well inside the Fermi sphere cannot scatter to higher levels inside the sphere because of the Pauli exclusion principle. The corresponding normal density of states, $N(E)$, is shown in Figure 2.9b. Here $N(E)dE$ is the number of electron states

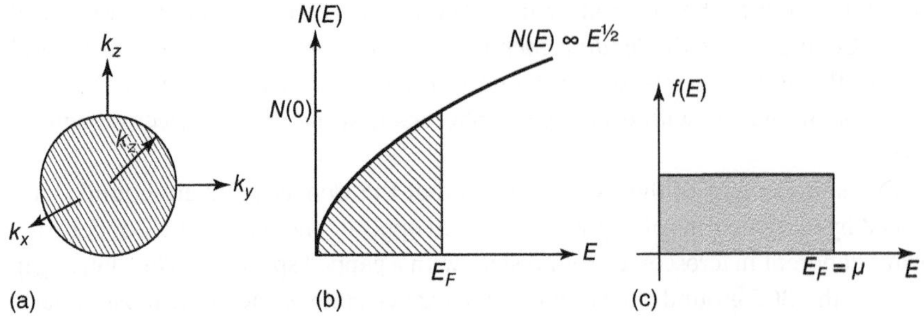

Figure 2.9 (a) Fermi sphere for the free electron gas for a normal metal at $T = 0$ K. (b) Electron density of states. (c) The Fermi distribution function showing the occupation of states. The shaded region represents occupied states at $T = 0$ K.

with energy between E and $E + dE$. $N(0)$ denotes the density of states at the Fermi surface at absolute zero. The *density of states* (DOS) *is* essentially the number of different states at a particular energy level that electrons *are* allowed to occupy.

The Fermi distribution function for $T \to 0$ is the θ-function, meaning that all the states are occupied up to the Fermi energy, $E_F = \mu$, where μ is the chemical potential, as shown in Figure 2.9c.

$$f(E, T) = \frac{1}{e^{(E-\mu)/T} + 1} \lim_{T \to 0} f(E, T) = \theta(\mu - E) \tag{2.53}$$

The phonon-induced attractive interaction affects only those electrons in the vicinity of the Fermi surface. As already pointed out by Pippard, the fraction of electrons that cooperate in superconductivity (i.e. in the formation of Cooper pairs) are those with energy $\kappa_B T_c \approx \hbar\omega_D$ around the Fermi level (E_F), where ω_D is the Debye frequency (when the wavelength of the phonon frequency reaches the smallest length, i.e. of the unit cell.). In BCS, electrons that are inside such an energy shell are pushed toward lower energies, and condense into Cooper pairs at the ground state with energy ($EF - \Delta_0$). The energy gap has the same symmetry as the Fermi surface of the normal state (Figure 2.10 a). To scatter electrons well below the Fermi level into unoccupied states requires phonon frequencies much larger than that generated from the electron interaction.

Formation of the gap substantially modifies the density of states $N(E)$ to introduce singularities at the gap edges (Figure 2.10b). Since Cooper pairs follow the Bose–Einstein distribution, a virtually infinite number of them can occupy the same energy level, hence the DOS of a superconductor takes the form:

$$N(E) = N(0)\frac{E}{\sqrt{E^2 - \Delta^2}} \tag{2.54}$$

where singularities arises when $|E| = \Delta$.

The DOS can be measured by point contact tunneling spectroscopy (PCTS) as presented in Sections 3.7.3 and 6.15. During the measurement, the tunneling current depends only on the charge density within eV below the Fermi surface, where ($-V$) is the applied bias voltage. Thus PCTS can measure the integrated density of full states below the Fermi level. By modulating the bias voltage by dV (typically a few mV) around the dc voltage V of interest, a current modulation dI

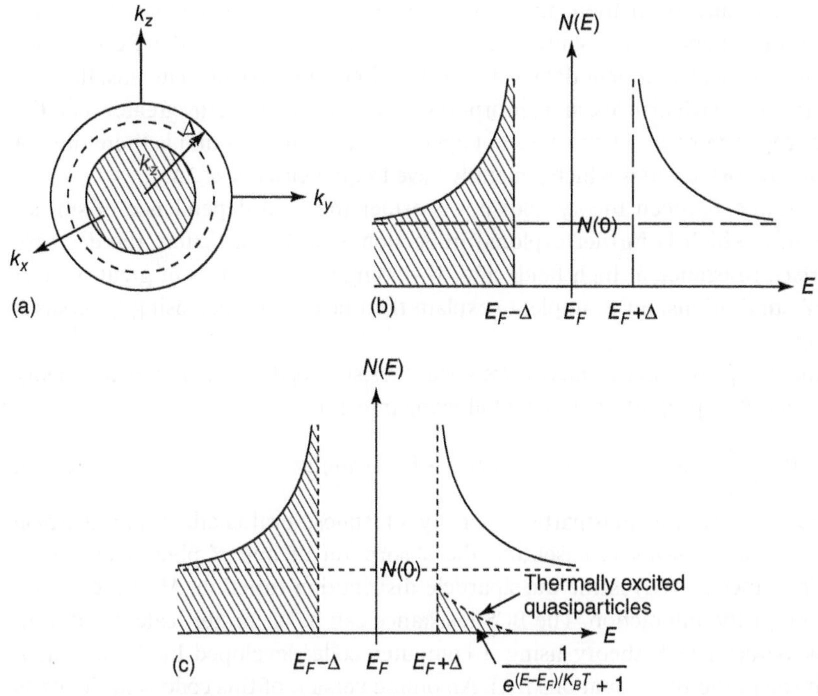

Figure 2.10 (a) Fermi surface (dotted line) for the ground state of a BCS superconductor at $T = 0$ K. Source: Adapted from Sonier [59]. (b) The density of states with singularities at the edges due to formation of the gap. The shaded region represents occupied states of the superconducting electrons [59]. (c) The density of states at $T > 0$ K. When Cooper pairs are broken the resulting normal electrons occupy states above $E_F + \Delta$. [59] following the Fermi–Dirac distribution as a function of the energy relative to the Fermi level.

can be measured to give the conductance $g(V) = dI/dV$, which is proportional to the DOS.

As shown in Figure 2.10, a superconductor is characterized by the density of states, and the distribution function of the electrons $f(E)$. The DOS exhibits an energy gap due to Cooper pair formation, and the distribution function is the Fermi–Dirac Distribution in thermal equilibrium.

With ω_c as the cutoff frequency for pair attraction, and $N(0)$ as the density of states at the Fermi level for electrons of one spin orientation, and V as the potential of the attractive interaction, BCS solved the self-consistent gap equation at zero Kelvin in the weak coupling limit, $k_bT_c \ll \hbar\omega_c\rangle$, $N(0)V \ll 1$, to obtain the relationships between the transition temperature, phonon frequency, and superconducting gap:

$$2\Delta_0 = 3.53k_bT_c = 4\hbar\omega_c e^{-\frac{1}{N_0V}} \tag{2.55}$$

Since $N(0)$ and V are material-dependent parameters, the energy gap varies with materials. Hence the ratio Δ_0/k_bT_c varies from superconductor to superconductor and is usually between 1.5 and 2.2 [21].

BCS addressed the electrodynamic properties of superconductors using the Pippard nonlocal approach. Mattis and Bardeen [60] derived expressions for the conductivity and from there the rf surface resistance and reactance. (The SRF community refers to the Mattis–Bardeen surface resistance as the BCS surface resistance.) Their treatment determines dissipation from two mechanisms: thermal excitation of quasiparticles, and absorption of photons with energy greater than the energy gap. The second term does not appear unless $\hbar\omega > 2\Delta$ and is therefore not relevant for SRF cavities which generally have frequencies $\hbar\omega \ll 2\Delta$.

The Mattis–Bardeen theory does not consider the field dependence of surface impedance which is further explored in Section 3.10. In particular, the decrease of surface resistance at high fields after N-doping (Chapter 3) is of great interest in SRF applications, for example, to explain the unexpected increasing Q versus E (anti-Q-slope).

In the BCS picture of the microwave surface resistance, the Mattis–Bardeen theory shows that R_s is proportional to the following integral:

$$R_s \propto \int_{\Delta}^{\infty} |M^2| N(\varepsilon) N(\varepsilon + \hbar\omega)[f(\varepsilon) - f(\varepsilon + \hbar\omega)] d\varepsilon \tag{2.56}$$

Here $N(E)$ is the quasiparticle density of states, evaluated at quasiparticle energies ε and $\varepsilon + \hbar\omega$ representing the absorption of one rf photon of energy $\hbar\omega$. The function $f(E)$ is the quasiparticle distribution function. M is the matrix element for the interaction. The BCS resistance can be accurately calculated from Mattis–Bardeen/BCS theory using a computer code developed by J. Halbritter, which solves the BCS equations [61]. An online version of this code is available as SRIMP [62]

For frequencies smaller than the gap frequency, energy from the electromagnetic field is absorbed only by thermally excited quasiparticles. The dissipated power is given by the net number of absorbed photons multiplied by the photon energy and by a matrix element which describes the absorption process and then integrated over available states and photons. For $T < \frac{T_c}{2}, \hbar\omega < kT, \hbar\omega < \Delta$, the matrix element can be considered constant, and the following approximate expression for the surface resistance is obtained

$$R_{BCS} = \frac{\mu_0^2 \omega^2 \lambda^3 \sigma_n \Delta}{kT} \ln\left(\frac{2.246kT}{\hbar\omega}\right) e^{-\frac{\Delta}{kT}} \tag{2.57}$$

which can be further approximated by

$$R_{BCS} = \frac{A}{T}\omega^2 e^{-\Delta/kT} \tag{2.58}$$

where A is a constant which depends on the following parameters: London penetration depth λL, coherence length ξ_0, Fermi velocity v_F, and mean free path ℓ. This relation is valid for $T < T_c/2$ [4].

These equations confirm (as before) that the surface resistance decreases exponentially with temperature, depends on the square of the rf frequency ω, and is directly proportional to the conductivity of the normal-conducting state σ_n. The surface resistance also depends on the mean free path via $\lambda^3\sigma_n$, in agreement with Eq. (2.38).

To extract the dependence of R_{BCS} with the mean free path, as it gets shorter with impurities, it is convenient to take into account the Pippard's definition of λ

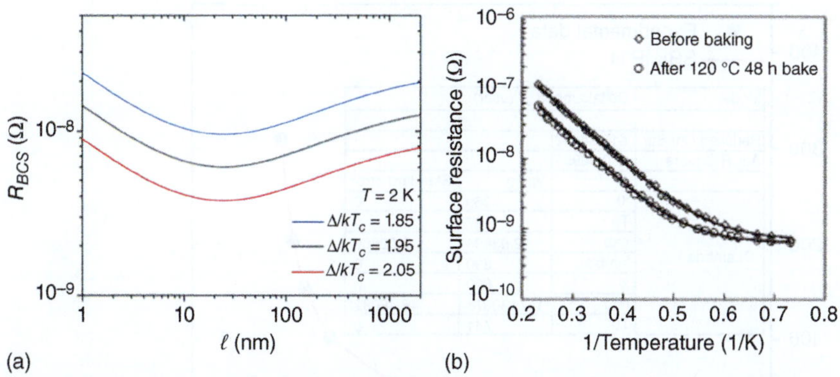

Figure 2.11 (a) BCS surface resistance as a function of the mean free path for different values of reduced energy gap Δ/kT_c, [49] Courtesy of M. Martinello, Fermilab. (b) Surface resistance versus $1/T$ showing how the BCS resistance changes for two different mfp values. Source: [63]/G. Ciovati, Jefferson Lab with permission of AIP Publishing LLC.

shown in Eq. (2.45). The conductivity σ_n is defined by the Drude formula (Eq. (2.7)). Therefore:

$$R_{BCS} \propto \lambda^3 \sigma_n \propto \ell \left(1 + \frac{\xi_0}{\ell}\right)^{3/2} \tag{2.59}$$

This relation has a minimum at $\ell = \xi/2$, meaning that the surface resistance is minimized when the mean free path is around $\xi/2$, as seen in Figure 2.11a.

In Figure 2.11a, curves of R_{BCS} as a function of ℓ calculated from SRIMP are shown for different values of the reduced energy gap Δ/kT_c, keeping all the other parameters fixed [49]. It is clear that with increasing Δ/kT_c the resistance decreases.

Figure 2.11b shows the surface resistance as a function of temperature for two different mean free paths achieved for Nb by 120 °C baking (Section 6.1). Measured values are compared with calculations from the BCS/Mattis and Bardeen theory using the SRIMP code [61, 64].

From experiments, the cavity mean free path can be determined with the help of the SRIMP code by fitting the measured R_s versus T data, or by fitting the resonance frequency change as a function of temperature during the cavity warm-up to T_c. During warm-up of a cavity to near T_c, the penetration depth increases along with the volume occupied by the rf magnetic field. This leads to a decrease of the resonance frequency, in agreement with the Slater's theorem [65].

$$\Delta\lambda = \frac{G\Delta f}{\mu_0 \pi f^2(T_0)} \tag{2.60}$$

where $\Delta f = f(T) - f(T_0)$, $\Delta\lambda = \lambda(T) - \lambda(T_0)$ and T_0 is the temperature at which the frequency becomes almost constant, usually 7.5 °K for Nb. The frequency variation can be approximated (according to Eq. [2.23]) as:

$$\frac{1}{\lambda^2(T)} = \frac{1}{\lambda_0^2}\left[1 - \left(\frac{T}{T_c}\right)^4\right] \tag{2.61}$$

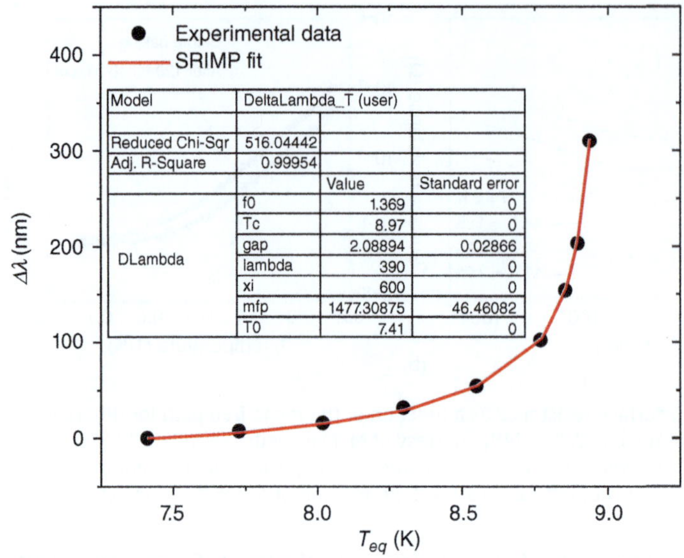

Figure 2.12 SRIMP interpolation of the penetration depth variation as a function of the temperature of a cavity to give the mfp and energy gap. Source: [49] Courtesy of M. Martinello, Fermilab.

For example, the frequency decreases about 5 kHz between liquid helium temperatures and T_c, for a 1.3 GHz Nb cavity. For dirty materials, the variation of the penetration depth is larger than for cleaner materials, as expected from Pippard's definition of the penetration depth:

$$\lambda_0 = \lambda_L \left(1 + \frac{\xi_0}{\ell}\right)^{1/2} \tag{2.62}$$

The SRIMP code is used to fit λ versus T data, with the following fixed parameters: critical temperature, T_c, coherence length ($\xi_0 = 38$ nm), and the London penetration depth ($\lambda_L = 39$ nm). The variable parameters used to fit the data are: mean free path ℓ and reduced energy gap (Δ/kT_c). Figure 2.12 shows an example of the SRIMP fit for frequency variation near T_c.

This procedure yields the variation of the penetration depth, as a function of the temperature close to T_c [49], and from there the mean free path.

2.3.6 Residual Resistance

When compared with experimental data, the theoretical behavior of the BCS surface resistance is generally good up to a certain temperature (see Figure 2.11b). Below this temperature, rather than continuing to decrease, the surface resistance approaches a constant value, called the residual surface resistance R_0 or R_{res} The complete expression of the surface resistance must therefore take into account not only the temperature-dependent term derived from the Mattis and Bardeen theory, R_{BCS}, but also a temperature-independent term, R_0:

$$R_s = R_{BCS} + R_0 \tag{2.63}$$

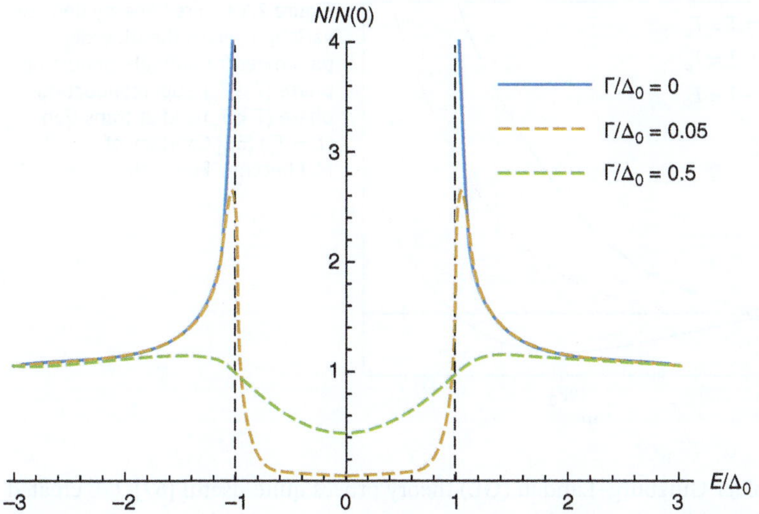

Figure 2.13 Superconducting density of states with various Dynes smearing parameter values. Source: [12] Courtesy of D. Bafia, Illinois Institute of Technology.

Possible causes for residual resistance may be the presence of defects, magnetic flux trapped in the superconductor, the presence of hydrides, the oxidized layer on the surface, residuals from chemical or mechanical surface treatment, and welding imperfections, and so on [4]. A fundamental contribution may also arise from quasiparticle states that exist within the gap as discussed below in Section 2.3.7.

2.3.7 Smearing of Density of States

Dynes introduced a phenomenological smearing parameter Γ [66]. This parameter describes a finite population of quasiparticle states that exist within the gap and modifies the density of states Eq. (2.54) to smear out the singularity. The height of the smeared coherence peak gives a phenomenological measure of the level of inelastic scattering. Figure 2.13 shows the density of states for different values of Γ. The Dynes smearing parameter causes the peak to decrease in height and the captured states in the sub-gap region.

$$\frac{N}{N_0} = \frac{E + i\Gamma}{\sqrt{(E + i\Gamma)^2 - \Delta^2}} \tag{2.64}$$

2.3.8 Ginzburg–Landau (GL) Theory

Note: Though formally H is the symbol for magnetic field strength, B (the magnetic flux density) is often used by a variety of authors, especially in Figures adopted from their work.

The BCS theory provides a thorough understanding of a large portion of the physics of superconductors. However, its quantum-mechanical formalism makes it difficult to use when addressing the macroscopic behavior of superconductors.

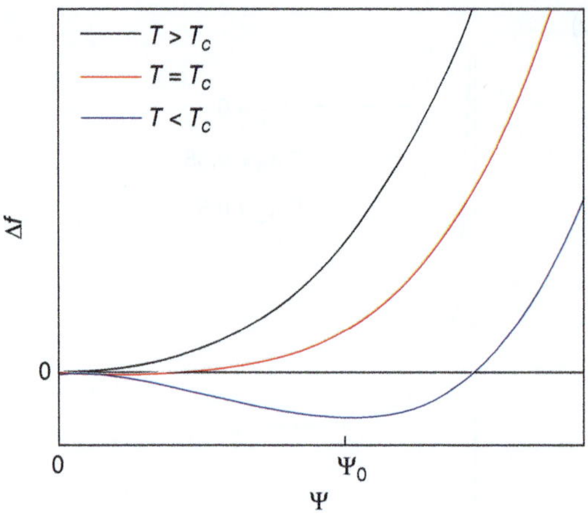

Figure 2.14 Free energy density variation versus the GL order parameter for normal-conducting phase ($T > T_c$), superconducting phase ($T < T_c$) and at transition ($T = T_c$) [68] Courtesy of M. Checchin, Fermilab.

Here the older Ginzburg–Landau (GL) theory proves quite useful [67]. GL created a theory describing the transition between the superconducting and the normal phases. Nowadays much of the SRF science for particle accelerators use the language of the GL theory.

The GL phenomenological model is a thermodynamic description of the superconducting transition as a second-order phase transition. The model uses a complex order parameter $\psi(r)$ that serves as a pseudo-wave function, where $n_s(\vec{r}) = |\psi(\vec{r})|^2$, and $n_s(\vec{r})$ is the local super-electron density, analogous to the superfluid density described in the two-fluid model. The order parameter goes to 0, as $T \to T_c$.

In the context of the BCS theory, ψ can be interpreted as the wave function describing the position of the center of mass of a Cooper pair, although, at the time the GL theory was developed, the nature of the superconducting carriers was not yet determined.

GL expresses the free energy density as a series expansion for small values of ψ, and for T near Tc. A second-order phase transition occurs when the energy of the superconducting state becomes lower than that of the normal-conducting state. The superconductive transition is described by a minimum in the free energy density as a function of the order parameter.

In Figure 2.14 the variation of free energy density is plotted against the order parameter. The value Ψ_0 corresponds to the minimum of Δf in the superconducting state.

Increasing the temperature above Tc increases the free energy above a certain threshold to drive the material normal-conducting. Increased energy due to the presence of electric currents and magnetic fields in the material also increases the free energy. Above a certain threshold, there is a phase transition to the normal-conducting state. GL derives a thermodynamic critical magnetic field H_c, above which the superconducting state is no longer energetically favorable:

$$\frac{H_c^2}{8\pi} = f_n - f_s \tag{2.65}$$

where f_n and f_s are the free energies in the normal and superconducting states.

The thermodynamic critical field is determined by equating the magnetic energy per unit of volume, associated with the magnetic pressure acting on the superconductor surface due to the expelled field, to the variation in free energy $f(T) = f_n(T) - f_s(T)$ related to the superconducting transition. The temperature dependence of the thermodynamic critical field closely follows an empirical quadratic relation:

$$H_c(T) = H_c(0)\left[1 - \left(\frac{T}{T_c}\right)^2\right] \tag{2.66}$$

where $H_c(T = 0)$ is the critical field at a temperature of 0 °K. Below $H_c(T)$, the material is superconductive and in the Meissner state.

By expanding the free energy density f_s in powers of $|\psi|^2$ and $|\nabla\psi|^2$, GL minimized the free energy density by a variational procedure with respect to the spatial variations of $\psi(r)$ and the vector potential $\boldsymbol{A}(r)$. Here $\boldsymbol{A}(r)$ describes the currents (and hence magnetic field) in the superconductor. The GL equations are two nonlinear second-order differential equations that describe the distribution of the magnetic field and currents in the material.

Under the assumption that $\psi(r)$ is small near T_c and varies slowly in space, GL expresses the free energy density f_s as a power series in $\psi(r)$ with coefficients α and β as phenomenological constants, together with an effective charge (e) and mass (m^*) of the electrons. Using the GL equations, they derive the temperature-dependent penetration depth and the coherence length in terms of the α and β.

$$\lambda_{GL} = \left(\frac{m^* c^2 \beta}{4\pi e^2 \alpha}\right)^{1/2} \tag{2.67}$$

$$\xi_{GL} = \left(\frac{\hbar^2}{2m^* \alpha}\right)^{1/2} \tag{2.68}$$

GL also calculates the temperature dependence of the penetration depth and coherence length. Near T_c both quantities vary as $(1 - T/T_c)^{-1/2}$, prompting the introduction of the Ginzburg–Landau parameter κ_{GL}, where:

$$\kappa_{GL} = \frac{\lambda(T)}{\xi(T)} \tag{2.69}$$

κ_{GL} is temperature independent to first order. An exact calculation from microscopic theory gives a weak temperature dependence, with increasing for decreasing temperature T.

κ_{GL} is related to the purity of the superconductor and varies with the electron mean free path. The relationship between λ_{GL} and ξ_{GL} governs the electrodynamics of the superconductor. κ_{GL} separates superconductors into two categories. Type-I superconductors are those with $\kappa_{GL} < 1/\sqrt{2}$ and Type-II are those with $\kappa_{GL} > 1/\sqrt{2}$. Type-I superconductors exhibit the complete Meissner effect up to the thermodynamic critical field H_c.

The difference between the two types of superconductors can also be understood in terms of an interface energy associated with a superconducting/normal-conducting boundary. The order parameter ψ does not vanish abruptly at such an interface, but approaches zero gradually within a distance ξ. Hence there is a loss in condensation energy per unit of area in the superconducting phase due to the variation of ξ at the boundary, approximately equal to $\left(H_c^2/\mu_0\right)\xi$. This term increases the superconducting/normal-conducting interface energy. When the magnetic field penetrates the superconducting region within a penetration depth λ there is a reduction of condensation energy per unit area of the order of $\left(H_c^2/\mu_0\right)\lambda$. Therefore, the interface energy per unit of area is:

$$\alpha = \frac{H_c^2}{2\mu_0}(\xi - \lambda) \tag{2.70}$$

where $\delta = \xi - \lambda$ is positive for Type-I superconductors (since $\xi > \lambda$) and negative for Type-II.

2.3.9 Critical Fields

When Shubnikov discovered Type-II superconductor alloys [69], he observed that the superconducting state survives up to larger magnetic fields than expected. Abrikosov [70] explained the observations when he predicted the existence of the "mixed state" by demonstrating that the GL equations allow solutions with periodical variations of the order parameter. The mixed state is often called the "Abrikosov state." He found that a superconductor with $\kappa_{GL} > 1/\sqrt{2}$ admits magnetic flux at a lower critical field H_{c1}, in the form of quantized vortices (see below), without becoming entirely normal. It remains in a mixed state up to an upper critical field H_{c2}, where it becomes entirely normal. This behavior contrasts with a Type-I superconductor with $\kappa_{GL} < 1/\sqrt{2}$ which becomes normal at a well-defined critical field H_c. Abrikosov derived the following relations between H_{c1}, H_{c2}, and H_c:

$$H_{c1} = \frac{H_c}{\kappa}(\ln \kappa - 0.27) \tag{2.71}$$

$$H_{c2} = \sqrt{2}\kappa H_c \tag{2.72}$$

The generation of the mixed state, for which magnetic flux structures (vortices) are stable in the superconductor, is a direct consequence of the negative surface energy of Type-II superconductors. H_{c1} and H_{c2} also show a quadratic dependence on the temperature (similar to H_c) in first approximation.

In particular, Abrikisov considered the case when the externally applied magnetic field is slightly below $H_{c2}(T)$. Approximate solutions for the linearized GL equations revealed the presence of a periodic microscopic magnetic field distribution, transverse to the applied field. Abrikosov [71] showed that the mixed state is populated by stable magnetic flux structures called vortices, carrying a single flux quantum

$$\Phi_0 = \frac{h}{2e} = 2.07 \cdot 10^{-15}\ \text{Wb} \tag{2.73}$$

and organized in a bi-dimensional hexagonal lattice. Magnetic fluxoid quantization is a consequence of the quantum-mechanical coherence of the wave function in closed superconducting contours. It is similar to Bohr–Sommerfeld quantization of normal electron wave function in atoms.

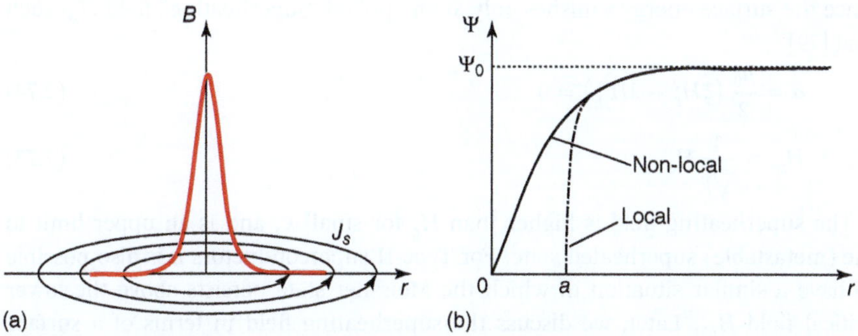

Figure 2.15 (a) Schematic representation of a vortex (b) Sketch of the order parameter variation as a function of the radius in the local and nonlocal models. Source: [68] Courtesy of M. Chechhin, Fermilab.

The supercurrent flows around the vortex to screen the magnetic field, which decays exponentially with constant λ. Independent of λ, the magnetic flux carried by a vortex is quantized and equal to Φ_0. Figure 2.15 gives a schematic representation of a vortex. The magnetic field profile is accompanied by a modulation of order parameter. In the local description [72], the vortex has normal conducting core with radius of the order of the coherence length ξ_0, with superconducting currents spinning outside the core and screening the magnetic flux confined within. In the nonlocal description (Figure 2.15b) the vortex represents a modulation of the order parameter that tends to zero at the center of the vortex, and approaches its finite value far from it [70].

Abrikosov predicted a periodic square array of thin filaments of magnetic flux in the mixed state. Subsequent solutions of the GL equations for magnetic fields just below the upper critical field H_{c2} (T), showed that a periodic triangular array of vortex lines has the lowest free energy of all possible periodic solutions, and hence is the most stable configuration [73–75]. In an isotropic Type-II superconductor, the vortex lines repel one another, to form an equilateral triangular lattice for energy minimization.

Decoration experiments that utilize small iron particles make the vortices visible through an electron microscope to provide confirmation of the triangular arrangement and verify the flux quantum of Eq. (2.73) [76]. The triangular vortex lattice was also imaged by scanning tunnel microscopy of $NbSe_2$ [77]. It is energetically more favorable to increase the number of vortices as one increases the magnetic field, rather than increasing the amount of flux in each vortex [78].

Returning to the boundary separating two phases, Type-I superconductors minimize the interphase boundary area, whereas Type-II superconductors maximize the boundary area, resulting in a maximum possible boundary area when the normal phase exists only in the cores of quantized vortices. In the actual process of magnetic phase transition from superconducting to normal state, a boundary between the two phases is created and so plays an important role, as discussed earlier.

In particular, the positive surface energy for Type-I superconductors suggests a possible persistence of Meissner state above the thermodynamic critical field, since the surface energy vanishes only at an applied "superheating" field H_{sh} such that [79]

$$\sigma = \frac{\mu_0}{2}\left(\xi H_c^2 - \lambda H_{sh}^2\right) = 0 \tag{2.74}$$

$$H_{sh} = \frac{1}{\sqrt{k}} H_c \tag{2.75}$$

The superheating field is higher than H_c for small κ, and is an upper limit to the (metastable) superheated state. For Type-II superconductors it is also possible to have a similar situation in which the Meissner state persists above the lower critical field H_{c1}. Later, we discuss the superheating field in terms of a surface barrier field opposing the entrance of magnetic flux at the surface. For both types of superconductors, the existence of the superheated state was first theoretically predicted by Ginzburg [80] and later observed experimentally for tin by Garfunkel and Serin [81].

From the GL theory, Saint-James and de Gennes [82] showed that in a magnetic field parallel to the surface, superconductivity will nucleate in a surface layer of thickness ξ at a field $H_{c3} = 1.695\ H_{c2}$, higher the field at which nucleation occurs in the volume of the material.

The lower critical field of a type-II superconductor is the field at which a single magnetic flux quanta (vortex) is stable in the bulk of the superconductor. It is also the lowest applied field at which the free energy is lowered by having a vortex in the bulk rather than outside the superconductor. Even though it is energetically favorable for flux to be inside the superconductor at H_{c1}, it does not penetrate at this field, unless there are defects present for nucleation. Just as pure water remains in the liquid state above its boiling point, a superconductor can remain in the flux-free state above H_{c1}. An energy barrier, called Bean–Livingston surface energy barrier, allows the superconductor to remain in this metastable state up to at most H_{sh}, the superheating field [83, 84]. The barrier can become lower at defects.

The superheating field corresponds to the highest field for which the Ginzburg–Landau free energy still possesses a local minimum as a function of the order parameter. In other words, H_{sh} corresponds to the highest field at which the superconductive Meissner state exists in a metastable condition above the lower critical field. The superheating field corresponds to the highest magnetic field for which there is an acceptable solution of the Ginzburg–Landau equations.

To emphasize the importance of H_{sh}, forming vortices on the surface and pushing them into the bulk requires a finite activation energy. As a result, if a superconductor is exposed to a magnetic field > H_{c1}, it will stay in a flux-free "superheated" metastable state until a higher "superheating field" H_{sh} at which the excess free energy is sufficient to form vortices and push them across the Bean–Livingston surface energy barrier.

The energy barrier is responsible for the occurrence of the metastable Meissner state. It represents the energy cost for a vortex to penetrate the surface of a type-II superconductor to reach the bulk. Superconductivity cannot survive above $H = H_{sh}$ because the screening surface current reaches the de-pairing value $J_d = n_s e\Delta/p_F$.

When the external field increases above H_{sh}, flux lines enter the superconductor, creating normal conducting volume, and sharply increasing rf losses. There is very large heat dissipation due to the motion of the vortex cores through the material [85], leading to a quench. Therefore, in a perfect cavity with no defects, the fundamental limit is H_{sh}.

However, in reality, defects in the material may alter the superconducting properties at the defect location, lowering the energy cost required for a flux line to enter at the defect. This can lead to vortex penetration at fields lower than H_{sh}. The size of the vortex cores sets the length scale at which defects become possible nucleation sites for flux penetration. Such defects could be understood as "magnetic defects." If normal conducting defects are present, excessive heating at the defect and its vicinity leads to a thermal runaway-induced quench well below H_{sh}. These defects are resistive defects that cause heating prior to the quench. A temperature mapping system can detect the heating to locate the defect and the corresponding quench.

Ginzburg [80] first analyzed the superheating field by solving the GL equations for the one-dimensional case where a superconductor occupies a half-space. He obtained the superheating field as a function of the GL parameter. He also examined two limiting behaviors for small and large κ and found the superheating field H_{sh} in terms of bulk critical field. His results improved by Brewer [86] are given below.

$$\text{For } \kappa \ll 1, H_{sh} \simeq \frac{0.84}{\sqrt{k}} H_c, \tag{2.76}$$

$$\text{For } \kappa \gg 1, H_{sh} \simeq H_c \tag{2.77}$$

Matricon and St. James [83]numerically treated the same problem later for better accuracy. Their results are shown in Figure 2.16 along with other H_{sh} calculations for comparison.

To determine the values of H_{c1} and H_{sh} more accurately without using a one-dimensional model, it is necessary to numerically solve the GL equations. Using freshly developed numerical codes Checchin [68] at Fermilab calculated the normalized lower critical field $h_{c1} = H_{c1}/(\sqrt{2}H_c)$ and the normalized superheating field $h_{sh} = H_{sh}/(\sqrt{2}H_c)$ for values of GL parameter in the range $0.2 \leq \kappa \leq 3$. In Figure 2.16, the numerical values he obtained are compared with other numerical calculations found in literature [83, 87–90]. The results are in good agreement with previous calculations. The numerical solutions can be fitted with two analytic formulas within the range $0.2 \leq \kappa \leq 3$:

$$h_{c1} \simeq 0.58\kappa^{-0.57}, \qquad h_{sh} \simeq 0.72 + 0.18\kappa^{-1} + 0.004\kappa^{-2} \tag{2.78}$$

H_{c1} has also been calculated numerically from the GL equations to cover a larger range for κ_{GL}, as shown in Table 2.1 [88].

Another approximation is useful for superconductors with $\kappa \gg 1$:

$$\frac{H_{c1}}{H_c} = \frac{\ln(\kappa)}{\sqrt{2\kappa}} \tag{2.79}$$

The superheating field has also been calculated numerically for $\kappa > 1$ and is very well approximated by the following, when $T \approx T_c$ [90]:

$$\frac{H_{sh}}{H_c} \approx \frac{\sqrt{5}}{3} + \frac{0.5448}{\sqrt{\kappa}} \tag{2.80}$$

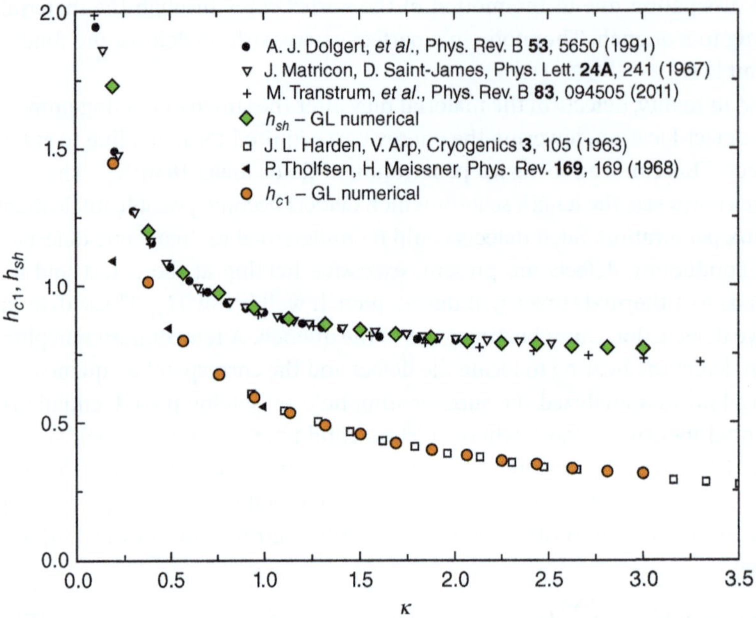

Figure 2.16 Numerically computed values from the GL equations for $h_{c1} = H_{c1}/(\sqrt{2}H_c)$ and $h_{sh} = H_{sh}/(\sqrt{2}H_c)$ as a function of the GL parameter. Source: Adapted from [68, 83, 87–90]. Courtesy of M. Checchin, Fermilab.

Table 2.1 B_{c1} versus κ_{GL} computed from GL theory in Harden and Arp [88].

κ_{GL}	B_{c1}/B_c via GL
$2^{-1/2}$	1.00
1	0.817
2	0.547
5	0.315
10	0.201
20	0.124
50	0.0622

Source: Adapted from [88].

The value of κ for the rf layer of a Nb cavity depends on the purity and the method of preparation. The value of $\kappa = 1$ is in common use, as determined by [96]. For pure Nb the value of $\kappa = 1.43$ has been determined [91]. For $\kappa = 1$, Eq. 2.80 gives: $H_{sh} = 1.29\ H_c$, which is close (1.278) to the numerical solution of GL equations, as from Eq. (2.78). This leads to a value of 260 mT for H_{sh}, which would translate to near 60 MV/m for E_{acc} for a TESLA shaped structure.

One source gives $\kappa = 1.43$ for pure Nb [91], for which $H_{sh} = 1.2\ H_c$, the value in customary use for Nb in the SRF community, which would translate to 240 mT or 56 MV/m.

2.3.10 Comparison Between Ginzburg–Landau and BCS

Gor'kov realized [92] that the GL theory is equivalent to the BCS theory around the critical temperature. He derived the GL theory as a limiting case of the BCS microscopic theory. Gor'kov proved that the fundamental quantities of the two theories were related so that ψ can be thought of as the Cooper pair wavefunction. He calculated the GL coherence length and the penetration depth from BCS theory using λ_L, ξ_0, and the mean free path [93] as:

$$\xi_{GL} = 0.739\left(\xi_0^{-2} + \frac{0.882}{\xi_0 \ell}\right)^{-1/2}\left(1 - \frac{T}{T_c}\right)^{-1/2} \tag{2.81}$$

$$\lambda_{GL} = \frac{1}{\sqrt{2}}\lambda_L\left(1 + 0.882\frac{\xi_0}{\ell}\right)^{1/2}\left(1 - \frac{T}{T_c}\right)^{-1/2} \tag{2.82}$$

The microscopic BCS theory gives H_c and H_{c2} as well as an approximation for H_{c1} [21, 94] in terms of penetration depth, coherence length, flux quantum, and penetration depth:

$$H_c = \frac{\Phi_0}{2\sqrt{2}\pi\lambda_{GL}\xi_{GL}} \tag{2.83}$$

$$H_{c1} = \frac{\Phi_0}{4\pi\lambda_{GL}^2}(\ln\kappa_{GL} + 0.5) \tag{2.84}$$

$$H_{c2} = \frac{\Phi_0}{2\pi\xi_{GL}^2} \tag{2.85}$$

Figure 2.17 shows the critical fields B_{c1}, B_c, B_{sh}, and B_{c2} of niobium versus mean free path. Low mean free path (high k_{GL}) leads to significantly lower B_{c1} and higher B_{c2}, more relevant for superconducting magnets than for cavities. B_c is relatively unchanged from changes in mfp, and B_{sh} decreases slightly with lower mean free paths. For this calculation $\xi_0 = 38$ nm and $\lambda_L = 39$ nm were used.

Before ending this chapter, Table 2.2 summarizes the key parameters for pure Nb

2.3.11 Derivation of R_s and X_s [98]

Review the definitions from Section 2.3.3

$$Z_s = \sqrt{\frac{i\omega\mu_0}{\sigma_n - i\sigma_s}},\ \sigma_n = \frac{n_n e^2\tau}{m},\ \sigma_s = \frac{1}{\omega\mu_0\lambda_L^2} = \frac{n_s e^2}{m\omega},\ \lambda_L = \sqrt{\frac{m}{n_s e^2\mu_0}}$$

Rationalize the denominator of Z_s^2

$$Z_s^2 = \frac{i\omega\mu_0}{\sigma_n - i\sigma_s} = \frac{i\omega\mu_0}{\sigma_n - i\sigma_s}\cdot\frac{\sigma_n + i\sigma_s}{\sigma_n + i\sigma_s} = \frac{\omega\mu_0(\sigma_n i - \sigma_s)}{\sigma_n^2 + \sigma_s^2}$$

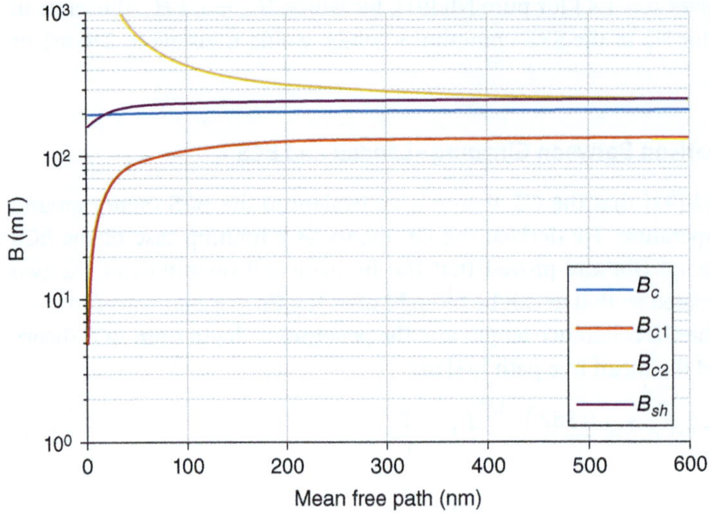

Figure 2.17 GL predicted critical fields of Nb versus mean free path. Source: [95] Courtesy of D. Gonnella, Cornell University.

Table 2.2 Summary of material parameters for Nb typically used for SRF Nb cavities.

Variable	Name	Clean Nb value[a]
T_c	Critical temperature	9.2 K
$\Delta/k_B T_c$	1/2 energy gap	1.81
λ_L	London penetration depth	39 nm
ξ_0	Clean coherence length	38 nm
ℓ	Mean free path	>1 um
k_{GL}	GL parameter	~1, [91]1.43
H_c	Thermodynamic critical field	200 mT
H_{c1}	Lower critical field	165 mT (Eq. [2.78]) 173 [91]
H_{sh}	Superheating field	2400 mT (Eq. [2.78])

a) Clean niobium means high RRR niobium with large mean free path.
Source: Data from Finnemore et al. [91], Maxfield and McLean [96], and Novotny and Meincke [97].

Translate the numerator from Cartesian into polar coordinates

$$-\sigma_s + \sigma_n i = re^{2i\theta}$$

where

$$r = \sqrt{\sigma_s^2 + \sigma_n^2}, r\cos(2\theta) = -\sigma_s \cos(2\theta) = -\frac{\sigma_s}{r}$$

$$Z_s^2 = \frac{\omega\mu_0 re^{2i\theta}}{r^2} = \frac{\omega\mu_0}{r}e^{2i\theta}$$

Taking the square root of both sides and using Euler's formula yields

$$Z_s = \sqrt{\frac{\omega\mu_0}{r}e^{2i\theta}} = \sqrt{\frac{\omega\mu_0}{r}}e^{i\theta} = \sqrt{\frac{\omega\mu_0}{r}}(\cos\theta + i\sin\theta)$$

We find the values of cos and sin using half-angle formulas:

$$\cos\theta = \pm\sqrt{\frac{1+\cos(2\theta)}{2}} = \pm\sqrt{\frac{1-\sigma_s/r}{2}} = \pm\sqrt{\frac{r-\sigma_s}{2r}}$$

$$\sin\theta = \pm\sqrt{\frac{1-\cos(2\theta)}{2}} = \pm\sqrt{\frac{1+\sigma_s/r}{2}} = \pm\sqrt{\frac{r+\sigma_s}{2r}}$$

Since σ_s, $\sigma_n \geq 0$, 2θ is in the second quadrant, θ is in the first quadrant, and thus $\cos\theta$ and $\sin\theta$ are nonnegative. We can drop the $\pm$ above.

Substituting into Z_s above yields

$$Z_s = \sqrt{\frac{\omega\mu_0}{r}}\left(\sqrt{\frac{r-\sigma_s}{2r}} + i\sqrt{\frac{r+\sigma_s}{2r}}\right) = \sqrt{\frac{\omega\mu_0}{r}\cdot\frac{r-\sigma_s}{2r}} + i\sqrt{\frac{\omega\mu_0}{r}\cdot\frac{r+\sigma_s}{2r}}$$

Which separates Z_s into real and imaginary parts as desired for

$$Z_s = R_s + iX_s$$

We use the approximation $\sigma_n \ll \sigma_s$, which gives $r = \sqrt{\sigma_n^2 + \sigma_s^2} \approx \sqrt{\sigma_s^2} = \sigma_s$

We will use the above approximation several times, $r \approx \sigma_s$

$$X_s = \sqrt{\frac{\omega\mu_0}{r}\cdot\frac{r+\sigma_s}{2r}} \approx \sqrt{\frac{\omega\mu_0}{\sigma_s}\cdot\frac{\sigma_s+\sigma_s}{2\sigma_s}} = \sqrt{\frac{\omega\mu_0}{\sigma_s}} = \sqrt{\omega\mu_0\left(\omega\mu_0\lambda_L^2\right)} = \omega\mu_0\lambda_L$$

As needed, $X_s = \omega\mu_0\lambda_L$

$$R_s = \sqrt{\frac{\omega\mu_0}{r}\cdot\frac{r-\sigma_s}{2r}}$$

If we reuse the approximation

$$r = \sqrt{\sigma_n^2 + \sigma_s^2} \approx \sqrt{\sigma_s^2} = \sigma_s$$

Then $R_s = \sqrt{\frac{\omega\mu_0}{r}\cdot\frac{r-\sigma_s}{2r}} = 0$

To obtain a more accurate approximation for R_s we use a Taylor approximation for

$$r = \sqrt{\sigma_n^2 + \sigma_s^2}$$

$$r = \sqrt{\sigma_s^2 + \sigma_n^2} \approx \sigma_s + \frac{\sigma_n^2}{2\sigma_s}$$

From which we obtain

$r - \sigma_s \approx \frac{\sigma_n^2}{2\sigma_s}$ which we substitute in the expression for

$$R_s = \sqrt{\frac{\omega\mu_0}{r}\cdot\frac{r-\sigma_s}{2r}}$$

$$R_s = \sqrt{\frac{\omega\mu_0}{r}\cdot\frac{r-\sigma_s}{2r}} = \sqrt{\frac{\omega\mu_0(r-\sigma_s)}{2r^2}} \approx \sqrt{\frac{\omega\mu_0\sigma_n^2}{4\sigma_s r^2}} = \frac{1}{2}\sigma_n\sqrt{\frac{\omega\mu_0}{\sigma_s r^2}}$$

Using $r \approx \sigma_s$ again,

$$R_s \approx \frac{1}{2}\sigma_n\sqrt{\frac{\omega\mu_0}{\sigma_s r^2}} \approx \frac{1}{2}\sigma_n\sqrt{\frac{\omega\mu_0}{\sigma_s^3}} = \frac{1}{2}\sigma_n\sqrt{\omega\mu_0\left(\omega\mu_0\lambda_L^2\right)^3} = \frac{1}{2}\sigma_n\sqrt{\omega^4\mu_0^4\lambda_L^6}$$

Finally, we reach the desired expressions:

$$R_s \approx \frac{1}{2}\sigma_n\omega^2\mu_0^2\lambda_L^3$$

$$X_s = \omega\mu_0\lambda_L$$

Part II

High Q Frontier: Performance Advances and Understanding

3

Nitrogen-Doping

3.1 Introduction

Nitrogen doping (N-doping) has been a surprising and exciting discovery, followed by substantial developments and applications. The phenomenon was discovered by Grassellino in 2012 at Fermilab [99]. N-doping is a surface treatment for niobium cavities to absorb nitrogen atoms as interstitial impurities into the niobium lattice. This treatment gives rise to an unexpected field-dependent Q *increase* at medium fields (15–25 MV/m, 60–100 mT), known as the "anti-Q-slope" or "positive Q-slope." The surprising aspect of the discovery is that standard cavity treatments normally show Q *decreases* with increasing field called the medium field Q-slope (MFQS). Interstitial N lowers the mean free path of the RF penetration layer of the niobium (Section 3.5). N-doping also lowers the residual resistance (Section 3.7). The N-doping process has proved to be highly reproducible, and is widely used for accelerators planning cw operation to take advantage of the higher Q and correspondingly lower cryogenic demands. In this chapter we attempt to cover and synthesize a very wide range of experimental and theoretical efforts.

Q enhancements of factors of 3–6 for 1.3 GHz cavities help decrease the cryogenic cost of accelerators operating in the continuous wave (cw) mode. The linear coherent light source (LCLS) at SLAC was first to embrace N-doping technology to incorporate into the planning and realization of the new accelerator LCLS-II [100], and later for LCLS-II-HE [101]. The LCLS-II baseline design required 1.3 GHz 9-cell TESLA-shaped cavities [25] with an average intrinsic quality factor $Q_0 \sim 2.75 \times 10^{10}$ at an accelerating gradient of 16 MV/m at 2.0 K. N-doping will also be used for a potential cw upgrade of the European XFEL [102]. Much of the technology for 1.3 GHz single and nine cells was developed at Fermilab, Jefferson Lab (JLab), and Cornell, with application to LCLS-II in mind.

A significant part of the research on N-doping focused on developing doping "recipes" that yield the best performance in terms of high Q, maximum quench field, topics to be covered in this chapter, and the lowest sensitivity of Q to trapped dc magnetic flux (Chapter 5). A variety of doping regimes have been explored at frequencies 1.3 GHz and 650 MHz, in the context of development for the LCLS-II and PIP-II projects [103, 104]. Many fundamental scientific questions have been addressed to

Superconducting Radiofrequency Technology for Accelerators: State of the Art and Emerging Trends, First Edition. Hasan Padamsee.

understand why the process works to give higher Q's, the physics of the anti-Q-slope, and its connections to the mfp. But many questions about the anti-Q-slope remain.

3.2 N-Doping Discovery

The N-doping technique was discovered during an attempt to form higher T_c niobium nitride (NbN) on the inner surface of the SRF 1.3 GHz cavity by reacting the bulk niobium cavity with nitrogen at 800 °C in a high-temperature UHV furnace [99]. The process resulted in the formation of lossy, metallic NbN phases on the surface, with a low Q (10^7–10^8) for the cavity. After removing five or more μm of material via electropolishing (EP), the cavity performances at 2 K (Figure 3.1) showed the surprising increase in Q with accelerating gradient. The removal of the NbN layer by EP revealed the underlying extraordinary results of doping Nb with interstitial N.

Figure 3.1 shows Q *versus* E results of several cavities treated with different doping and post-doping EP parameters [99]. For each treatment, the doping and EP parameters are shown in the accompanying table (Table 3.1). Note that cavities doped at 1000 °C for 1 hour needed higher (80 μm) surface removal after doping, whereas cavities doped for 10 minutes needed smaller (10 μm) surface removal. The one cavity doped at 800 °C for 10 minutes needed a removal of 5 μm [99]. For comparison, Figure 3.1 also shows the performance of one cavity prepared with EP plus 120 °C baking, the standard preparation technique evolved over the previous decade for high gradients (Section 2.2.2) and applied extensively for EXFEL [105]. Recall also that SRF cavities prepared with this standard preparation technique show a MFQS

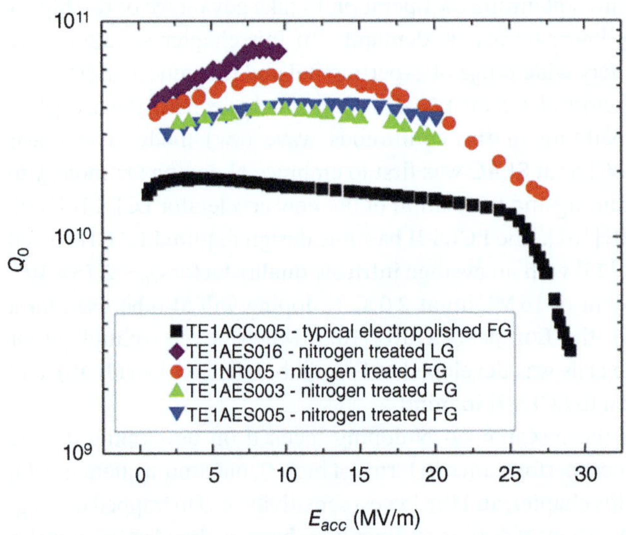

Figure 3.1 First discovery results of N-doping. Source: [99]/A. Grassellino/with permission of IOP Publishing. Several Q *versus* E cavity performance curves shown after N-doping, along with parameters for doping.

Table 3.1 The amount (in bold) of material removed after doping, and before RF test [99].

CAVITY ID	Treatment	Subsequent cumulative material removal via EP for each RF test [μm]
TE1AES016	1000 °C 1 hour with 2×10^{-2} Torr p.p. nitrogen	**80**
TE1AES003	1000 °C 10 min with 2×10^{-2} Torr p.p. nitrogen	**10**,60
TE1AES005	1000 °C 1 hour with 2×10^{-2} Torr p.p. nitrogen	20, 40, **80**
TE1NR005	800 °C 3 hours in UHV. followed by 800 °C 10 min with -2×10^{-2} Torr p.p. nitrogen	**5**,15

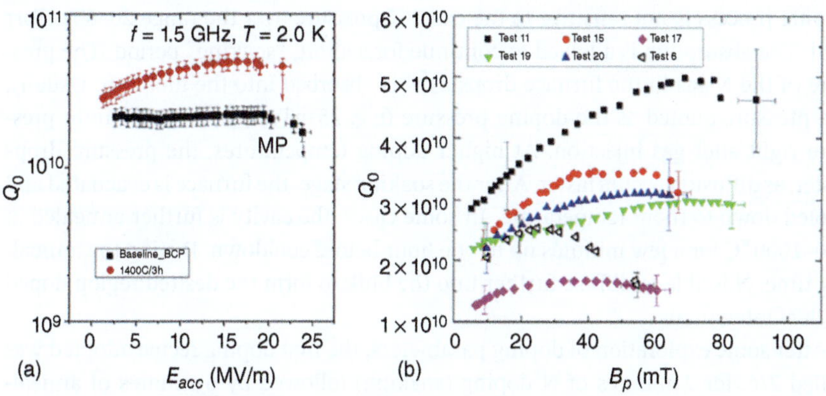

Figure 3.2 First discovery of titanium doping. Performance of a 1.5 GHz single-cell cavity made from large grain Nb. Source: [106]/with permission of Pashupati Dhakal. The cavity was treated first with BCP, then heat treated at 1400 °C for 3 hours. (b) Ti-doping results from several cavities. Source: [109]/G. Ciovati et al./with permission of IOP Publishing.

between 5 and 25 MV/m, as seen in Figure 3.1 in the curve with the lowest Q. Here, the Q drops slowly with field. N-doping is one of the first techniques found to replace the MFQS (falling Q) with the opposite behavior of the anti-Q-slope (rising Q).

In the same year, and a few months earlier than the discovery of N-doping, another serendipitous discovery of enhanced Q with positive slope (Figure 3.2a) was realized by Dhakal at JLab by annealing a 1500 MHz Nb cavity at high temperature (~1400 °C), but without any post-annealing chemistry [106, 107]. Later it was established that Ti impurities from the end flanges were accidentally introduced during the annealing, leading to about 1% at Ti within the first 2 μm. It is suspected that these impurities play a similar role as interstitial N introduced via N-doping.

Ti-doping was repeated on several 1500 MHz Nb cavities [109] by annealing in a high-vacuum induction furnace above 1200 °C for a few hours, resulting in a Ti concentration of ~1 at% within the top ~100 nm. The source of Ti again was from the Ti45Nb cavity flanges. Results are shown in Figure 3.2b.

Between the two methods of obtaining higher *Q*s, N-doping became the popular choice due to the ease of introducing the interstitial N impurity. Also, N diffusion takes place at a lower temperature of 800 °C compared to Ti diffusion, which needs temperatures above 1200 °C to achieve the effective diffusion length. Most of this chapter is devoted to N-doping since Ti-doping has not been explored very much.

N-doping is generally performed just after the 800 °C baking, normally carried out to degas dissolved hydrogen from the bulk Nb. If this H is not removed it leads to a falling *Q* at low fields (*Q*-disease) [4] (Section 2.2.2 and 6.11). Typically, the temperature of the furnace is ramped up to 800–1000 °C and held for few hours (~3 hours). At the end of the H degas cycle, high purity N gas is injected through a particle filter, with a partial pressure ~ 25 mTorr for a soaking time (2–60 minutes). At this stage, niobium has almost no surface oxide (which normally serves as an absorption barrier) so it readily absorbs the nitrogen into its surface. Studies show that the doping process is not sensitive to the exact N pressure over the range 20–60 mTorr [95]. The absorption is allowed to continue for a short, "soaking" period. The pressure of the N gas in the furnace drops as N is absorbed into the niobium. Usually, the pressure quoted as the doping pressure (e.g. 25 mTorr) is the maximum pressure right after gas injection. At higher doping temperatures, the pressure drops faster, as diffusion occurs faster. After the soaking stage, the furnace is evacuated and cooled down to room temperature. In some cases, the cavity is further annealed at 800–1000 °C for a few minutes up to one hour before cooldown. During the annealing time, N is able to diffuse further into the bulk to form the desired region doped with N interstitials.

After some exploration of doping parameters, the first doping recipe adopted was called 2/6, for 2 minutes of N doping (soaking) followed by 6 minutes of annealing, followed by 5–10 μm of EP to remove the high-loss nitride layer [110–112]. Doping recipes are typically named to indicate the doping time/annealing time, followed by statement about the amount of EP.

A typical N-doping temperature and pressure profile for a doping run is shown in Figure 3.3 [108]. A clean high-vacuum furnace is necessary to avoid contamination by hydrocarbons that can be detrimental to the cavity's performance. All cavities are loosely capped with niobium foil on each flange during the heat treatment to prevent cavity contamination. The small openings between the loose caps and the cavity flanges allow pumping.

3.3 Surface Nitride

The N-doping process leaves a thin surface layer (1–3 μm) of lossy niobium nitride [113]. The EP step, which follows the furnace treatment, removes this layer and exposes the doped layer underneath. SEM surface image with EDX analysis shows triangular or star-shaped structures (Figure 3.4a) identified as normal-conducting nitride, mainly the β-Nb_2N phase [113] consistent with the phase diagram.

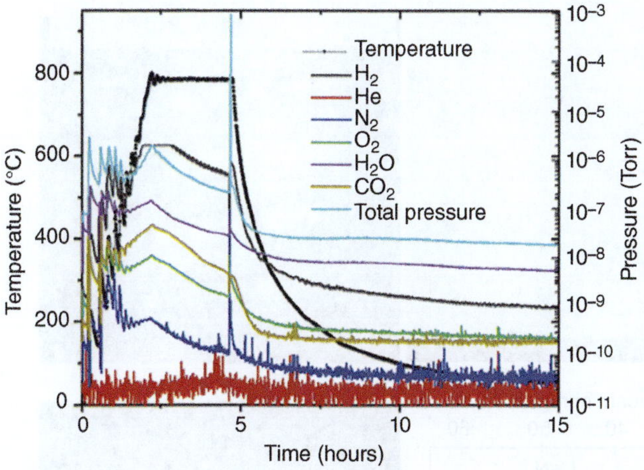

Figure 3.3 Residual gas analyzer pressure profile and temperature profile during nitrogen doping at JLab. Nitrogen was injected into the furnace at 800 °C for 2 minutes at a partial pressure of 2×10^{-3} Torr. The cavity was further annealed in UHV for 6 minutes before cooldown. Source: [108]/P. Dhakal with permission of Elsevier.

The Nb_2N phase is identified by TEM analysis (Figure 3.4b) as well as XPS measurements [114–116]. According to the equilibrium phase diagram of Nb–N, (Figure 3.4c) in the low N concentration regime below 10 at% N, the main α-phase is that of interstitially dissolved N in Nb. As the N concentration increases, a two-phase mixture forms of α-Nb (N) and a stoichiometric β-Nb_2N [117]. After 5 μm Nb removal by EP, the clean TEM image (Figure 3.4d) shows the absence of any β-Nb_2N phase.

3.4 Interstitial N

Secondary ion mass spectroscopy (SIMS) measures the diffusion depth of nitrogen. SIMS measures the elemental composition of a material by sputtering the surface of a metal with a focused ion beam and analyzing the ejected secondary ions. Figure 3.5 shows SIMS results for two samples, one doped at 800 °C for 20 minutes followed by 30 minutes annealing, and the other doped at 990 °C for 5 minutes. Results show that the diffusion depth of nitrogen is ten to a hundred micrometers [119–122] depending on the temperature. The nitride layer is about 1–3 μm thick. The interstitial concentrations range from 10^{19} to 10^{20} atom/cm^3, corresponding to a level of 200–2000 at ppm, 0.02–0.2 at%, equivalent to 30–300 wt ppm.

As one way to calibrate the level of the N impurity, if a pure Nb sample, starting with RRR = 300 is loaded with 300 wt ppm of N, the RRR would drop to about 15, which is quite heavy doping. According to the N profile from Cornell SIMS data of Figure 3.5, cavities doped for longer times and higher temperatures require more material removal via EP to reach optimal RF performance compared with cavities that are doped for shorter times. The levels of N concentration for 800 °C N-doping

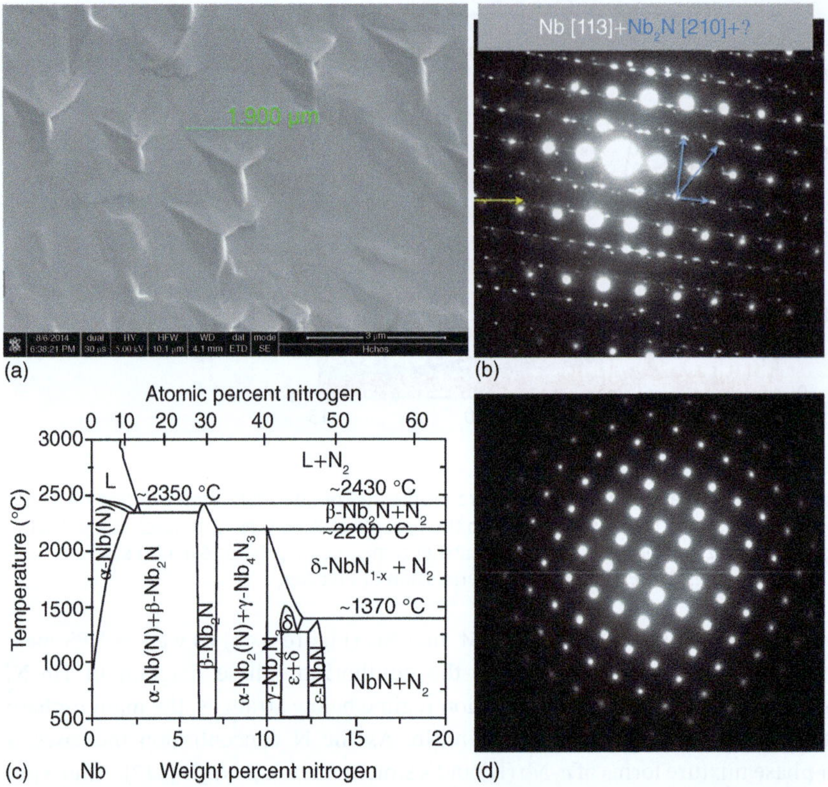

Figure 3.4 (a) SEM image of Nb surface after N-doping with 2 N6 recipe at 800 °C shows triangular Nb_2N regions [113] Y. Trenikhina et al., Fermilab/JACoW/CC BY 3.0 (b) TEM image showing Nb_2N phase. [113]/Y. Trenikhina et al./JACoW/CC BY 3.0. (c) Phase diagram of Nb–N system. [118] W. Mayr et al. with permission from Elsevier via CCC. (d) TEM image of clean Nb after 5 µm EP, showing absence of Nb_2N. [113]/ Y. Trenikhina et al./JACoW/CC BY 3.0.

from Cornell SIMS data are in agreement with similar SIMS concentration measurements after doping at other labs.

Gonnella from Cornell used a diffusion model [119] to include the formation of the nitride layer and the doped layer, along with the effect of final annealing. The concentration of nitrogen just below the nitride layer only depends on the temperature. Concentration profiles measured on N-doped samples typically follow the simulation results using Fick's laws [121]. The maximum concentration does not change with time. The N concentration observed depends on the temperature of diffusion [123], and the duration of N-doping [124].

We mentioned that for the range of doping pressures used (tens of mTorr of N), the concentration of N in niobium is independent of the pressure. This is due to the process being diffusion limited rather than source limited. The nitride layer acts as the source. The diffusion equations are solved numerically. For reference, the diffusivity of N in Nb at 800 C is $1.34 \times 10^{-13} m^2/s$. The temperature dependence of the

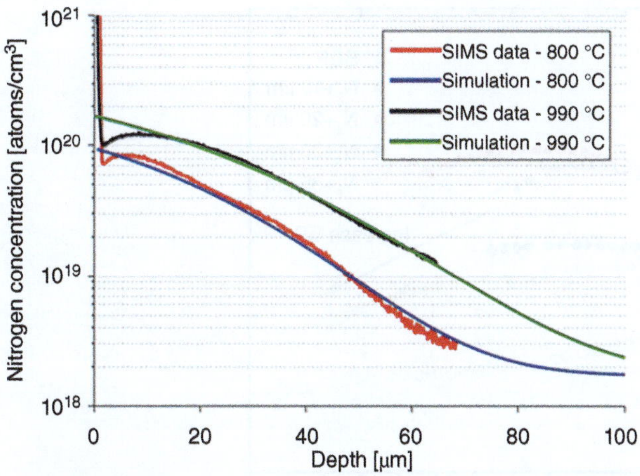

Figure 3.5 Nitrogen concentration in two N-doped samples measured with SIMS. The red curve was from a sample doped at 800 °C while the black was from a sample doped at 990 °C. Also shown are the predictions from the diffusion simulation. Note that 10^{20} atoms/cc correspond to 2000 at ppm The atomic density of pure Nb is 5.4×10^{22} atoms/cm^3. Source: [95] Courtesy of D. Gonnella, Cornell University.

diffusion constant is $4.8 \times 10^{-9} \exp(-1.12 \times 10^4/T)$. The change of the pressure in the furnace can also be determined by calculating the amount of nitrogen uptake by the cavity and found to agree with measurements [95].

There is good agreement between measured and simulated results except for the large N spike in the nitride layer, and the small dip just under the nitride layer. Of course, the non-monotonic behavior of N presence under the NbN layer cannot be explained by a linear diffusion model. But adding a sink term for N depletion near the interface helps account for the dip due to post-dope annealing to improve the prediction around the nitride–metal interface [125].

To obtain optimal RF performance after doping, it is necessary to remove by EP not only the Nb_2N layer but also an appropriate amount of doped Nb so that the final RF layer has the optimal concentration of N. Figure 3.6 shows cavity performance results upon removing increasing amounts of Nb after doping. The level of N concentration affects R_{BCS}, R_0, the anti-Q-slope strength, and the quench field. At very low fields, the Q is 50% higher due to the lower electron mean free path from a large amount of interstitial N, as expected from the BCS theory (Section 2.4.5). 10 μm material removal after doping results in a strong anti-Q-slope, but does not give the best Q, or the best quench field. At this stage, we estimate that the material is in the "over-doped" regime with about 8×10^{19} atoms/cc of N concentration. Note that the diffusion length for 800 °C for 20 minutes doping plus 30 minutes anneal is about 20 μm (for a total 50 minutes diffusion). Going further into the Nb, 20 μm removal by EP is appropriate to give the highest Q (4×10^{10}), highest quench field (105 mT), as well as a strong anti-Q-slope. At this depth, the N concentration is 5×10^{19} atom/cc (corresponding to local RRR = 30). Another 10 μm material removal leads to a lower Q at medium fields, as the N concentration in the RF layer drops below the optimum.

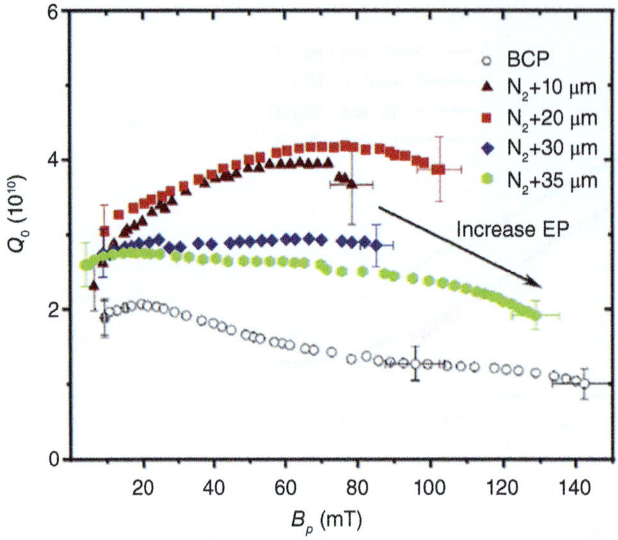

Figure 3.6 Changes in performance after successive EP on a single cell cavity after N-doping [108] at 800 °C for 20 minutes at 25 mTor followed by annealing for 30 minutes. The accelerating gradient increases with more EP removal, but the overall Q decreases. Source: [126]/P. Dhakal et al./JACoW/CC BY 3.0.

This is the "under-doped" regime. Eventually, after 35 μm removal (when the N concentration falls to between 1 and 2×10^{19} atoms/cc), the Q *versus* E performance moves toward the baseline before N-doping. Remarkably, even at the low N concentration (2×10^{19}), the Q at medium fields shows improvement over the baseline result with standard BCP. Thus even a small quantity of residual N has a beneficial effect on the Q.

The anti-Q-slope is strongest in the over-doped regime (brown curve in Figure 3.6), where the N concentration is close to the peak, but the quench field is lowest (Section 3.11). After 20 μm removal (red curve), the N concentration has fallen by about 60% without significantly changing the strength of the anti-Q-slope, but the quench field has significantly increased. A N concentration near 5×10^{19} seems effective for high Q gains and appears to be in the optimal doping regime.

3.5 Electron Mean Free Path Dependence

An important material property affected by doping is the electron mean free path (mfp). As discussed in Section 3.5 the mfp (l) can be used to calculate the "dirty" penetration depth and coherence length from Eqs. (2.81) and (2.82). The mfp for the cavity can be determined experimentally in a number of different ways. It is related to the normal conducting electrical resistivity (ρ) via the ρl product. For Nb, [127].

$$\rho l = 0.37 \times 10^{-15} \Omega \cdot \text{m}^2 \tag{3.1}$$

(Note: Other values of the product have been given [128])

The N impurity concentration determines the resistivity [5] via

$$\rho = [5.2 \times 10^{-8}\Omega \cdot \text{m}] \cdot c \tag{3.2}$$

where c is the N impurity concentration in at %. Thus the nitrogen concentration can be used directly to calculate the mean free path.

For a second way to determine mfp, the BCS resistance depends on the mfp as discussed in Section 2.4.5. Surface resistance results versus temperature from cavity Q measurements can be fit with Halbritter's BCS resistance calculations using the SRIMP program to yield the mfp, and the reduced energy gap.

In a third method, also discussed in Section 2.4.5, by closely monitoring the cavity frequency near T_c, the frequency shift Δf *versus* T can also be used to extract mfp by fitting the data with SRIMP.

Figure 3.7a shows the mfp calculated (smooth curve) from the N concentration, as discussed above, as well as discrete mfp data points calculated self-consistently from the change in penetration depth near T_c, and from the BCS resistance results [95]. Figure 3.7b shows the calculated BCS resistance versus mfp along with the data for the doped cavities. Note the minimum obtained in R_{BCS} near $\ell = \xi_0/2$ (20 nm) as expected from BCS theory (Section 2.4.5). The Ti-doped cavity from JLab has a mfp of about 75 nm, extracted from the BCS resistance change with temperature (Figure 3.7c). The cavity falls in the lighter doped region, with the mfp > the BCS resistance minimum.

The doping treatments explored by Gonnella in the Cornell study cover a wide range of mfp on either side of the minimum, from heavily doped cavities with mfp of 5 nm to lightly doped cavities with mfp of 200 nm. The study found a strong dependence of the anti-Q-slope strength on the doping level, with stronger Q-slopes at lower mfp, as shown in Figure 3.8a. The anti-Q-slope has an approximate logarithmic dependence (not shown) on the peak magnetic field in the medium field region: 5–20 MV/m [129, 130]. Figure 3.8 b shows how stronger doping (with lower mean free path) leads to a stronger anti-Q slope, where the dependence on mean free path is roughly logarithmic. As we will discuss in Section 3.11, stronger doping also leads to lower quench fields, so the higher Q's are not useful.

3.5.1 LE-μSR Measurements of Mean Free path

The mfp can also be estimated from low energy muon spin rotation (LE-μSR) measurements [131, 132]. (For a discussion of LE-μSR, see Section 6.4.3) The technique is useful for relative comparisons of penetration depths (and mfps) for different treatments, rather than for accurate measurements of absolute penetrations depths. Figure 3.9 shows penetration depth data via LE-μSR on several samples cut out from cavities treated by a variety of commonly used methods, such as EP, BCP, EP/BCP plus 120C bake, and N-doping. The heavier doped treatment studied here has the surface mfp of about 40 nm with penetration depth about 28 nm. The N-doping was carried out for 1 hour N (at 1000 °C), followed by 60 μm EP [49].

The cleanest samples prepared by EP/BCP (without any baking) give the shortest penetration depth (about 25 nm), and longest mean free path >400 nm since the purity of the Nb surface layer is not polluted by any means. The famous 120 °C bake treatment used to cure the HFQS (Section 6.4.3) strongly decreases the mfp

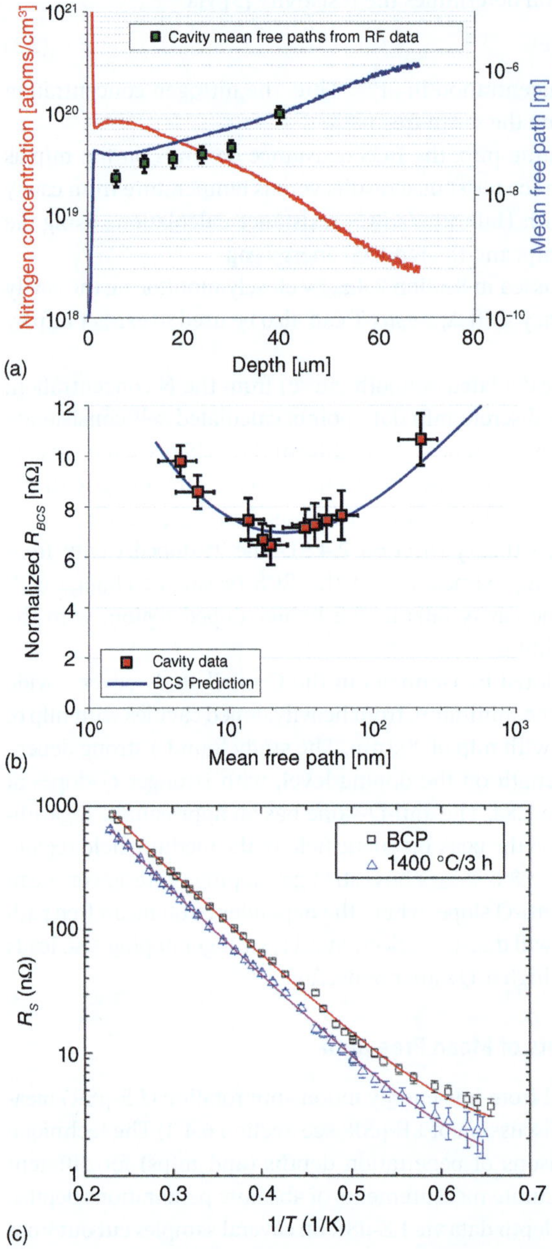

Figure 3.7 (a) Nitrogen concentration from SIMS (red) and the computed mean free path (blue) for the 800 °C N-doped sample. Discrete data points are mean free paths for single-cell N-doped cavities. Source: [95] Courtesy of D. Gonnella, Cornell University (b) 2.0 K R_{BCS} at low fields for the ten nitrogen-doped cavities. Experimental data agrees well with the BCS prediction. To vary the mfp, N-doping recipes followed were: 990 °C in 30 mTorr of N for 5 minutes, 900 °C in 60 mTorr of N for 20 minutes plus 900 °C anneal in vacuum for 30 minutes, 800 °C in 60 mTorr of N for 20 minutes plus 800 °C annealing in vacuum for 30 minutes [95] Courtesy of D. Gonnella, Cornell University. (c) Mean free path determination $\ell = (76 \pm 17)$nm for the Ti-doped cavity by fitting the surface resistance versus 1/T with the BCS theory. Source: [107]/P. Dhakal et al./American Physical Society/CC BY 3.0.

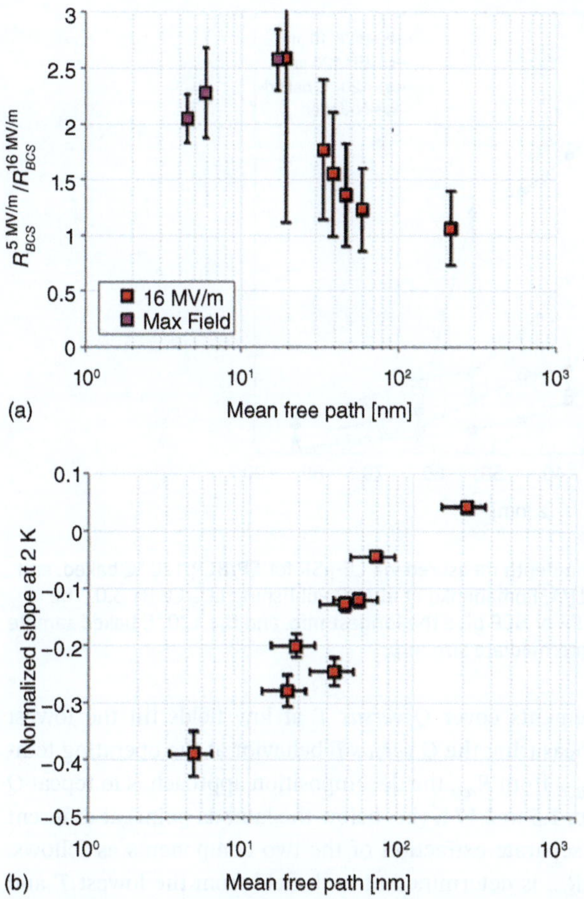

Figure 3.8 (a) Q-slope strength at 2.0 K for the ten doped cavities. The strength of the anti-Q slope increases with stronger doping, and lower mfp. For three cavities that did not reach 16 MV/m due to quench limitations, R_{BCS} at the maximum field was used. (b) Logarithmic fit of the anti-Q slope (at 2.0 K) versus mean free path. Larger mean free path (less doping) weakens the anti-Q-slope, which eventually disappears to manifest as the medium field Q slope. Source: [95] Courtesy of D. Gonnella, Cornell University.

for EP/BCP cavities to <2 nm (very near the surface) and to about 16 nm deeper in the RF layer (50 nm), corresponding to the long penetration depths. The 2–6 doping recipe nm after 5 μm EP (commonly used) yields an intermediate mfp, 90–180 nm.

3.6 Anti-Q-Slope Origins from BCS Resistance

When studying the dependence of surface resistance on the RF field, the technique of decoupling the temperature-independent residual resistance (R_{res}) from the temperature-dependent BCS resistance (R_{BCS}) was developed [133] to understand in a simple way which of the two surface resistance component is responsible for changes with field.

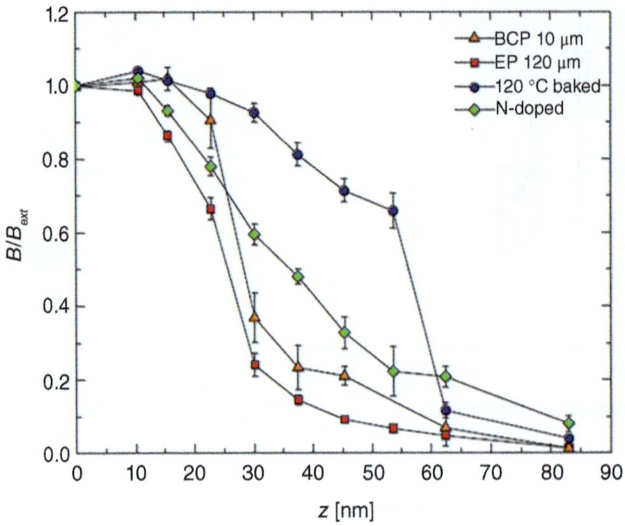

Figure 3.9 London penetration depth measured via LE-μSR for EP/BCP, 120 °C baked, and N doped samples. Source: [131]/A. Romanenko et al./AIP Publishing LLC/CC BY 3.0. The cleanest samples prepared by EP or BCP give the longest mfp, and the 120 °C baked sample the shortest. N-doping gives intermediate size mfp.

Normally cavity measurements cover Q *versus* T at low fields till the lowest temperature, followed by measuring the Q *versus* B behavior at the operating temperature. To deconvolute R_{BCS} from R_{res}, the decomposition approach is to repeat Q *versus* T measurements from 1.5 to 2.17 K (i.e. below the lambda point) at different field levels B, followed by separate extraction of the two components as follows: (a) the residual resistance R_{res} is determined at each field from the lowest T and (b) the temperature-dependent surface resistance is measured at each field after subtracting the result from (a).

$$R_{res}(E_{acc}) = \frac{G}{Q_0(E_{acc}, T < 1.5\,\text{K})} \tag{3.3}$$

$$R_{BCS}(E_{acc}, T = 2K) = \frac{G}{Q_0(E_{acc}, T = 2K)} - \frac{G}{Q_0(E_{acc}, T < 1.5\,\text{K})}. \tag{3.4}$$

Here G is the cavity geometry factor, $G = QRs$. For example, Figure 3.10 compares the R_{BCS} and R_{res} results versus surface B via deconvolution for various standard treatments, such as BCP, EP, BCP plus 120 °C bake, and EP plus 120 °C bake. We will discuss the implications of the observed changes in the surface resistance components for the 120 bake treatments in Chapter 6 with HFQS. Here we focus on the results for N-doping treatment.

Figure 3.10a shows that the surprising anti-Q-slope result for N-doping arises primarily from the BCS resistance, which *decreases* with increasing field. The decomposition analysis clearly shows that the residual resistance does not play a role in the field-dependent resistance drop. The deconvolution result confirms that at low field, the BCS resistance decreases due to the decrease of mfp from N interstitials, as expected from the BCS theory (Section 2.4.5). Interestingly, the

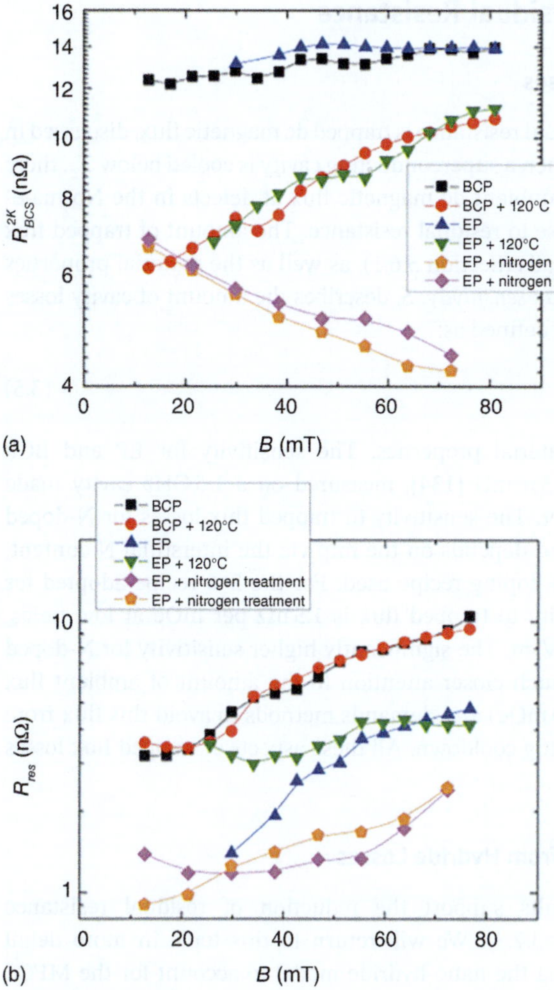

Figure 3.10 (a) Temperature-dependent (BCS resistance) and (b) temperature-independent (residual resistance) components of the microwave surface resistance. A comparison is shown for various treatments: standard treatments EP, BCP, and 120 °C bake and N-doping. Source: A. Grassellino et al. [99]/with permission from IOP Publishing.

decomposition analysis shows that N-doping also lowers residual resistance (Figure 3.10b). Section 3.7.2 discuss possible causes for the impact of interstitial N on the residual component.

These performance improvements by N-doping raise important questions: (i) Why does the BCS resistance fall with increasing field? (ii) How does the presence of interstitial impurity N (or Ti) induce an anti-*Q*-slope? and (iii) How does N-doping lower the residual resistance? The deconvolution analysis shows that the field-dependent decrease needs to be understood from a fundamental perspective, needing an extension of the BCS theory for high fields. Theoretical efforts to understand the anti-*Q*-slope behavior at medium fields will be discussed in Section 3.10 in this chapter.

3.7 N-Doping and Residual Resistance

3.7.1 Trapped DC Flux Losses

A familiar component of residual resistance is trapped dc magnetic flux, discussed in greater detail in Chapter 5. When a superconducting cavity is cooled below T_c, there is the potential to trap local ambient dc magnetic flux at defects in the Nb material. The trapped flux gives rise to residual resistance. The amount of trapped flux depends on the cooldown details (Section 5.6.1), as well as the material properties (Section 5.11). The trapped flux *sensitivity*, *S*, describes the amount of cavity losses per unit of trapped flux and is defined as:

$$S = \frac{R_{fl}}{B_{trap}} \tag{3.5}$$

Sensitivity depends on material properties. The sensitivity for EP and BCP cavities with long mfp is 0.35 n/mG [134], measured on a 1.5 GHz cavity made out high-purity niobium sheet. The sensitivity to trapped flux losses for N-doped cavities is generally higher and depends on the mfp via the interstitial N content, and therefore on the detailed doping recipe used. For the 2–6 recipe adopted for LCLS-II cavities, the sensitivity to trapped flux is 1.5 nΩ per mOe at low fields, rising to 2.5 nΩ/mOe at 15 MV/m. The significantly higher sensitivity for N-doped cavities, therefore requires much closer attention to the amount of ambient flux in the cryostat (typically 5–10 mOe) and demands methods to avoid this flux from getting trapped in the Nb during cooldown. All these aspects of trapped flux losses will be discussed in Chapter 5.

3.7.2 Residual Resistance from Hydride Losses

Point-contact tunneling results support the reduction of residual resistance with N-doping (next Section 3.7.3). We will return to this topic in more detail (Section 6.13) when discussing the nano-hydride model to account for the MFQS and HFQS as well as the 120 °C baking effect that reduces HFQS [135]. For now, we point out that the reduction in residual resistance from N-doping is possible due to N interstitials inhibiting the formation of lossy niobium hydrides by providing trapping centers for H, just as the 120 °C baking provides O interstitials to inhibit hydride formation to ameliorate the HFQS. TEM results (Figure 3.11) support the reduction of hydrides as the possible explanation, showing (a) absence of hydrides in some of the doped samples compared to hydrides in non-doped samples.

Using GI (grazing incidence)-synchrotron XRD at APS, a structural and phase composition study on cutouts from cavities processed with N-doping show further supporting results [137, 138]. A Nb (NHx) solid solution is found, where the N atoms occupy the interstitial octahedral sites of Nb, trapping hydrogen atoms located in the tetrahedral sites. This formation shows how N prevents the formation of Nb hydrides.

Theoretical studies show that N presence inhibits formation of hydrides. The introduction of a thin N-rich layer could have a significant impact on hydride

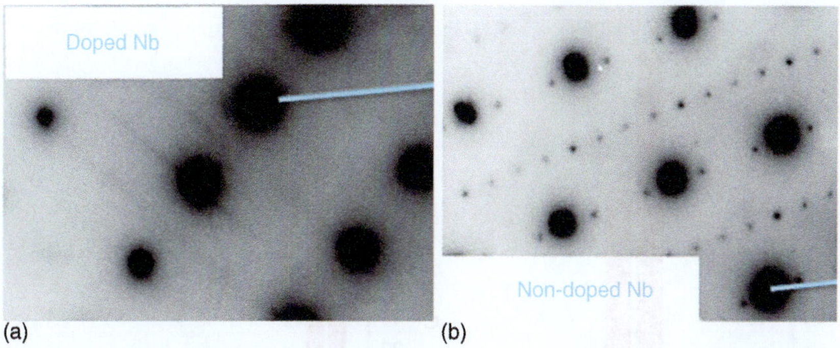

Figure 3.11 TEM diffraction patterns at 94 K show (a) Nb lattice without hydrides for a N-doped Nb sample. Source: [136]/Courtesy of Y. Trenikhina, Fermilab, U.S. Department of Energy Office of Science. (b) Nb lattice with secondary diffraction peaks due to Nb-hydrides for a non-doped sample.

nucleation behavior by forcing nano-hydrides to nucleate further away from the surface owing to the energetic disadvantage of nucleating in the surface region [139–143].

3.7.3 Tunneling Measurements

As discussed in Section 2.3.5, point contact tunneling spectroscopy (PCTS) is a versatile technique that uses the substrate native oxide as the tunnel barrier to measure the surface density of states (DOS), the superconducting gap Δ, the critical temperature T_c and the parameter Γ/Δ, called the quasi-particle lifetime broadening parameter, connected to the residual resistance R_{res}. These parameters shed light on the performance of SRF cavities at low RF fields. As discussed in Chapter 2, R_S depends on the DOS at the Fermi level, which in turn depends on Δ, Γ, and the temperature.

Upon etching 5 μm of material after N-doping, the tunneling junctions measured by Proslier on the cut out sample of a N-doped cavity exhibit (Figure 3.12) high-quality superconducting DOS with peaked gap around 1.49 meV and Γ/Δ values of 4.6% [144]. The maps reveal homogeneous superconducting gap values with a small spread (mostly 1.4–1.6 meV) over a significant area of 350 × 168 μm. The corresponding ratio $2\Delta/k_B \cdot T_C = 4$ is in agreement with the ratio value of 4 extracted from the temperature dependence of the surface impedance, R_S (T), measured on N-doped cavities, as discussed in Section 3.6.

Niobium cavities subject to a 1400 °C annealing in UHV for 1 hour (with Ti interstitial impurities of about 1 at% within the first micron) show a similar increase of the quality factor with increasing accelerating RF field. The PCTS measurements done on Nb coupons exhibit consistently near-ideal BCS DOS.

For comparison to the behavior without doping, similar measurements conducted on an EP sample without N-doping or without 120 °C baking (presented in Section 6.15) show a large spread in gap values (from 1.0–1.7), lower $2\Delta/k_B T_C = 3.5$, and larger $\Gamma/\Delta = 6.5\%$, along with broader DOS curves. This behavior is likely connected

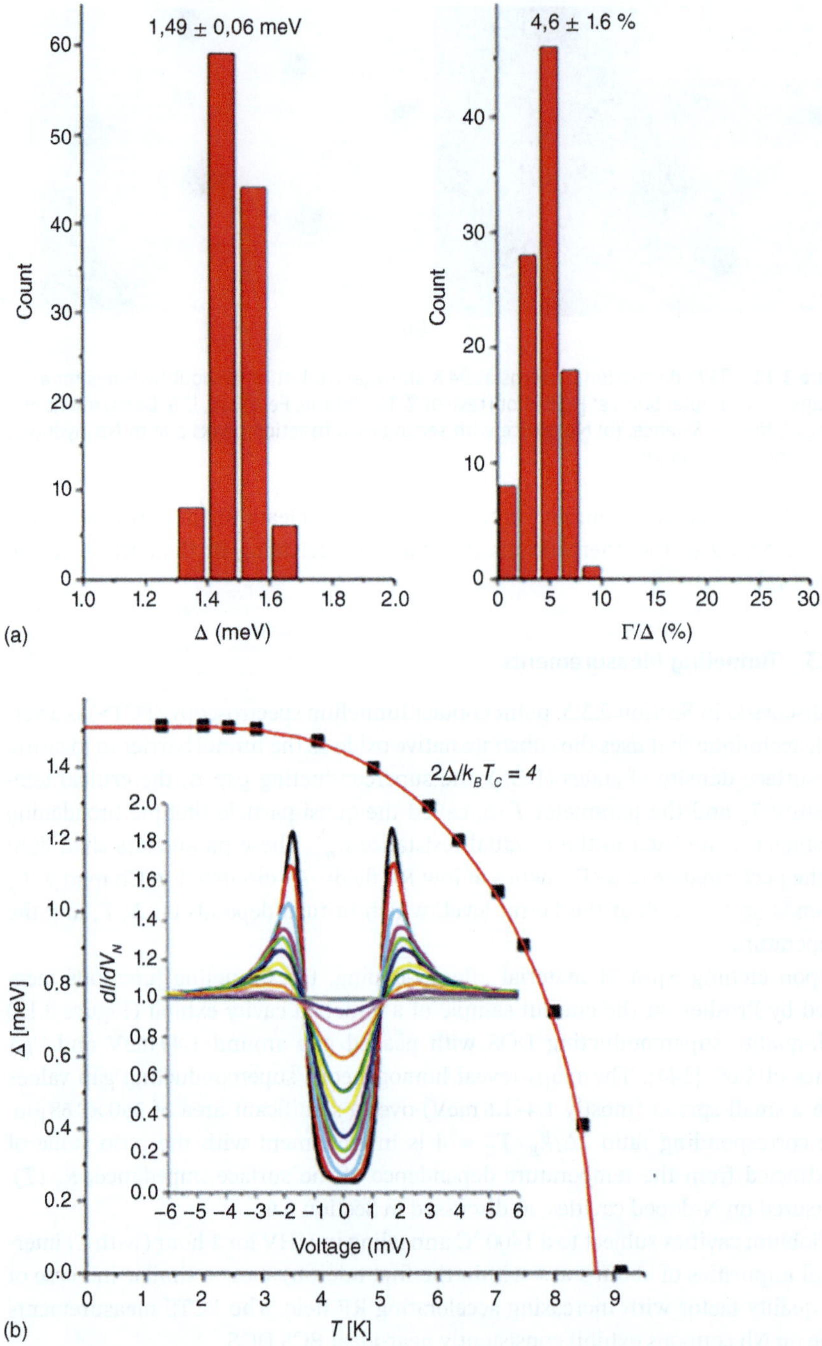

Figure 3.12 (a) Summary of the PCTS measurements on N-doped samples cut out from a N-doped cavity after 5 μm EP. Statistics of Δ in meV show small spreads around 1.5 meV, and a smaller Γ/Δ compared with non-doped samples to be presented in Section 6.15 (b) Temperature dependence of Δ for typical tunnel junctions measured from sharp DOS curves shown in the insert. [144] N. Groll et al/arXiv.

to the presence of niobium hydrides. Both N-doping (or Ti-doping) and 120 °C baking inhibit the formation of hydrides to give better PCTS properties and lower residual resistance.

3.8 RF Field Dependence of the Energy Gap

The remarkable new feature of N-doping is the anti-Q-slope in the medium field region. Both Martinello at Fermilab [49] and Gonnella at Cornell [95] present data analysis that suggests that the reason for the anti-Q-slope is an increase in the gap with increasing field. At Cornell, the SRIMP program is used to fit R_{BCS} data for all nitrogen-doped cavities tested by allowing Δ/k_BT_C to vary with field. Figure 3.13a shows R_{BCS} *versus* E_{acc} for one cavity (LT1–3) along with fits via SRIMP with adjusted Δ/k_BT_C. Figure 3.13b shows how Δ/k_BT_C increases with field over

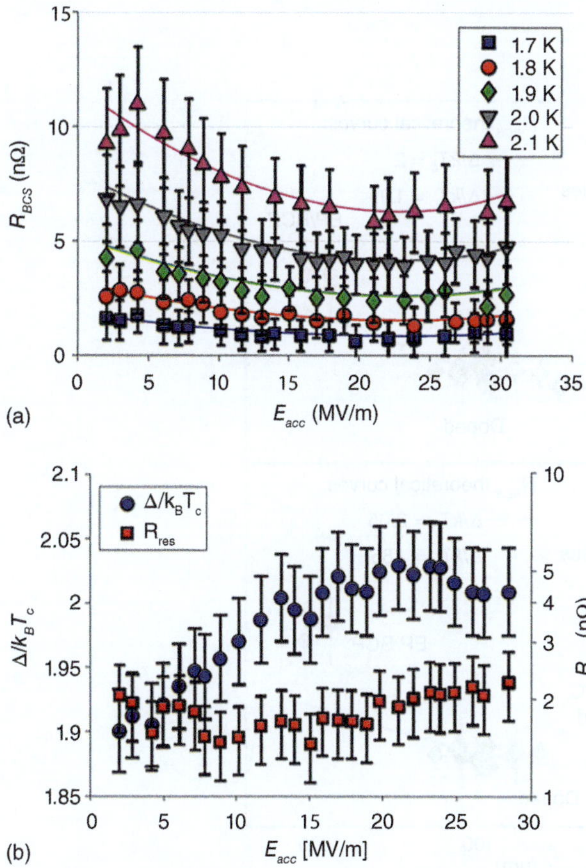

Figure 3.13 (a) R_{BCS} *versus* E_{acc} for Cornell N-doped cavity, with fits to the data via SRIMP by varying Δ/k_BT_C. (b) Corresponding results for Δ/k_BT_C with increasing RF field. Source: [95] Courtesy of D. Gonnella, Cornell University.

the medium field range. Thus an increasing field-dependent energy gap in the medium field region could account for the anti-Q slope. Section 3.10.3 discusses microwave-driven nonequilibrium effects as a possible explanation for the change in the energy gap with field [145].

A similar conclusion was reached from Fermilab data on N-doped cavities. Figure 3.14 shows the BCS surface resistance at 2 K as a function of the mean free path. The upper part of Figure 3.14 shows the results obtained at low field (5 MV/m), while the lower part is at medium field (16 MV/m). The BCS resistance values can be fit with SRIMP by selecting the best-reduced energy gap Δ/k_BT_C. Doped cavities are best fit by setting $\Delta/k_BT_C = 2$ at low fields and 2.05 at medium fields. In contrast, a similar analysis for 120 °C baked, BCP, and EP cavities show a drop in Δ/k_BT_C from 1.95 to 1.85. For the 120 °C bake cavities the mean free path was measured with LE-μSR (Section 6.4.3).

For completeness, we add that the mfp values of the Ti-doped cavities at JLab (Figure 3.2b) fall in the 25–80 nm range [109], close to the minimum in the R_{BCS} curves of Figure 3.14.

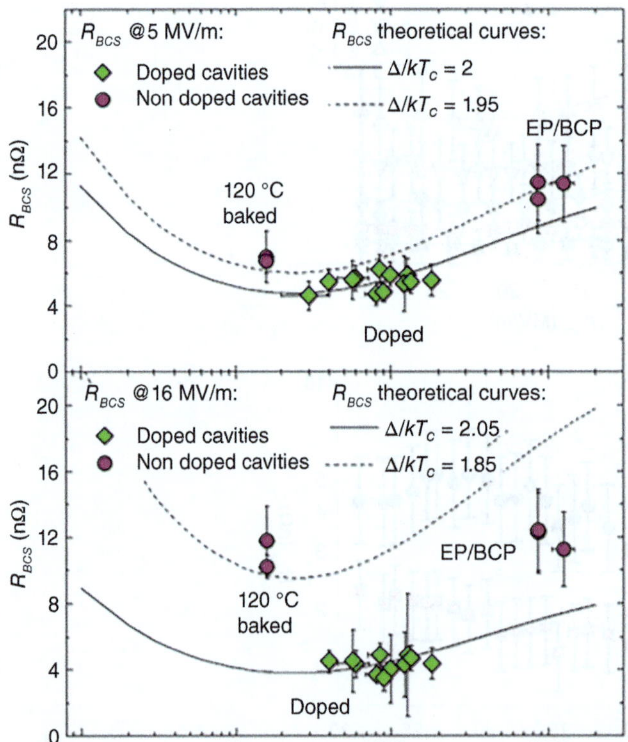

Figure 3.14 2 K BCS surface resistance *versus* mean free path, at 5 MV/m (upper graph) and at 16 MV/m (lower graph). The green diamonds represent doped cavities, the pink circles represent non-doped cavities. The black curves are theoretical curves of R_{BCS} *versus* l (mean free path) estimated using SRIMP for different reduced energy gap values. Source: [49] Courtesy of M. Martinello, Fermilab.

3.9 Frequency dependence of Anti-Q-Slope

The behavior of the anti-Q-slope for cavities of different frequencies gives important clues for the underlying reasons for the rise of Q with field. Fermilab studies [49, 146] by Martinello show that the anti-Q-slope is stronger at frequencies higher than 1.3 GHz, and weaker at lower frequencies. Figure 3.15a,b, and c compare the strength of the anti-Q-slope for cavities with different frequencies when given a variety of treatments, N-doping, 120 C bake, EP, and BCP. At 2.6 and 3.9 GHz, an anti-Q-slope is present even without N-doping, as shown in Figure 3.15c. Most of the 120 °C baked cavities show absence of the anti-Q-slope, except at the highest frequency, 3.9 GHz. The anti-Q-slope appears at 1.3 GHz with N-doping and gets stronger for higher frequencies, as evident from Figure 3.15a. At 650 MHz the anti-Q-slope is absent, but doping nevertheless reduces the Q-slope of 650 MHz cavities. Also, Cornell studies of a 500 MHz cavity treated with the 2/6 recipe showed an improved overall R_{BCS}, but no anti-Q-slope at all [147].

The frequency dependence of the anti-Q-slope lends support to the nonequilibrium explanation for the anti-Q-slope of N-doped cavities as discussed in Section 3.10.3.

3.10 Theories for Anti-Q-Slope

Several models have been proposed to understand the decreasing microwave resistance for increasing RF fields with N-doping, as well as to explain the surprising anti-Q-slope that develops with higher RF frequencies, without N-doping. The theory extended by Mattis and Bardeen to calculate the BCS surface resistance (Section 2.4.5) includes the effects of mean free path, and energy gap, but for zero RF fields only. Some of the proposed explanations for the field dependence of Q and the anti-Q-slope have been reviewed by Maniscalco [148, 149].

As discussed in Section 2.4.5. a superconductor is characterized by the density of states (DOS), and the distribution function of the electrons $f(E)$. The DOS exhibits an energy gap due to Cooper pair formation, and the distribution function is the Fermi–Dirac distribution in thermal equilibrium. The RF field can drive nonlinear effects in a superconductor by changing the density of states, or by changing the distribution function. In either case, there is a redistribution of quasiparticles which is used to account for R_{BCS} changes with increasing RF field. Thus the models to be discussed modify the Mattis–Bardeen calculations by incorporating changes in the DOS, by changes in the distribution function, or both. An additional model is based on the two-fluid theory.

Among the models discussed, none are completely satisfactory. Some account for the mfp dependence, but not for the frequency dependence, and others have the contrary feature. An unanimously accepted explanation is missing. Nevertheless, it is worthwhile to examine the physics of the models for future understanding.

3.10.1 Xiao Theory

In 2014, Xiao at Jefferson Lab extended the Mattis–Bardeen BCS theory in the presence of a nonzero RF field based on coherently moving Cooper pairs [150, 151]. In

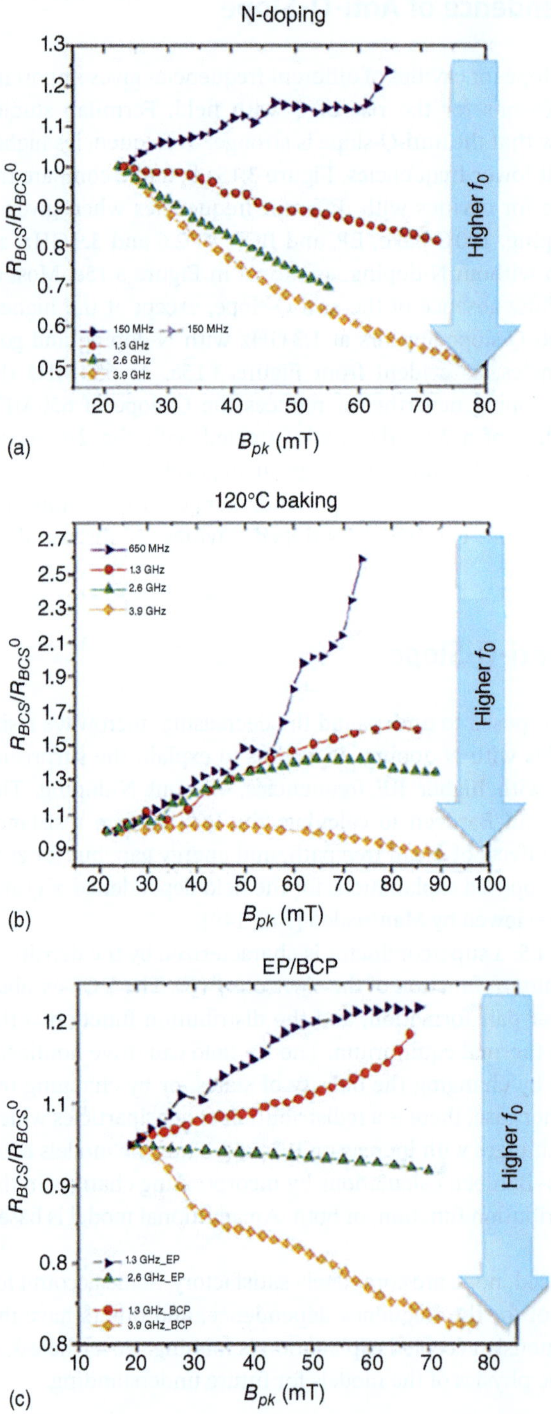

Figure 3.15 (a) Behavior of normalized surface resistance with increasing RF field for cavities at various frequencies for (a) N-doped (b) 120 °C baked and (c) EP/BCP cavities. Source: [146]/M. Martinello et al/American Physical Society/CC BY 4.0.

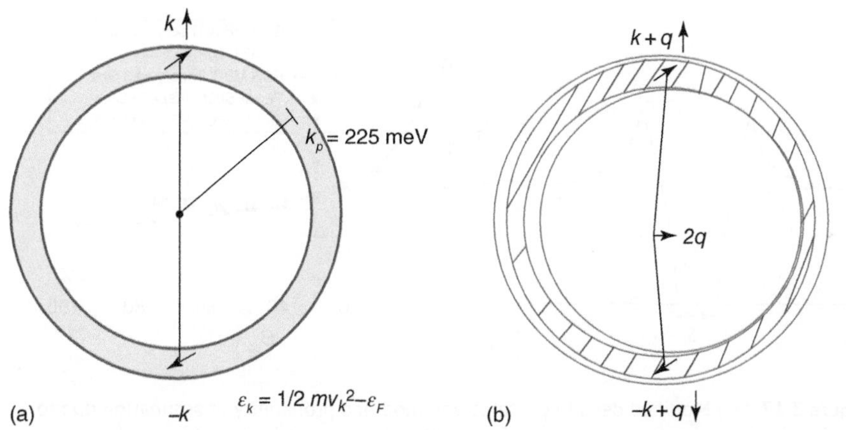

Figure 3.16 (a) In the zero field limit of the BCS theory, electrons in the shaded region of the Fermi sphere of the superconductor form Cooper pairs leading to superconductivity. (b) At high RF fields, electrons in the active slice of the Fermi sphere gain a net momentum *2q*, the same for all Cooper pairs, distorting the active region to shift the DOS. Source: Xiao and Reece [150]/B. Xiao and C. Reece/Jefferson Lab/U.S Department of Energy/JACoW/CC BY 3.0.

the BCS theory, paired particles in the ground state, with total mass $2m$ and zero total momentum occupy states ($\boldsymbol{k}\uparrow$, $-\boldsymbol{k}\downarrow$), with velocity $\boldsymbol{V_k}$ in random direction (Figure 3.16a). Xiao's theory, extended for high RF fields, states that net flow in a certain direction can be obtained by pairing ($\boldsymbol{k}+\boldsymbol{q}\uparrow$, $-\boldsymbol{k}+\boldsymbol{q}\downarrow$), with total momentum $2\boldsymbol{q}$ the same for all Cooper pairs, corresponding to net velocity $\boldsymbol{V}_s = \hbar\boldsymbol{q}/m$. This change is illustrated by a shift in the slice of the Fermi sphere (Figure 3.16a), as shown in Figure 3.16b.

The coherently moving Cooper pairs shift the DOS and $f\ (E)$, as shown in Figure 3.17a. These changes cause a significant change in single particle scattering, which leads to a field dependence of R_s. The main result predicts the surprising decreasing R_{BCS} with increasing Cooper pair velocity (proportional to the strength of the RF field) as observed in N-doped and Ti-doped cavities. Initial comparison with experimental data from JLab and Fermilab show reasonable agreement with measurements on Ti-doped and N-doped cavities [150], as shown in Figure 3.17b.

The theory correctly predicts that lower frequency cavities at 2 K will show a weaker anti-*Q*-slope (Figure 3.18a), in rough agreement with the measurements (Figure 3.15). There is no prediction for frequencies higher than 1.5 GHz, a gap that needs to filled. The theory tends in the right direction for lower frequency, showing weaker anti-*Q*-slope at 700 MHz in Figure 3.18a, although the data at 650 MHz show no anti-*Q*-slope (Figure 3.15). Theoretical predictions do show deviations from measured values at the highest fields, possibly due to thermal effects, which the theory does not take into account. Aspects of thermal effects are included in the following theory by Gurevich via an overheating parameter.

However, the mfp dependence predicted by the Xiao theory (Figure 3.18b) is in contradiction to the data for the strength of the anti-*Q*-slope, as shown in Figure 3.8 above. The theoretical anti-*Q*-slope does not disappear when the mfp increases to 1000 nm, but the data of Figure 3.8a and b show the strength of the anti-*Q*-slope goes to zero at long mfps.

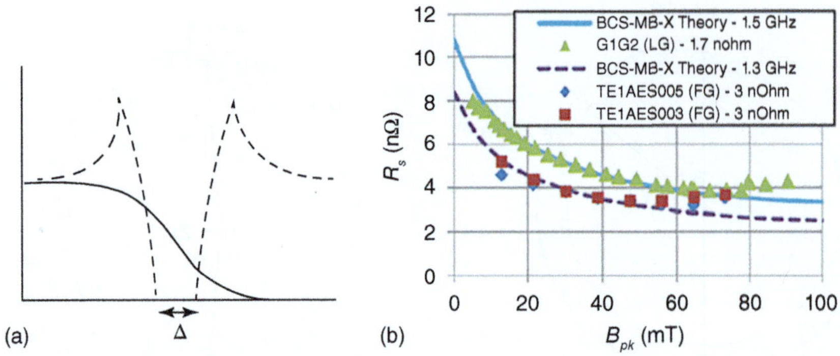

Figure 3.17 (a) Modified density of states and modified probability of occupation due to moving Cooper pairs. (b) Comparison of calculated field-dependent surface resistance at 2.0 K from Xiao's theory with data from JLab at 1.5 GHz for a Ti-doped single cell Nb cavity and 1.3 GHz N-doped cavities. Source: [151] B. Xiao and C. Reece/JACoW/CC BY 3.0.

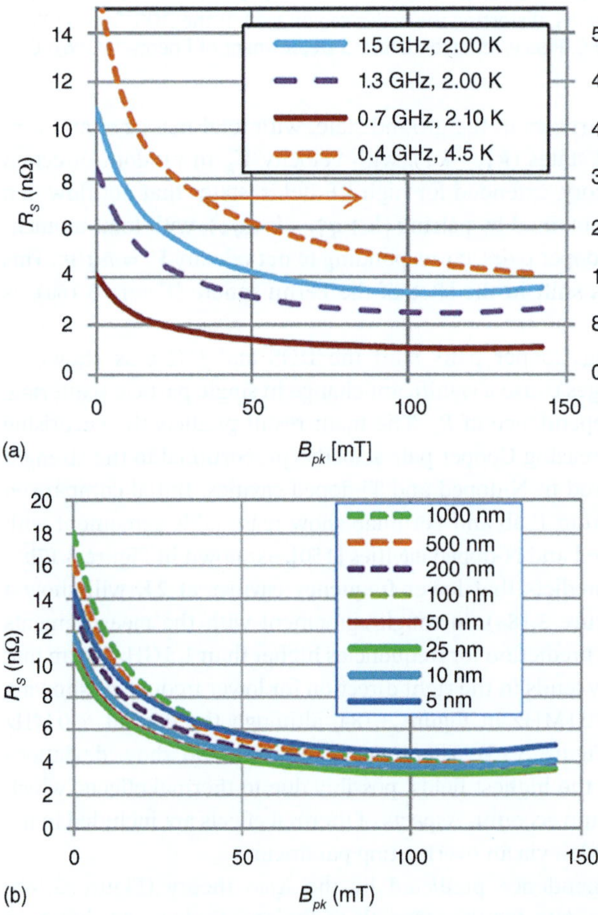

Figure 3.18 (a) Surface resistance versus peak RF magnetic field calculated from Xiao's theory for Nb for (a) frequency dependence and (b) mean free path dependence. Source: [150] B. Xiao and C. Reece/JACoW/CC BY 3.0.

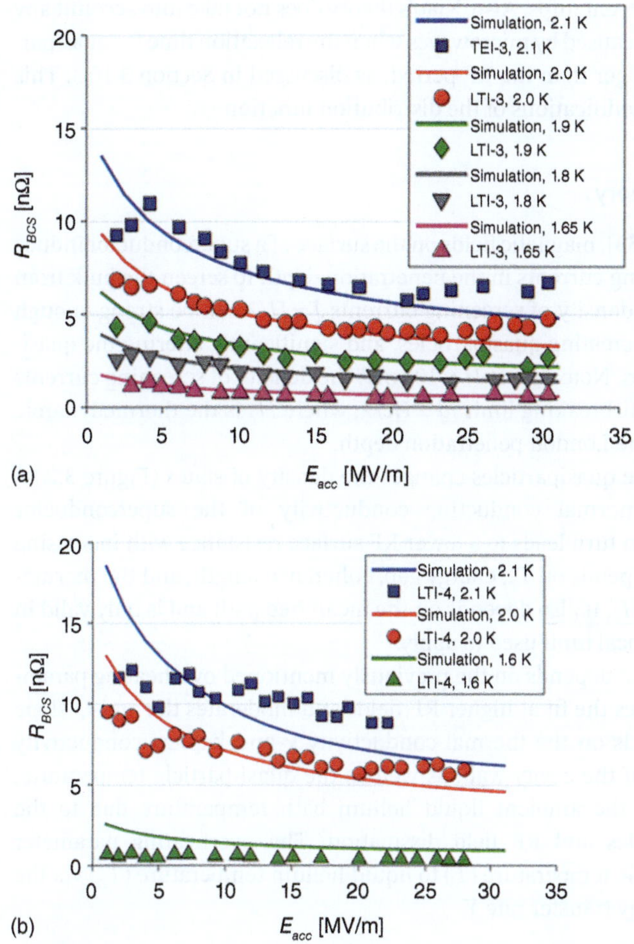

Figure 3.19 (a) R_{BCS} versus E_{acc} at different temperatures for LT1–3, mfp = 34 nm, compared with predictions from Xiao theory [95]. (b) R_{BCS} versus E_{acc} at different temperatures for LT1–4 mfp = 47 nm, compared with predictions from Xiao theory. Source: [95] Courtesy of D. Gonnella, Cornell University.

For a more detail comparison, R_{BCS} versus E_{acc} was calculated from the Xiao theory and compared to the data for two 1.3 GHz N-doped cavities at Cornell (Figure 3.19a and b). There is good agreement in the medium field region where the anti-*Q* slope is strong for a cavity with a short (34 nm) mfp. But the fit degrades for the cavity with higher mfp of 47 nm, Figure 3.19b. Here the observed anti-*Q* slope gets weaker, but the Xiao theory continues to show a strong anti-*Q*-slope for longer mfps. Thus the Xiao theory does well for a cavity with short mfp and strong *Q*-slope. But the theory contradicts data for long mfps and weaker *Q*-slopes.

Some of the deficiencies of the Xiao theory may be explained by features of the other theories that follow [152]. At high fields, pair-breaking due to higher flowing currents becomes intertwined with impurity scattering that also affects both the density of states and the distribution function. Thus, there is disagreement between theory and data for different mfps. Also, Xiao's theory does not take into account any nonequilibrium effects caused by microwaves when the relaxation time for quasiparticles could become longer than the RF period, as discussed in Section 3.10.3. This could lead to further modifications of the distribution function.

3.10.2 Gurevich Theory

In Gurevich's theory [153], magnetic fields on the surface of a superconductor induce Meissner effect screening currents in the penetration depth to screen the bulk from the external field. The density of screening currents $J \simeq H/\lambda$ can be strong enough to break Cooper pairs, creating quasiparticles, and significantly altering the quasiparticle density of states. Note that at $H \simeq 100$ mT, the density of screening currents reaches $\simeq$ 50% of the pairbreaking limit $Jd \simeq Hc/\lambda$, where H_c is the thermodynamic critical field, and λ is the London penetration depth.

The presence of these quasiparticles changes the density of states (Figure 3.20a) so as to lower the normal conducting conductivity of the superconductor (Figure 3.20b), which in turn leads to a lower RF surface resistance with increasing RF field. The theory depends on T_c, energy gap, coherence length, and the thermodynamic critical field, H_c. It also depends on the mean free path and is only valid in the dirty limit for the local limit used to apply.

Gurevich's theory also depends on the previously mentioned overheating parameter (α), which improves the fit at higher RF fields and moderates the anti-Q-slope for long mfp. α depends on the thermal conductivity κ and kapitza conductivity h_K, the thickness (d) of the cavity wall, as well as the quasi-particle temperature, which is higher than the ambient liquid helium bath temperature due to the motion of quasiparticles and RF field dissipation. The overheating parameter relates the quasi-particle temperature (T) to liquid helium temperature (T_0) via the electron–phonon energy transfer rate Y

$$\alpha = \frac{R_{\mathrm{BCS},0} H_{\mathrm{c}}^2}{2T_0} \left(\frac{1}{Y} + \frac{d}{\kappa} + \frac{1}{h_{\mathrm{K}}} \right) \tag{3.6}$$

The overheating $T - T_0$ between the quasiparticle temperature and the bath temperature is linearly proportional to α and to R_{BCS}, and scales quadratically with H_{pk}:

$$T - T_0 = \frac{\alpha T_0}{R_{BCS,0}} \left(\frac{H_{pk}}{H_c} \right)^2 R_{BCS}(H_{pk}, T) \tag{3.7}$$

As a property of the RF surface layer, Y changes with the strength of impurity doping, measured by the electron mean free path. Shorter mean free paths mean more efficient transfer of energy from electrons to phonons and thus a higher Y and lower α. More impurities (shorter mean free path) lead to more collisions and more efficient energy transfer. Cavities with shorter mean free paths, therefore, exhibit stronger anti-Q-slopes as seen in Figure 3.8b for Cornell doped cavities.

Longer mean free paths correspond to weaker anti-Q-slopes higher α, as observed, as well as higher overheating at higher fields (Figure 3.20b). Stronger fields dissipate

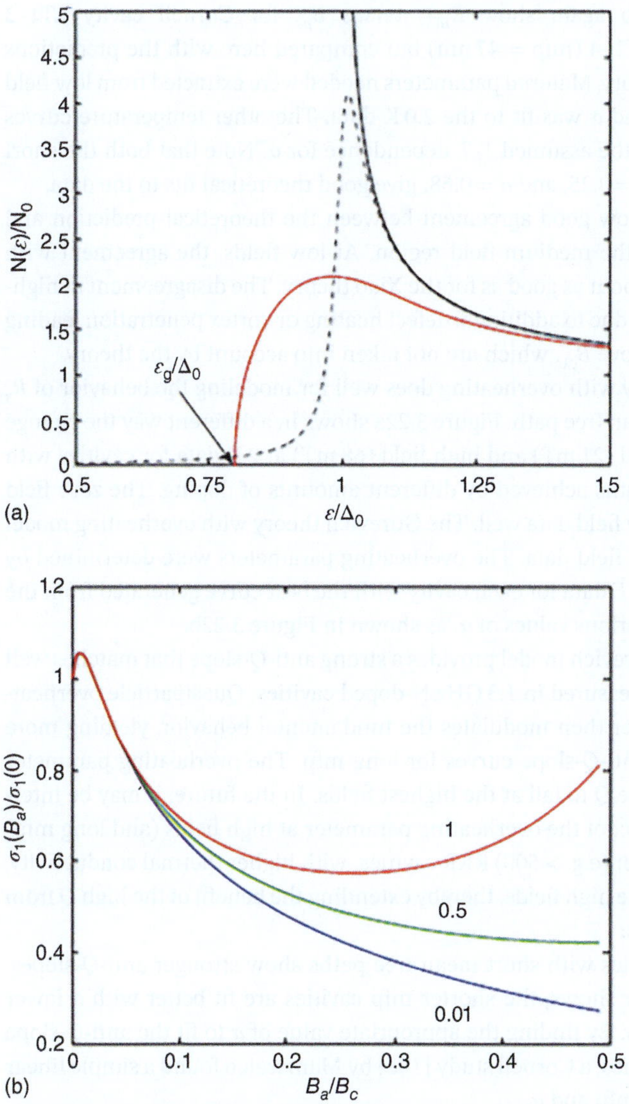

Figure 3.20 RF currents smear the density of states. For high currents, the density of states oscillates between two solid curves; the peak gets averaged over the RF period resulting in the dotted curve. (b) RF surface resistance is proportional to the normal conductivity $\sigma_1(B_a)$ calculated here for $T/\Delta_0 = 0.1$, $\omega/\Delta_0 = 0.0025$, where T is the temperature and ω the rf frequency, and Δ_0 is the gap; for low (0.01), medium (0.5) and high (1) overheating parameters (α). Source: [153]/A. Gurevich with permission of American Physical Society.

more power, and the inefficiency in the transfer of that power away from the quasiparticles raises the temperature of the quasiparticles and raises the surface resistance in a feedback effect. Stronger overheating thereby leads to a weaker reduction in R_{BCS} and thus a weaker anti-*Q*-slope. At high RF fields, the overheating becomes strong enough to give an increasing R_{BCS} with RF field, as seen in the data.

Figure 3.21a and b again show R_{BCS} *versus* B_{pk} for Cornell cavity LT1–3 (mfp = 34 nm) and LT1–4 (mfp = 47 nm) but compared here with the predictions from the Gurevich theory. Material parameters needed were extracted from low field fitting with SRIMP, and α was fit to the 2.0 K data. The other temperature curves were generated using the assumed $1/T$ dependence for α. Note that both the short and long mfp cases, $\alpha = 0.25$, and $\alpha = 0.68$, give good theoretical fits to the data.

All temperatures show good agreement between the theoretical prediction and experimental data in the medium field region. At low fields, the agreement with experimental data is about as good as for the Xiao theory. The disagreement at highest fields is most likely due to additional defect heating or vortex penetration leading to additional losses above B_{c1}, which are not taken into account by the theory.

The Gurevich theory with overheating does well for modeling the behavior of R_s (H) with changing mean free path. Figure 3.22a shows in a different way the change in R_s between low field (21 mT) and high field (68 mT) at 2 K data for cavities with different mean free paths achieved by different amounts of doping. The zero field BCS theory fits the low field data well. The Gurevich theory with overheating model fits well with the high field data. The overheating parameters were determined by matching the Q *versus* E data for each cavity with the best curve generated from the Gurevich theory for various values of α, as shown in Figure 3.22b.

In summary, the Gurevich model provides a strong anti-Q-slope that matches well the strongest slopes measured in 1.3 GHz N-doped cavities. Quasiparticle overheating via the α parameter then modulates the fundamental behavior, yielding more moderate or absent anti-Q-slope curves for long mfp. The overheating parameter for long mfp causes the Q to fall at the highest fields. In the future, it may be interesting to test the impact of the overheating parameter at high fields (and long mfp) by lightly doping higher (e.g. > 500) RRR cavities, with higher thermal conductivity, to arrest the Q fall at the high fields, thereby extending the benefit of the high Q from doping to higher fields.

Heavier doped cavities with short mean free paths show stronger anti-Q-slopes. As Figure 3.21 further shows, the shorter mfp cavities are fit better with a lower overheating parameter. By finding the appropriate value of α to fit the anti-Q-slope data from several cavities, a Cornell study [148] by Maniscalco found a simple linear relationship between mfp and α.

For the physics, it appears that N-doping works by introducing interstitial impurities that improve the electron–phonon energy transfer mechanism. Without these impurities, quasiparticles become heated above the lattice temperature by RF dissipation, releasing that energy to the lattice during inelastic collisions with impurities. As the concentration of impurities increases (i.e. as α decreases), the scattering rate increases, improving the heat transfer rate from the electrons to the lattice phonons, and mitigating the quasiparticle overheating effects.

The Gurevich theory suffers from an important deficiency for the frequency dependence. The minimum in $R_{BCS} \propto \sigma_1(H_0)$ becomes more pronounced (Figure 3.23a, b) as the RF frequency decreases. This is opposite to experimental data of Figure 3.15, which shows a weaker anti-Q-slope at low frequencies.

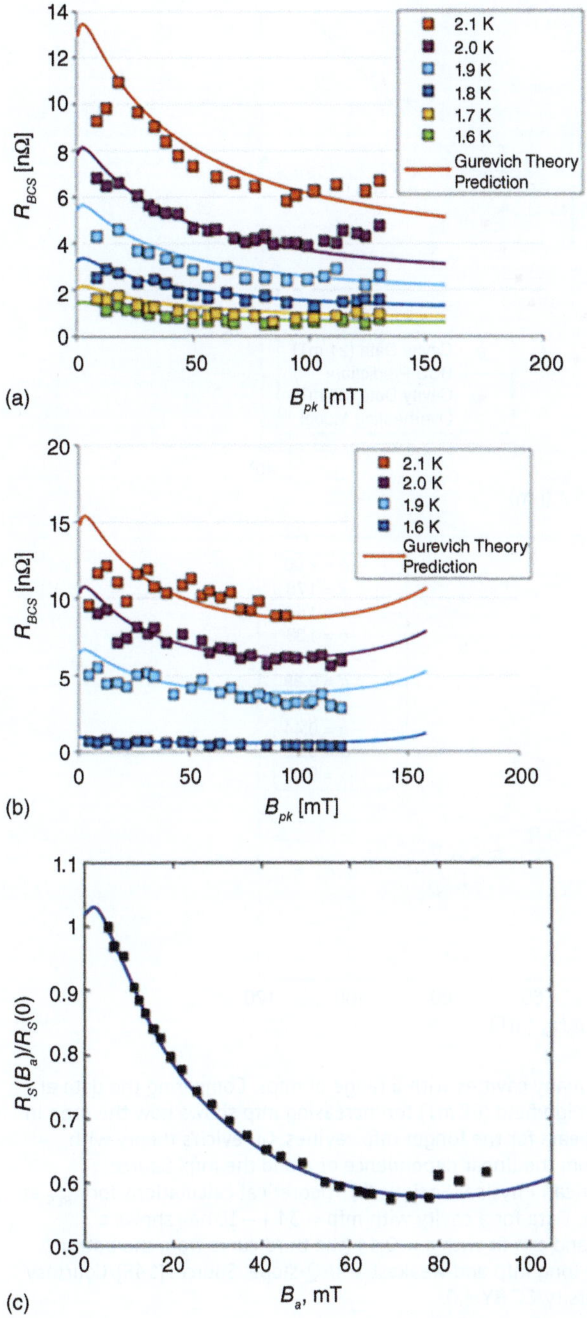

Figure 3.21 (a) The short mfp case: R_{BCS} *versus Bpk* at different temperatures for LT1-3 compared with predictions from the Gurevich theory. $\alpha = 0.25$ was fit to the 2.0 K data. Other curves assumed the *1/T* dependence for α [95] Courtesy of D. Gonnella. Cornell University. (b) The longer mfp case: R_{BCS} *versus Bpk* at different temperatures for cavity LT1-4 compared with predictions from the Gurevich theory, $\alpha = 0.68$ [95] Courtesy of D. Gonnella. Cornell University (c) *Rs* calculated for longer mfp case with $\alpha = 0.91$, from Gurevich [153] and compared with data from a Nb cavity doped with Ti at 1.75 GHz [104], corresponding to a large mfp, weak *Q*-slope case. Source: [153]/A. Gurevich with permission of American Physical Society.

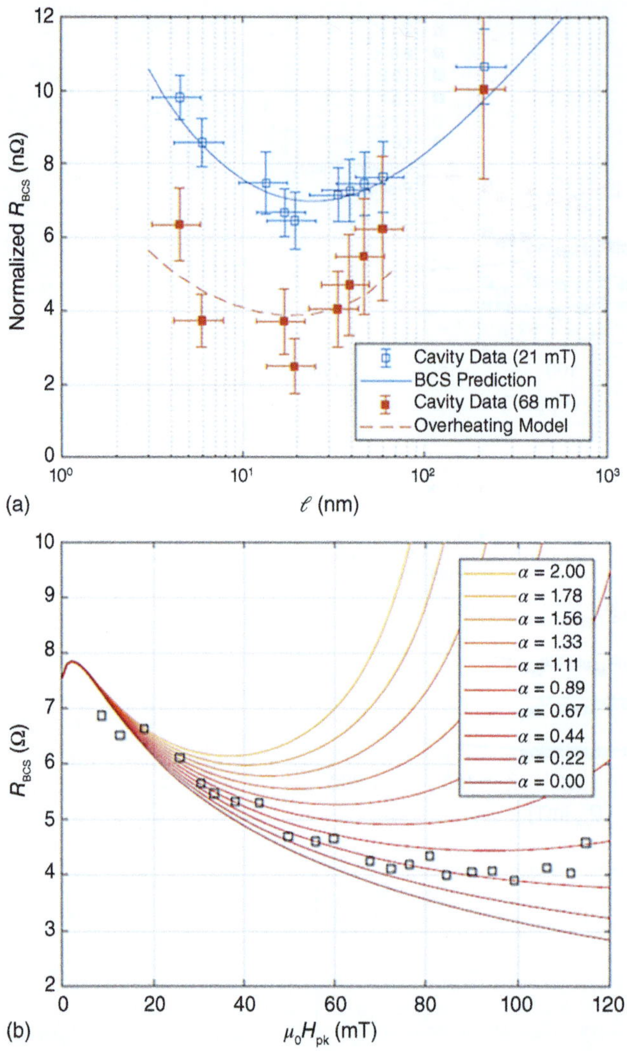

Figure 3.22 (a) R_{BCS} data for many cavities with a range of mfps. Comparing the data at low field (21 mT) with data at high field (68 mT) for increasing mfp shows how the drop in R_{BCS} for the higher field disappears for the longer mfp cavities. Gurevich's theory with overheating tracks the data from the linear dependence of α and the mfp. Source: [148]/with permission of American Physical Society. (b) Theoretical calculations for R_{BCS} at 2 K with a range of values of α. Data for a cavity with mfp = 34 + −10 nm, shows a medium-strong anti-Q-slope and can fit with $a = 0.44$ (the third curve from the bottom). $\alpha = 2$ is the topmost curve, for long mfp and weakest anti-Q-slope. Source: [148]/Courtesy of J. Maniscalco, Cornell University/CC BY 4.0.

3.10.3 Nonequilibrium Superconductivity

Nonequilibrium effects [154] lead to changes in the distribution function [155, 156] that stimulate superconductivity leading to Δ increases and R_{BCS} decreases with the RF field amplitude. The hypothesis that the anti-Q-slope of N-doped cavities

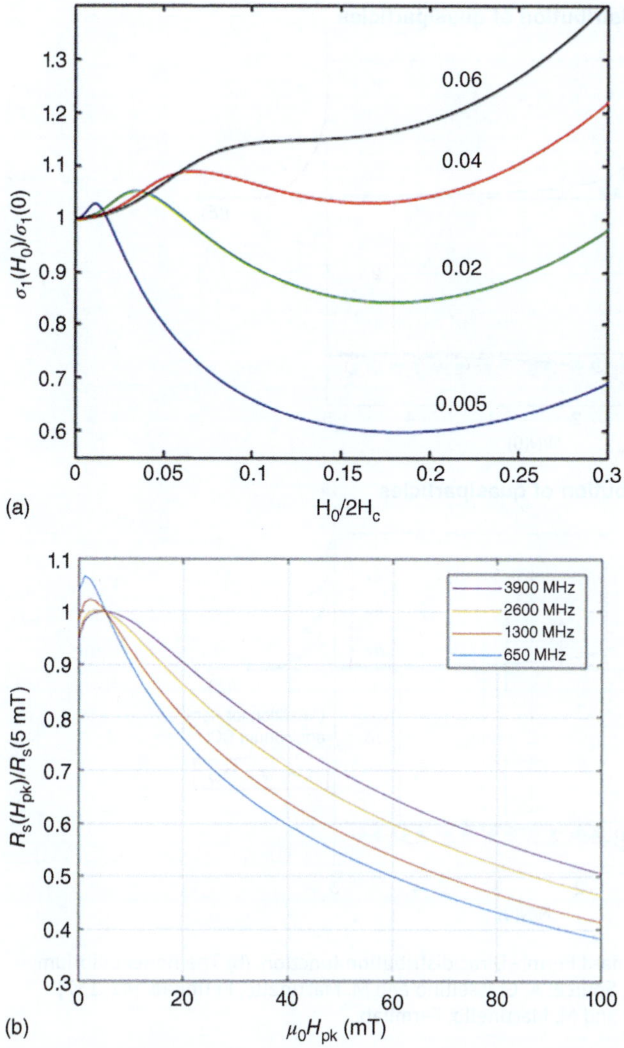

Figure 3.23 RF frequency dependence predictions of the Gurevich theory. Linear conductivity $\sigma_1(H_0) \propto R_{BCS}$ calculated at low temperature for ω/Δ: 0.005, 0.02, 0.04, 0.06. Source: Gurevich [153]/American Physical Society, where ω is the frequency and Δ is the energy gap. (b) Frequency-dependent results of the relative field-dependent reduction in surface resistance predicted by the Gurevich model, calculated with $T_c = 9:1$ K, $\Delta/k_BT_c = 1:886$, $\ell = 9$ nm, and $T_{bath} = 2$ K. Source: [148]/Courtesy of J. Maniscalco, Cornell University/CC BY 4.0.

may originate from a deviation of the quasiparticle energy distribution from thermal equilibrium was proposed by Romanenko [145].

Absorption of a microwave photon with energy $\hbar\omega$ causes quasiparticles at an energy E to move to an energy $E + \hbar\omega$. Figure 3.24a compares the standard Fermi–Dirac distribution function with the nonequilibrium quasiparticle

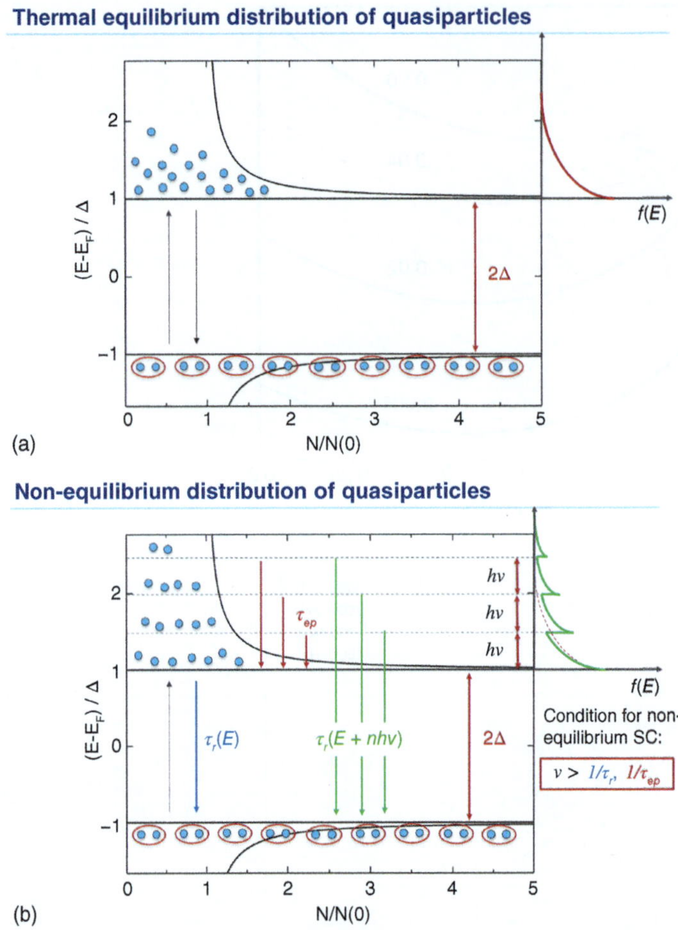

Figure 3.24 (a) The standard Fermi–Dirac distribution function. (b) The nonequilibrium quasiparticle distribution. Source: A. Grassellino and M. Martinello, Fermilab. [42, 157] Courtesy of A. Grassellino and M. Martinello, Fermilab.

distribution [157]. Under non-equilibrium conditions, a structure with sharp peaks at multiples of $\hbar\omega/\Delta$ shows up due to microwave photon absorption. Sequential single-photon absorption processes introduce spikes spaced at $\hbar\omega$ intervals. The effect increases with RF field as more photons are absorbed, driving the system further out of the standard thermal distribution. As the microwave field excites quasiparticles to higher energies, states near the gap edge free up, thereby increasing the effective gap. These effects lead to gap enhancement despite the creation of excess quasiparticles [146]. Both Cornell and Fermilab data (Figures 3.13 and 3.14) detect supporting gap increases.

An open question for applying the nonequilibrium model is whether 1.3 GHz frequency may be too low compared to kT (at 2 K) for the photons to drive the quasiparticles out of thermal equilibrium.

Eliashberg [158] also proposed it is possible, above a certain RF frequency, for the gap to *increase* with field. Thermally excited quasiparticles at the upper gap edge can be promoted to higher energies faster than they can recombine back down into Cooper pairs, or relax via inelastic scattering processes with phonons. This causes an increase in the number of available quasi-particle states existing near the gap edge, resulting in a decrease in the probability of photon absorption. As a result, the superconducting gap may increase with field, leading to a decrease in the surface resistance with field. According to Eliashberg, the minimum frequency (15 GHz for Nb [158, 159]) at which nonequilibrium effects may be visible depends on the inelastic collision time of quasi-particles scattering with phonons, and the quasi-particle recombination time. The introduction of interstitial impurities may alter the scattering and recombination times to cause nonequilibrium effects to become visible at lower frequencies (e.g. 1.3 GHz) than in the case of pure niobium, where the gap effect on R_s is visible without interstitials at higher frequencies (e.g. 3.9 GHz). The same effects may also strengthen the anti-Q-slope effect at higher frequencies as shown in Figure 3.15a.

Increase of the energy gap with the RF field has been measured in the past [160], and attributed to nonequilibrium effects [158]. De Vissser et al. [156] presents measurements of the electrodynamic response, quality factor, and resonant frequency, of an Al superconducting resonator (at 5.3 GHz) as a function of temperature and microwave power at low temperatures. They explain response and quasiparticle recombination time measurements by microwave-induced nonequilibrium effects on the distribution function, $f(E)$. Above 200 mK, the quality factor increases with increasing power. The microwave suppression of surface resistance was also observed in thin films and explained due to the nonequilibrium quasi-particle distribution function leading to the enhancement in supercurrent [161]. These are many promising hints but much more experimental work is needed to strengthen the non-equilibrium effects explanations.

3.10.4 Two-Fluid Model-Based on Weak Defects

The Weingarten [162] model is based on the two-fluid description of the surface resistance and the postulated presence of weak superconducting defects. The model uses the conductivities of the normal conducting current components of the superconductor and of weak superconducting defects when in the normal state. Other major features of the model are the superconducting proximity effect, and percolation behavior.

Weingarten considers nitrogen-doped niobium as a disordered [163] composite of small pockets of material with poor superconducting properties (like T_c, H_c, or Δ) that remain proximity-coupled to the superconducting Nb due to their small size [164, 165]. As the strength of the RF field increases, these pockets transition to the normal conducting state. This decreases the overall quasi-particle conductivity (and thereby lowers the BCS surface resistance) and is dependent on both RF field strength and frequency. Although the model contains many finely tuned parameters, it finds good agreement between the theory and recent experiments exhibiting anti-Q-slope, especially as it varies with frequency (Figure 3.25a).

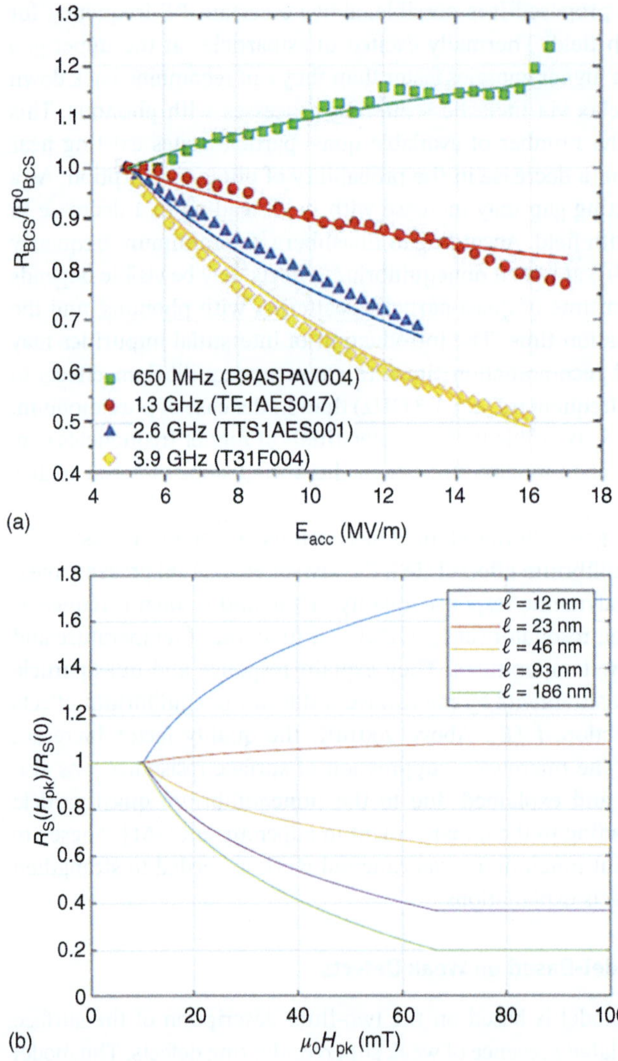

Figure 3.25 (a) Data for the normalized surface resistance R_s versus E_{acc} at 2 K, at various frequencies. Source: [146]/M. Martinello et al/APS/CC BY 4.0. Superimposed are the calculations from the Weingarten two-fluid model showing good agreement with data [162] with permission via CCC from IEEE. (b) Theoretical predictions of the field-dependent surface resistance from the Weingarten model for varying electron mean free path ℓ. Long mfps show stronger anti-*Q*-slopes, contradictory to data. Results calculated with $\sigma_{Nb} = \ell \cdot 1.94 \times 10^7$ Sv/m, $T = 2$K, and $f = 1.3$GHz. Source: [148]/Courtesy of J. Maniscalco, Cornell University/CC BY 4.0.

However, further calculations [148] using the model find that its prediction for the electron mean free path dependence of the strength of the anti-*Q*-slope is in disagreement with experiment: as modified by [148] the model predicts that cleaner cavities (i.e. with longer electron mean free path) should exhibit stronger anti-*Q*-slopes (Figure 3.25b). This is in contradiction with experimental observations (Figure 3.8).

3.11 Quench Field of N-Doped Cavities

Section 2.4.9 discusses the critical fields H_{c1} and H_{sh} for different κ values (different mean free paths) to understand how N-doping affects the quench field. As discussed in Chapter 2, H_{c1} depends on penetration depth, coherence length, and κ_{GL}. These parameters all depend on the mfp. Since doping with nitrogen lowers mfps, N-doping reduces H_{c1}. At fields greater that H_{c1}, it becomes energetically favorable for a vortex to be inside the bulk. However, there is an energy barrier to flux entry, delaying flux entry, keeping the superconductor in the metastable Meissner state. Therefore, reduction of H_{c1} with mfp does not necessarily mean that the cavity quench field will fall to H_{c1}. In a perfect cavity with no defects, the fundamental limit is the higher field, H_{sh}. However, the presence of defects can lower the barrier for flux entry, pushing the quench field down, closer to H_{c1}. Defects effectively lower the energy cost required for a flux line to enter at the defect. This can lead to earlier vortex penetration, and quench follows with increasing field. Surface roughness leading to local field enhancements may also contribute to early flux entry, and quenching near H_{c1}.

Figure 3.26a shows the quench field for N-doped cavities at Cornell compared to calculations for H_{c1}. Most of the cavities quench at or above H_{c1}. Several doped cavities exceed H_{c1}, some by a significant amount. Two cavities quenched below H_{c1} likely due to major defects, such as normal conducting niobium nitride inclusions that heat up in the RF field, and lead to thermal runaway. It is possible that poor phases of niobium nitride [113] are present even after significant EP due to deeper diffusion of the nitrides at the grain boundaries, or perhaps the inefficiency of EP that took place at higher temperatures in the early days of N-doping and EP evolution. Magnetic field enhancement due to surface roughness could also be responsible. One cavity quenched significantly below H_{c1}, likely due to an insurmountable multipacting barrier faced by the cavity performance.

Clearly, the data of Figure 3.26 from both Cornell and Fermilab show that lowering H_{c1} by lowering the mfp with N-doping lowers the average quench field, and that there is a correlation between longer mean free paths and higher average quench fields, especially for short mfp cavities, with the additional burden of defects for flux entry.

Figure 3.27a shows the quench field data for N-doped cavities from Fermilab, prepared by early doping recipes, (as well as early EP at warmer temperatures) and compared to cavities prepared by standard preparation methods, such as EP or BCP, as well as the high-performance cavities prepared by EP followed by 120 °C baking. The data here are plotted as a function of κ instead of mfp. The Ginzburg–Landau parameters were calculated using:

$$\kappa = \frac{\lambda}{\xi}; \quad \lambda = \lambda_0\sqrt{1+\frac{\xi_0}{l}}; \quad \xi = \left(\frac{1}{\xi_0}+\frac{1}{l}\right)^{-1} \tag{3.8}$$

with $\xi_0 = 38\,\text{nm}$ and $\lambda_0 = 39\,\text{nm}$ [96].

The earlier developed doping recipes were 10/0 and 20/0 (followed by 4–15 μm EP). The later recipe 2/6 (plus 5 μm EP) was finally used for accelerator cavities bound for LCLS-II. During the LCLS-II cavity production, the average accelerating

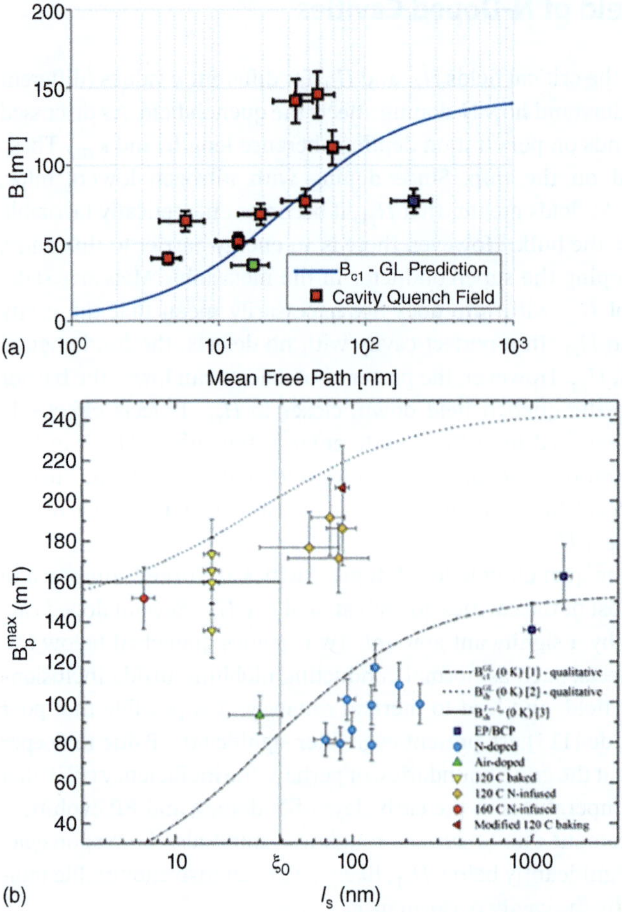

Figure 3.26 (a) Predicted B_{c1} from GL theory versus mean free path and single-cell cavity quench fields from Cornell data. The low field point in blue is most likely limited by multipacting and the point in green by normal conducting NbN islands on the surface due to only a small amount of final EP after nitrogen-doping [95] Courtesy of D. Gonnella, Cornell University. (b) Fermilab data: Comparing quench fields with calculated B_{c1} and B_{sh}. Note how other treatments such as 120 °C bake, and N-infusion (Section 7.3) give quench field results close to B_{sh}. Source: [68] Courtesy of M. Checchin, Fermilab.

gradient of about 370 cavities was 22.0 ± 3.6 MV/m (94 mT) [167]. The mfp for the 2–6 recipe is near 120 nm, leading to $\kappa = 1.55$.

Note that most of the Fermilab results in Figure 3.27a are near or below H_{c1}. Also shown are the calculated curves for H_{sh} discussed in Section 2.4.9. N-doped cavities do not approach this value. However, as we will discuss in Chapter 7, treatments such as 120 °C baking, two-step baking, and N-infusion do come closer to H_{sh}.

Figure 3.27b shows the data on several N-doped cavities from Fermilab prepared by the newer recipes 2/0 and 3/60, along with doped cavities from the older recipes. Here the quench fields from the newer recipes exceed H_{c1}. The incentive to explore the newer recipes came from higher gradient requirements of LCLS-HE

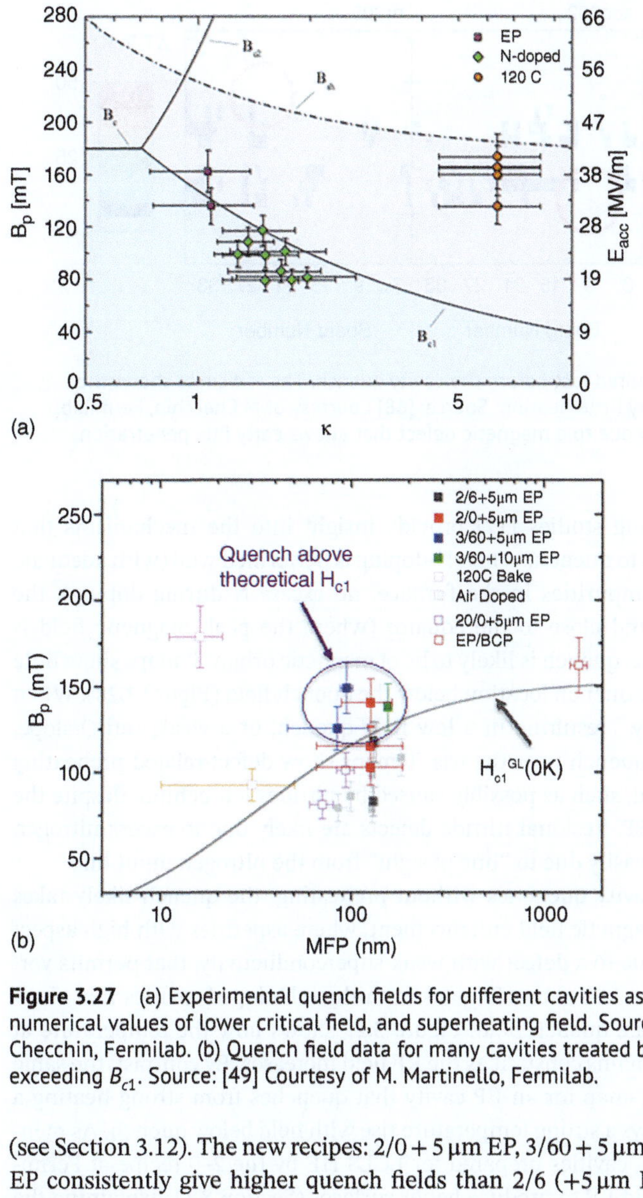

Figure 3.27 (a) Experimental quench fields for different cavities as a function of κ with numerical values of lower critical field, and superheating field. Source: [68] Courtesy of M. Checchin, Fermilab. (b) Quench field data for many cavities treated by 2/0 and 3/60 recipes exceeding B_{c1}. Source: [49] Courtesy of M. Martinello, Fermilab.

(see Section 3.12). The new recipes: 2/0 + 5 μm EP, 3/60 + 5 μm EP, and 3/60 + 10 μm EP consistently give higher quench fields than 2/6 (+5 μm EP) and heavy doped cavities.

Another important improvement that was implemented along with the new doping recipes at Fermilab was colder EP (Section 11.4.2) after doping. It is possible that cold EP achieves a smoother surface with less field enhancement for premature flux penetration to play an important role in reaching higher fields that exceed H_{c1}. It is also possible that cold EP is more effective in removing NbN precipitates [168]. One SEM study suggested that cold EP slows down preferential etching around the nitrides, eventually leading to a smoother surface. It is also possible that cold EP applied to the older recipes, such as 2/6, would give values that exceed H_{c1}.

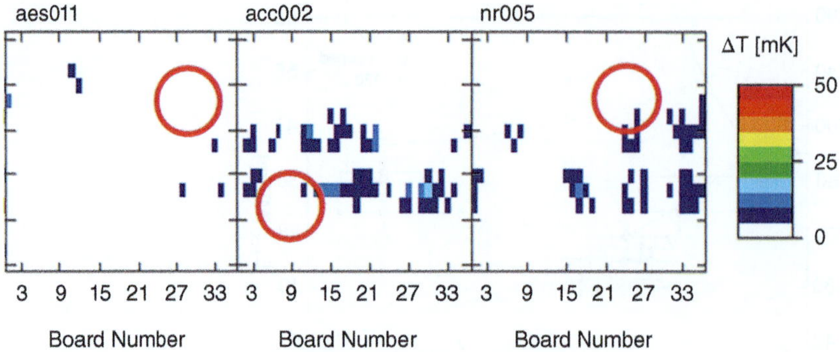

Figure 3.28 T-map acquired just before the cavity quench. The red circle shows the quench location with very little heating. Source: [68] Courtesy of M.Checchin, Fermilab. Here the quench is likely due to a magnetic defect that allows early flux penetration.

Temperature mapping studies [12] provide insight into the mechanisms that drive N-doped cavities to quench. When N-doping is performed well (with adequate EP after doping, no impurities in the furnace, no excess N during doping), the quench is usually found close to the equator (where the peak magnetic field is maximum). In this case, quench is likely to be of magnetic origin. T-maps show little or no preheating at the quench location below the quench field (Figure 3.28). When N-doping goes "poorly," resulting in a low field quench, or a weak anti-Q-slope, the cavity is likely to quench near the iris. T-maps show defect-related preheating below the quench field, such as possibly caused by nitrides left behind, despite the material removal by EP. Residual nitride defects are likely due to excess nitrogen deposited there more easily due to "line of sight" from the nitrogen input line.

When a N-doped cavity quenches without preheating, the quench likely takes place due to a local magnetic field enhancement, where asperities with high aspect ratio are present, or due to a defect with weak superconductivity, that permits vortex penetration. Figure 3.28 shows T-maps from three N-doped cavities that show *no preheating* below the quench field. Figure 3.29 shows how the temperature at these quench spots remains constant as the RF field increases. By contrast, the same Figure 3.29 shows a T-map for an EP cavity that quenches from strong heating a defect. The data displays a strong temperature rise with field below quench. As mentioned in Section 3.12, cavities prepared for LCLS-HE by the 2–0 recipe at Fermilab were treated by cold EP to produce better surfaces (Section 8.1), explaining the higher quench fields achieved.

At Cornell [95] Gonnella studied the quench of two strongly doped N-doped cavities using short RF pulses (100 μs) of high power (1.5 MW). (Descriptions of the high power pulsed technique can be found in [169, 170].) For a quench triggered by heating at a defect, the short pulses avoid thermal effects which grow with pulse length (time) to lead to thermally induced quenches. Short pulses allow higher fields by reducing the heating due to penetrated vortices at magnetic defects that lower the surface energy barrier, or the ohmic heating caused by highly resistive defects. The pulsed measurements show quench fields well above H_{c1}.

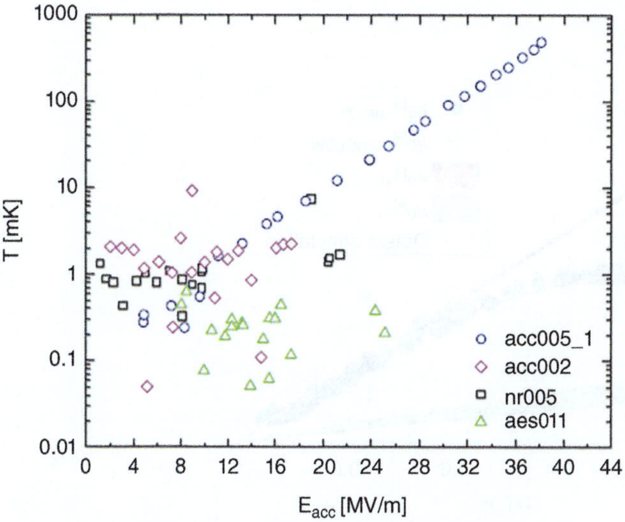

Figure 3.29 Temperature changes at the quench location as a function of the accelerating field. For well-prepared N-doped cavities, the temperature at the spot where the quench takes place remains constant as the RF field increases. In the case of a quench induced by heating at a defect, seen in the cavity (acc005_1), the temperature increases almost exponentially. This cavity was prepared by EP and 120 °C bake. Source: [68] Courtesy of M. Checchin, Fermilab.

Thus, the maximum gradient in pulsed operation well exceeds the maximum in cw operation.

One cavity quenched at 15 MV/m during cw operation at 2.0 K corresponding to H_{pk} = 64 mT. Short, high-power pulses yielded a significantly higher quench field near 100 mT (24 MV/m) at the same temperature. Figure 3.30a compares the short pulsed (100 μs) operation results with calculated values of H_{c1} and H_{sh} as well as with the cw results. Near T_c, the quench field with pulsed RF closely follows the superheating field, confirming theoretical expectations. At lower temperatures (and higher RF fields), the quench field departs from H_{sh}, but is still higher than H_{c1}, and higher than the cw quench field. This behavior indicates that quench takes place at a defect with ohmic losses, or that the quench is due to flux entry at a magnetic defect with lower surface barrier for flux penetration. It is not possible to distinguish between the two mechanisms from data of Figure 3.30a. The calculated field from the two defect mechanisms is the same.

The T-map [95] in Figure 3.30b pinpoints the location of the defect to be just above the equator, and helps to differentiate between the type of defect present. Below the quench field, there is no preheating at the quenching site. This result shows that quench at the defect is due to flux entry rather than ohmic defect heating. Another more heavily doped Cornell cavity showed a similar behavior, but at lower fields.

In summary, N-doping impacts the material properties of niobium, through the presence of interstitial N, lowering the mean free path of the RF penetration layer. This change in mean free path leads to a significant decrease of the lower critical field, H_{c1}, and to a small reduction in the superheating field, H_{sh} (Figure 2.16).

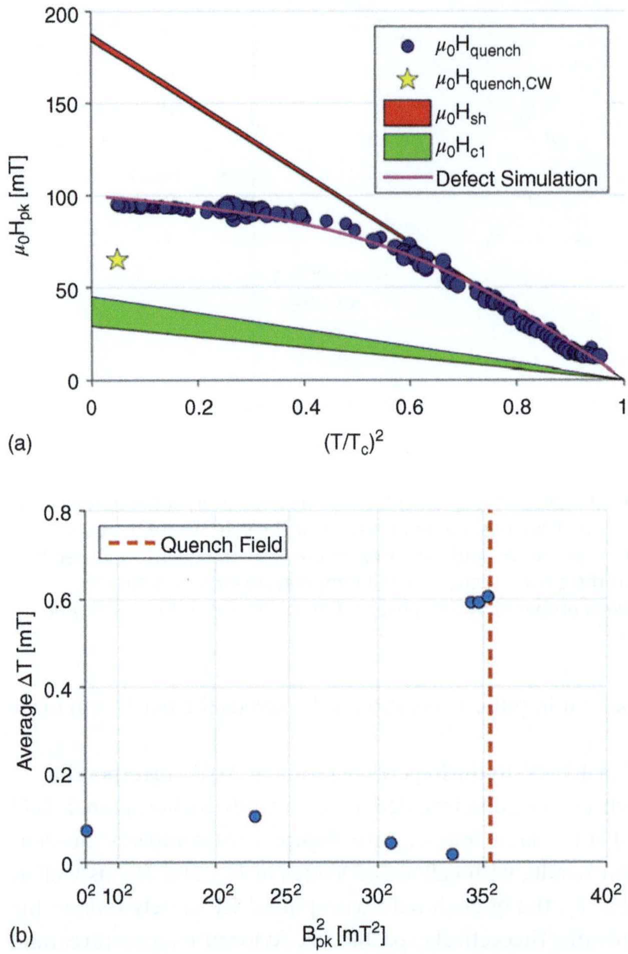

Figure 3.30 (a) B_{pk} *versus* $(T/T_c)^2$ for Cornell cavity LT1-2. Near T_c, the quench field follows the superheating field H_{sh}. However, at lower temperatures, the quench field is significantly lower than H_{sh} but still well above H_{c1}, consistent with a quench at a thermal or magnetic defect. CW quench field is also shown and can be seen to be higher than H_{c1} for this cavity. The (pink) smooth curve through the data points shows the expected drop in quench field due to heating at the defect from higher field, or from heating due to vortices that enter at the defect. The calculations do not distinguish between the two defect possibilities (thermal or magnetic). Source: [95] Courtesy of D. Gonnella, Cornell University. (b) Average temperature increase at the quench location versus $B^2{}_{pk}$. Very little preheating is present as the field in the cavity is increased until just prior to the quench field. This is indicative of abrupt flux entry at a defect, and not thermally triggered breakdown, which helps distinguish between thermally induced quench or one caused by abrupt flux entry. This quench was most likely caused by flux entry at a magnetic defect. Source: [95] Courtesy of D. Gonnella, Cornell University.

Quench studies under both cw and pulsed operation suggest that N-doped cavities are limited by magnetic defects more strongly than un-doped cavities. The quench field is most likely due to early vortex penetration below H_{sh} as evident from the absence of preheating at the quench location. The lower quench fields typically observed in moderate to heavily N-doped cavities are therefore likely due to lowering of H_{c1}, followed by vortex penetration at the field at which the energy barrier drops at a magnetic defect.

The similar results from Cornell and Fermilab on lack of preheating show that quench in N-doped cavities is more often triggered by flux entry when the RF field exceeds the critical magnetic field. Local magnetic field enhancements, such as at geometrical asperities with high aspect ratio will allow the magnetic field to penetrate at fields lower than H_{c1}. This behavior is quite different from quench in 120 °C baked cavities as discussed in Section 6.1.

3.12 Evolution and Comparison of N-doping Recipes

During the early period of the high Q R&D for LCLS-2 (2014), heavy doping recipes, such as 20/30 led to 40–60 nm mean free path, and gave quench fields around 16 MV/m. Later recipes such as 2/6 increased the mfp to >100 nm, along with higher H_{c1}, and also the cavities achieved quench fields around 22–23 MV/m.

The next big step was to move from longer to shorter duration of doping, from 20/30 to 2/6. The 2/6 doping treatment produces cavities with very high-quality factors [99, 171]. Besides the higher mfp, the 2/6 recipe tends to give a thinner doped layer, and smaller nitride precipitates.

For LCLS-II, more than 370 9-cell cavities were commercially produced at two different vendor sites, from raw Nb materials supplied by two different vendors. The cavities were doped with the 2/6 recipe followed by 5–7 μm EP at vendor sites [124]. The average gradient performance in vertical tests of 22 MV/m exceeded LCLS-II quench field specifications >19 MV/m, to allow reliable operation at an average gradient of 16 MV/m. According to single cell studies [49] the resulting mfp was between 90 and 200 nm, corresponding to H_{c1} values of 115–130 mT (see Figure 3.27b) equivalent to 27–30 MV/m. As a result, most 9-cell cavities for LCLS-II prepared by the 2-6 recipe (plus 5–7 μm EP) fall at or below H_{c1}, probably due to a significant presence of defects due to the large surface areas of 9-cells. These results were achieved before development of cold EP, discussed in Section 8.1.

3.13 High Q and Gradient R&D Program for LCLS-HE

For the 8 GeV upgrade, the specifications for LCLS-HE call for quench fields >23 MV/m (97 mT) to reach average operating field of 21 MV/m to accommodate space considerations in the tunnel. The Q_0 specified was also 2.7×10^{10}, same as for LCLS-II. For the higher gradient, FNAL pursued 2/0 doping, in which the cavity is doped for two minutes at 800 °C and immediately cooled with no anneal.

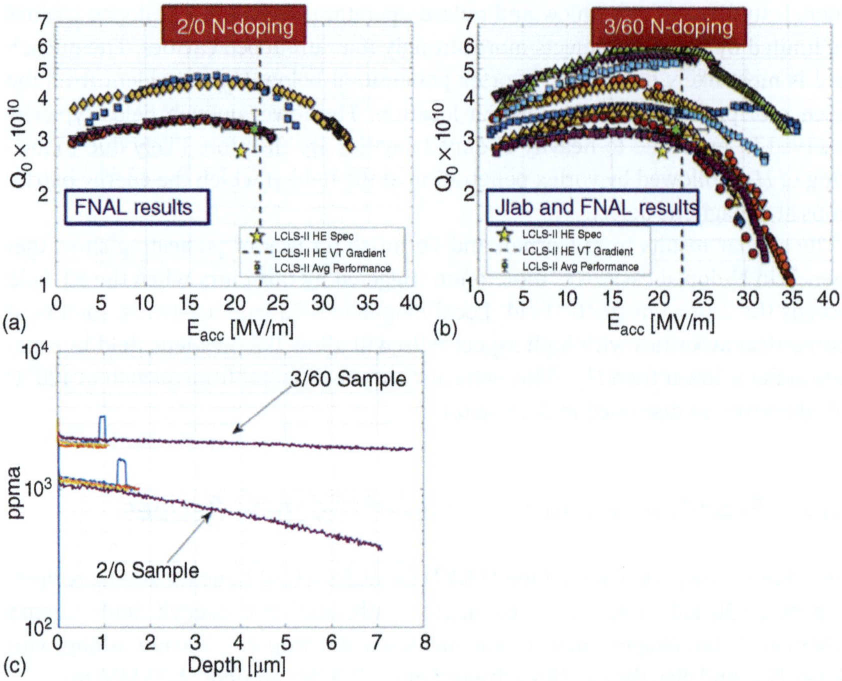

Figure 3.31 (a) Q_0 *versus* E_{acc} for single-cell cavities prepared with the 2/0 recipe [168] Figure provided Courtesy of D. Gonnella, SLAC. The stars show the LCLS-HE specifications and the average LCLS-II performance. All single-cell cavities exceed the LCLS-II HE specification. (b) Q_0 *versus* E_{acc} performance for single-cell cavities treated with the 3/60 recipe. Excellent performance was achieved, with some Q_s as high as 6×10^{10} at 2 K Source: [168] Figure provided Courtesy of D. Gonnella, SLAC. (c) Comparison of SIMS profiles for the N- concentration for the two doping recipes for LCLS-HE. Source: [172]/D. Gonnella, SLAC/JACoW/CC BY 3.0.

The choice of this recipe was based on expectations of increasing the mean free path of the doped layer (to reach higher H_{c1}) leading to higher quench fields based on the results of Figure 3.27. Results from single-cell cavities prepared with the 2/0 recipe are shown in Figure 3.31a [168]. As can be seen, all single-cell cavities exceeded LCLS-II HE specifications.

JLab pursued the 3/60 doping recipe [173], in which the cavity is doped for three minutes at 800 °C and then annealed in vacuum for 60 minutes. Table 3.2 compares the recipes for LCLS-II and LCLS-HE, i.e. 2/6, 2/0, and 3/60. The 3/60 recipe was motivated by the desire to create a uniform and deep doped layer, which could potentially reduce nanohydrides – that may be responsible for quench (Section 6.13) – and to remove the sensitivity of performance to the exact final EP amount [124]. Later, FNAL also investigated the 3/60 recipe. Figure 3.31b shows the combined results from JLab and Fermilab for the 3–60 recipe. Performance of these cavities was excellent, with nearly all cavities exceeding gradients of 30 MV/m. Q_0 performance was also stellar, with one cavity reaching more than 6×10^{10} at 21 MV/m and 2 K. Cold EP was a major improvement followed.

Table 3.2 N-doping recipes for LCLS-II and for LCLS HE [12].

2/6 Doping-baseline	2/0 Doping-FNAL	3/60 Doping-JLab
800 °C 3 h in UHV	800 °C 3 h in UHV	800 °C 3 h in UHV
800 °C 2 min 25 °mTorr N	800 °C 2 min 25 °mTorr N	800 °C 3 min 25 mTorr N
800 °C 6 min UHV	N/A	800 °C 60 min UHV
5 °μm EP	5 °μm EP	5 μm EP

In an effort [12] to systematically compare the performance of the three recipes shown in Table 3.1, Bafia at Fermilab carried out sequential treatments with each recipe using three single cell 1.3 GHz cavities. After each treatment and test, the cavities received a 40 μm EP reset to create a clean surface for the next treatment. First, the cavities were treated by the 2/6 + 5 μm EP recipe, which served as a baseline process for LCLS-II cavities. After a 40 μm EP reset, the cavities were treated by the FNAL developed 2/0 + 5 μm EP. Finally, the cavities were treated with the 3/60 + 5 μm EP recipe [173].

Figure 3.32 summarizes the performance of the three treatments as histograms of Q and quench fields. The solid lines represent average values. After 2/6 doping, the average quench field was 24 MV/m with a Q_0 at 16 MV/m of 3.61×10^{10}. After 2/0 doping, these averages substantially increased to 30 MV/m and 4.37×10^{10}. After 3/60 doping, the average quench field and Q_0 at 16 MV/m increased even further, up to 32.5 MV/m and 4.67×10^{10}, respectively. The additional 5 μm EP showed that the average values decreased to 30 MV/m and 4.53 10^{10}, which is comparable to the values obtained post 2/0 doping. Clearly, the optimized 2/0 and 3/60 nitrogen doping treatments improve both the quench field and Q_0 when compared to the baseline 2/6 treatment.

The mfp for the 2–6 (+5μm EP) recipe is between 120 and 130 nm. The new recipe 2/0 (+5μm EP) has similar mfp about 125 nm, and the 3/60 (+5μm) falls between 85 and 95 nm, while the 3/60 (+10μm) falls 90–155 nm [12]. It is reasonable to expect a somewhat shorter mfp for the 3/60 (+5μm) recipe due to the higher N content as visible from the SIMS analysis comparison [124].

The new recipes (2/0 and 3/60) appear not to raise quench fields by increasing the mfp, since these are comparable to the mfp of the (2/6) recipe. However, the older recipe has quench fields that fall at or below H_{c1}, whereas the new recipes appear to exceed H_{c1}. For example, the best single cells prepared with the new recipes reach 150 mT (35 MV/m). Hence the success of the new recipes may be due to additional improvements (such as cold EP) resulting in fewer defects for vortices to enter, rather than a higher H_{c1}.

First attempts to translate the better results to 9-cell cavities proved to be challenging, especially for the 3/60 recipe. At Fermilab, it was realized that the EP process for the single cells discussed above had changed to cold EP with electrolyte temperature reduced from 32 °C to <16 °C (Section 8.1), but not yet applied to the 9-cells. When cold EP was applied to one of the 9-cells, the gradient improved from 25 to

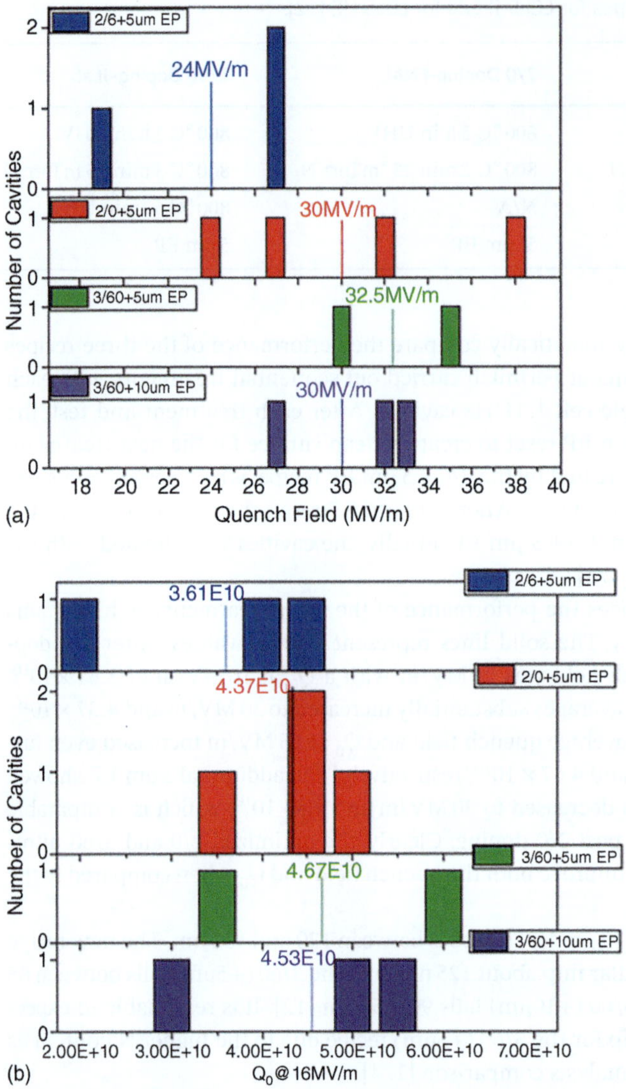

Figure 3.32 (a) Quench fields at 2 K of cavities after each treatment. (b) Q_0 measured at 16 MV/m and 2 K. Source: [12]/Courtesy of D.Bafia, Fermilab.

32 MV/m [168]. Subsequently, the average gradient for ten 9-cell cavities reached 25.6 MV/m (Figure 3.32a) as compared to the 9-cell average of 22 MV/m for the 2/6 recipe.

It appears that cold EP after N-doping is effective in reducing the defect population. A possible explanation is based on studies of EP on the nitride layer. During the higher temperature EP to remove the NbN layer, the HF etches the Nb around the nitride islands to effectively carve out the precipitates, leaving behind pits as shown in Figure 3.33 with possibly sharp edges [174]. By slowing down the EP process with

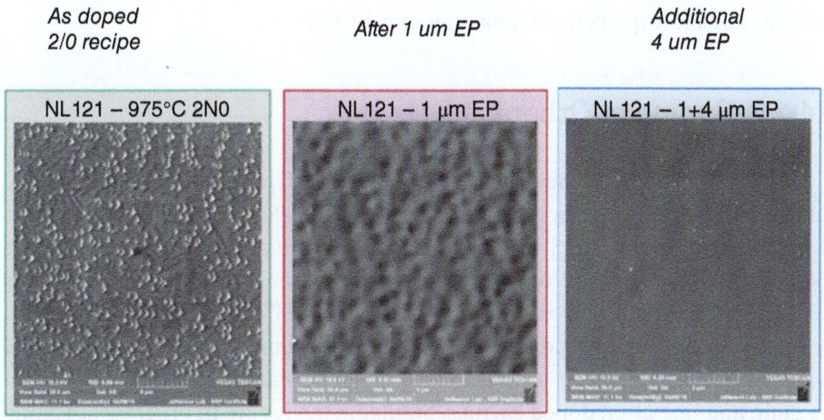

Figure 3.33 SEM photographs of N-doped sample before and after EP showing how nitride precipitates end up as pits due to a Nb excavation process of the warm EP. Source: Spradlin et al. Source: [174]/J. Spradlin et al. JACoW/CC BY 3.0.

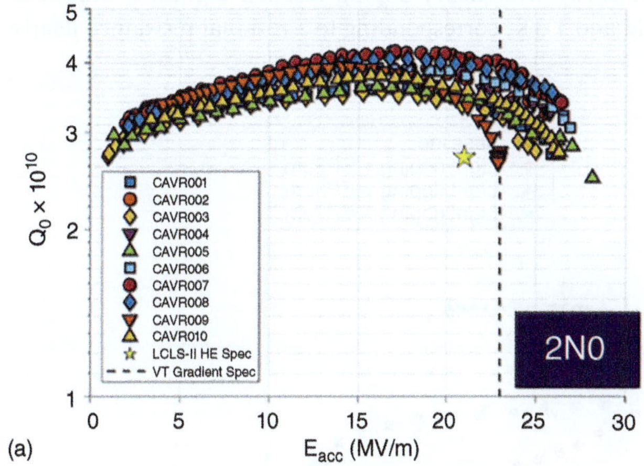

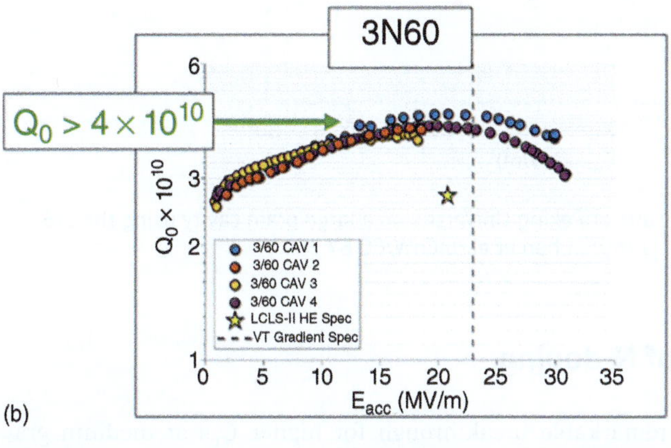

Figure 3.34 (a) Q_0 *versus* E_{acc} for ten 9-cell cavities processed with 2/0 N-doping recipe followed by cold EP. Average $Q_0 = 3.6 \times 10^{10}$ and gradient 25.6 MV/m, compared to the 9-cell average of 22 MV/m for the 2/0 recipe [175] Courtesy of D. Gonnella, SLAC. (b) Best early 9-cell results for 3/60 recipe. Source: [167] Courtesy of D. Gonnella, SLAC.

lower temperatures, the pits may possibly be avoided. Further EP reduces the number of pits, but a few quench-causing ones may be left over.

3.14 N-Doping at Other Labs

The Chinese Academy of Sciences (IHEP), Peking University (PKU), and the Shanghai laboratories explore N-doping for the SHINE (Section 11.4.3) accelerator, and for other projects. The SHINE laboratories have worked with heavy doping 20/30 and LCLS-II recipes (2/6), [177]. With heavy doping, they reached gradients of 16–20.3 MV/m and best Q values of 4.7×10^{10} at 16 MV/m. With the 2/6 recipe, they reached a bit higher gradient (22.5 MV/m), but somewhat lower Q values of 3.7×10^{10} at 16 MV/m. They are also studying performance differences between large grain and fine grain doped cavities. Similar activities are in progress at Peking University [176]. Figure 3.35 shows an example of a typical doping result on a large grain cavity. Good Q values with anti-Q-slope could be reached after sufficient EP.

At KEK [178], 1-cell and 3-cell cavities have been N-doped with preliminary encouraging results. One cavity with heavy doping reached highest Q values of nearly 2×10^{11} at low field and 1.5 K, corresponding to a residual resistance nearly as low as one nano-Ohm.

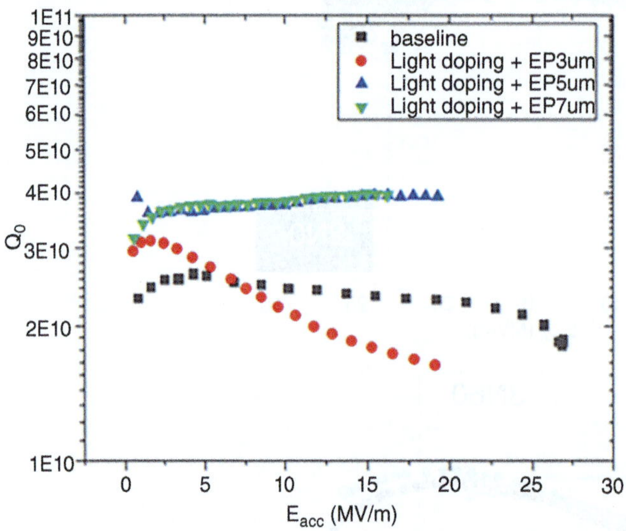

Figure 3.35 N-doping results at Peking University on a large grain cavity using the 2/6 recipe. Source: Chen et al. [176]/S. Chen et al./JACoW/CC BY 3.0.

3.15 Summary of N-doping

N-doping has been a remarkable breakthrough for higher Q_0s at medium gradients (15–20 MV/m). The treatment shows an extraordinary behavior in that the Q_0 increases as the field is raised. Substantial Q_0 enhancements have been demonstrated. N-doping impacts the final quench field. For carefully tailored recipes quench fields as high as 35 MV/m have been achieved.

Substantial theoretical and experimental R&D has been carried out to understand the nitrogen-doping effect and the origin of the anti-Q_0-slope. Decomposition analysis [99, 133] of the Q_0 vs. E curves shows that the rising Q_0 stems from the temperature-dependent part of the surface resistance showing that the rising Q_0 lies within the BCS theory. After doping, the residual resistance also falls. Several models have been proposed to understand these improvements. It is clear that the electron mean free path decreases due to the interstitial nitrogen content, Studies suggest [68] that the quench field for doped cavities is close to H_{c1} which falls due to increasing *kappa* (Ginzburg–Landau parameter) with lower electron mean free path due to the interstitial nitrogen. With superior surface treatment such as cold EP it is possible to exceed H_{c1} by avoiding defects that are prone to magnetic quench.

N-doping opens significant cryogenic benefits to accelerators planning CW operation, such as LCLS-II or future ERLs.

4

High Q via 300 °C Bake (Mid-T-Bake)

4.1 A Surprise Discovery

Another major discovery with major Q benefits has been the Fermilab invention [179] by Posen of a new procedure, called "mid-T baking, or 300 °C baking." The treatment raises the Q for medium gradients and cw operation. Figure 4.1 shows several benefits, some unexpected. Despite the absence of N exposure, the *Q versus E* behavior is similar to N-doping, especially the occurrence of the anti-Q-slope. The low-field residual resistance reaches a very low value to yield a record high Q at 1.5 K. The low field Q-slope (LFQS) disappears. Q values at extremely low fields become higher to open the door to new applications in quantum computing (Chapter 13). A further simplification of the 300 C bake treatment explored at IHEP (China) [180] and KEK (Japan) [181] leads to a more straightforward procedure than N-doping plus EP, and again yields results similar to the N-doping.

In the discovery stage of the 300 °C mid-T bake, 1.3 GHz, 1-cell cavities, prepared by EP, were completely sealed under their own vacuum to prevent exposure to a possibly lower quality furnace vacuum. The baking was carried out under vacuum at 300 °C for 2.5 hours while connected to a turbo pump. To prevent oxidation on the outside wall, the cavity was placed in a high-purity Argon atmosphere inside a stainless-steel chamber. The vacuum flanges with the usual Al–Mg joints were water cooled, and heating bands around the beam pipes were added to achieve 300 °C at the cavity cell. After baking, the cavity was kept under vacuum till the RF test. Later tests explored the consequences of exposure to clean air, and high-pressure water rinsing, opening the possibilities for simplifying the implementation of mid-T baking, as demonstrated later by KEK and IHEP.

4.2 Similarities to N-Doping

Surprisingly, at 2.0 K, the Mid-T baked cavities show an anti-Q-slope typically associated with N-doping [99]. The *Q increases* with E_{acc} from 5 to 25 MV/m reaching 3–5 × 10^{10}. For the best 1-cell cavity, the Q arrives at 5 × 10^{10} at 30 MV/m. Decomposition analysis of the surface resistance (Section 3.6) shows that the anti-Q-slope behavior originates from the BCS resistance, similar to N-doped

Superconducting Radiofrequency Technology for Accelerators: State of the Art and Emerging Trends, First Edition. Hasan Padamsee.

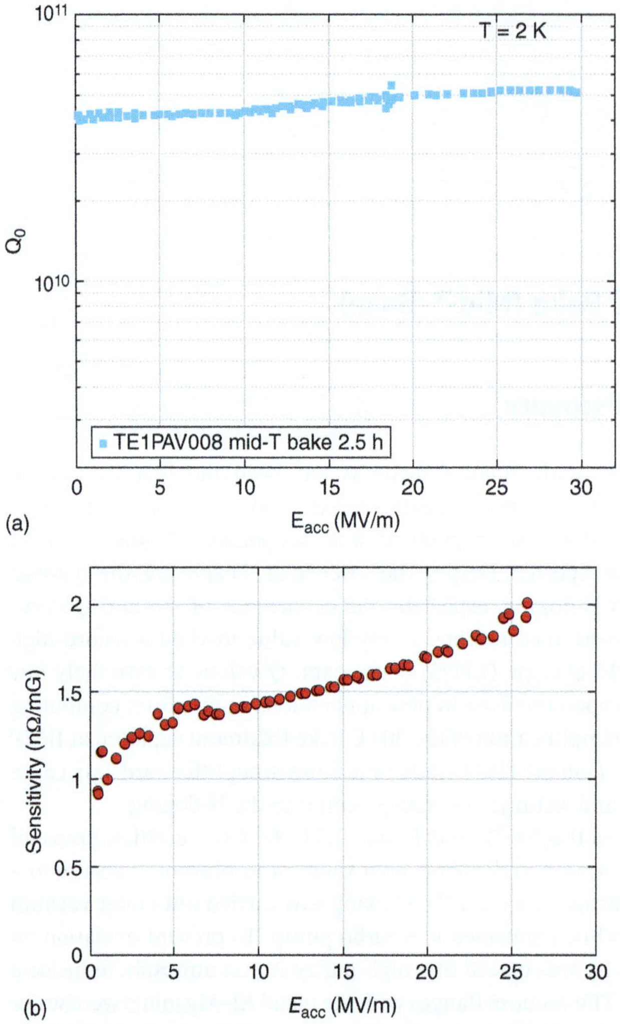

Figure 4.1 (a) Very high $Q = 5 \times 10^{10}$ at 30 MV/m by baking at 300 °C to dissolve the natural oxide (and other surface layers) into the bulk, but not exposing the cavity to air or water before RF measurements. (b) Sensitivity to trapped flux as a function of accelerating field after mid-T bake, prior to venting. The sensitivity is similar to that of N-doped samples. Source: [179]/S. Posen et al./American Physical Society/CC BY 4.0.

cavities. These properties suggest the presence of interstitial N, to be discussed further below.

When the gradient after treatment exceeds 25 MV/m, mid-T baking also shows elimination of the high field *Q*-slope (Chapter 6), which usually starts above 20–25 MV/m for the standard EP treatment.

The final quench field for most cavities after mid-T bake is in the range of 25–30 MV/m (H_{pk} = 106–128 mT), again similar to the best nitrogen-doped cavities [110]. The sensitivity of residual resistance to the trapped dc magnetic field is higher than that of an EP-treated cavity, as shown in Figure 4.1b, resembling that of a N-doped cavity [182]. At low fields, the sensitivity to trapped flux is 1.5 nΩ per mOe rising to 2.5 nΩ/mOe at 15 MV/m, consistent with the behavior of N-doped cavities (Section 3.7.1).

The mfp extracted from the frequency increase near T_c is about 50 nm, corresponding to a heavier N-doped case (Section 5.1). Yet the higher quench field (128 mT) is well above H_{c1} (90 mT for 50 nm mfp), suggesting that the mid-T bake preparation is low in defects that lead to quenching at or below H_{c1}. Note that a cold EP treatment carried out prior to 300 °C bake could be an important contributing factor to the higher gradient.

In the temperature range 1.4–1.5 K, governed by the residual resistance, Q_0 values reach $3–4 \times 10^{11}$ even for gradients as high as 30 MV/m. While Q's of 2×10^{11} have been reported previously [171, 183, 184], a Q value above 3×10^{11} at a gradient ~20 MV/m is unprecedented. An upper bound to the corresponding residual resistance is 0.63 ± 0.06 nΩ at 16 MV/m. To minimize residual resistance from trapped flux dissipation, magnetic field compensation was used to provide a low dc magnetic field environment. As discussed in Chapter 5, the best treatments available were applied to minimize trapped flux. These included heating the cavity at high temperatures (900 °C) to improve flux expulsion from the material, and fast cooling to He temperature with strong thermal gradients for best flux expulsion (Chapter 5).

To study the impact of mid-T baking on the composition of the surface, electropolished niobium samples were analyzed by a SIMS facility equipped with an in situ heating stage. Measurements were carried out on the EP baseline samples, on samples after mid-T bake, and after exposure to air. Figure 4.2a shows how the mid-T bake drops the Nb_2O_5- signal to the noise level over a very short sputter time in the SIMS. Therefore 300 °C bake completely eliminates the pentoxide layer. There is a small residue of NbO and NbO_2. After reexposure to air, the pentoxide regrows on the sample, though it appears thinner, due to growing the oxide in air, rather than in water. Previous surface studies of niobium samples via XPS [185–187] or X-ray reflectivity have shown that pentoxide can be mostly dissociated by vacuum heat treatment at temperatures above ~300 °C, driving O into Nb, and leaving a small amount of residual suboxides on the surface.

Figure 4.2b shows an unexpected increase in the intensity of the NbN signal after the mid-T bake, even though there was no direct N exposure to the cavity surface. A bump appears close to the surface, some tens of nm deep. It is similar in depth to a near-surface N-bump found for a 120 °C nitrogen "infused" sample (discussed in Section 7.3). The N-bump persists after the oxide is regrown.

The unexpected SIMS result, showing the presence of nitrogen over the depth of tens of nm accounts for the anti-Q-slope behavior. Even though no N gas was injected during the mid-T bake, N may have come from surface contamination. Other labs that explored the mid-T bake have also found N-doped behavior, which shows that the source of N is not a singular accident, but a general feature.

The presence of interstitial N and interstitial O from the 300 °C baking could be responsible for curing the HFQS (which normally has an onset field of 20–25 MV/m) by inhibiting the formation of hydrides, as will be discussed in

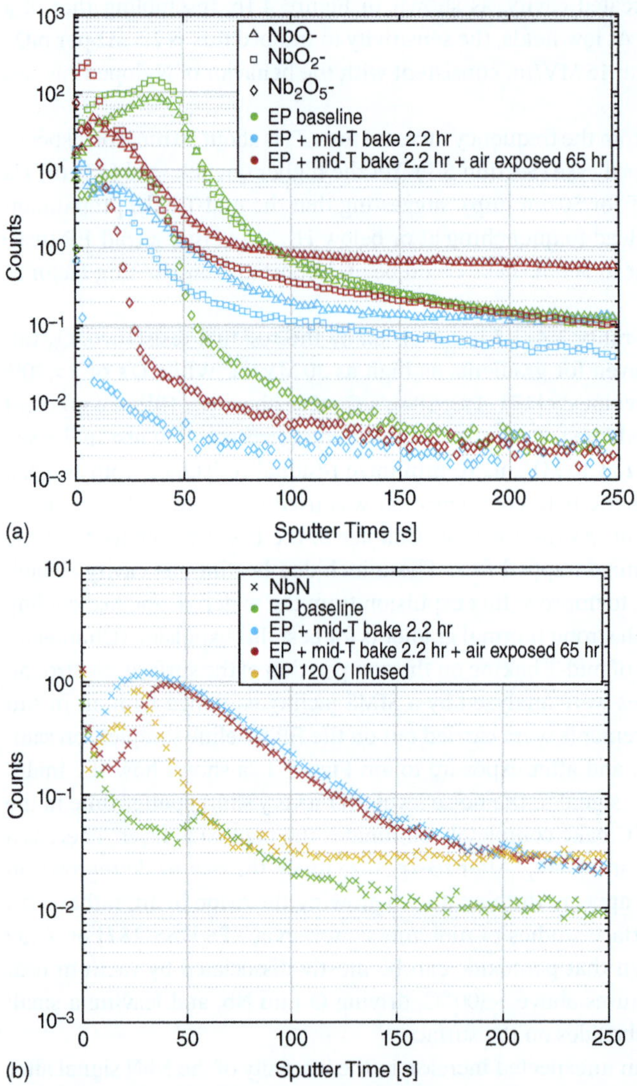

Figure 4.2 (a) SIMS depth profiles for various Nb oxides for a baseline EP sample, then after mid-T bake, then exposed to air, all inside the SIMS system. (b) Comparison of nitrogen signal between EP baseline, mid-T bake (with and without air exposure), and separately 120 °C N-infused sample. Source: [179]/S. Posen et al./American Physical Society/CC BY 4.0.

Section 6.6. However, the maximum fields are not as high as achieved with 120 °C bake or N-infusion, as we will see in subsequent chapters. The N interstitial profile (Figure 4.2b) shows a slightly higher concentration of N and goes out to a greater depth than for the 120 °C N-infused, high gradient case. It opens the possibility that reduced baking time may push the behavior of mid-T baked cavities toward the N-infusion case- a possibility worth future exploration.

To study the effect of reoxidizing the surface on the RF performance, two cavities were vented with nitrogen gas, followed by a high-pressure water rinse before reassembly, followed by another RF test. The Q_0 degraded some but remained higher than the EP baseline at 2 K. The HFQS reduction was preserved after oxidation and HPR, to reach similar quench fields. Some of the lost Q could be regained by 2X HF rinsing to grow a fresh pentoxide layer. It may be interesting to explore the possibility of preservation of the oxide-free surface, and its superior behavior by exposing the cavity only to N or Ar, without water.

4.3 Mid-T Baking at Other Labs

Both KEK and IHEP [180, 181] have successfully followed up on the mid-T bake, but with a simpler baking procedure to realize similar benefits. They used a regular furnace with a double vacuum system instead of baking self-sealed cavities. Nb foil endcaps offered protection for the cavity surface.

IHEP explored a wider baking temperature range from 250 to 400 °C for three hours with ten 1.3 GHz single cells. After mid-T furnace bake, these cavities were exposed to air, received high-pressure water rinse (HPR) and reassembly in a clean room, followed by vertical tests, all standard procedures. All the cavities reached high Q in the range of $3–5 \times 10^{10}$ at medium field and exceeded 25 MV/m. The anti-Q-slope appears between 5 and 18 MV/m, as observed with N-doped cavities (Figures 3.1, 3.31, 3.32). The average Q was 2.9×10^{10} at 24 MV/m. The highest Q is 4.9×10^{10}@16 MV/m. One cavity treated at 250 °C reached 1.2×10^{10}@37 MV/m, which is the highest gradient among cavities treated with mid-T furnace bake at IHEP. A likely explanation for the high gradients is that O and N absorbed from the oxide during baking serve to prevent formation of niobium hydrides by trapping H (Section 6.6). It is likely that the slower diffusing N plays the dominant role because the O interstitial content becomes too dilute at 300 °C. Overall, 300 °C for three hours was judged to be the best temperature for reaching high Q at medium field, as desired for upcoming applications such as SHINE (Section 11.4.3).

Using a bigger furnace, IHEP treated six 9-cell cavities. The cavities were first prequalified with standard treatment to reach 30 MV/m. Figure 4.3 shows all 9-cell units demonstrate high Q in the range of $3.5–4.4 \times 10^{10}$ at the gradient between $16 \sim 24$ MV/m. The average Q of all the six 9-cell cavities is 3.8×10^{10}@16 MV/m. The highest gradient is 26.5 MV/m. At 1.5 K, where residual resistance dominates, the best Q was 1×10^{11}, corresponding to residual resistance of 2.7 nΩ. Due to regrowth of the oxide, the residual resistance is not as low as the mid-T baked results when there is no exposure to air.

Thus mid-T baking can accomplish the main benefits of N-doping in a simpler way. It is furthur encouraging for practical implementation that mid-T baking can be carried out in a regular furnace (using additional good protection against contamination) and the benefits are preserved after HPR. Another important simplification over high-temperature N-doping is that there is no need for post-doping EP since there are no nitrides formed, and post-furnace treatment EP can be avoided since the N interstitials are only present in the first tens of nm. The absence of nitrides is also favorable for higher quench fields due to fewer magnetic or thermal defects.

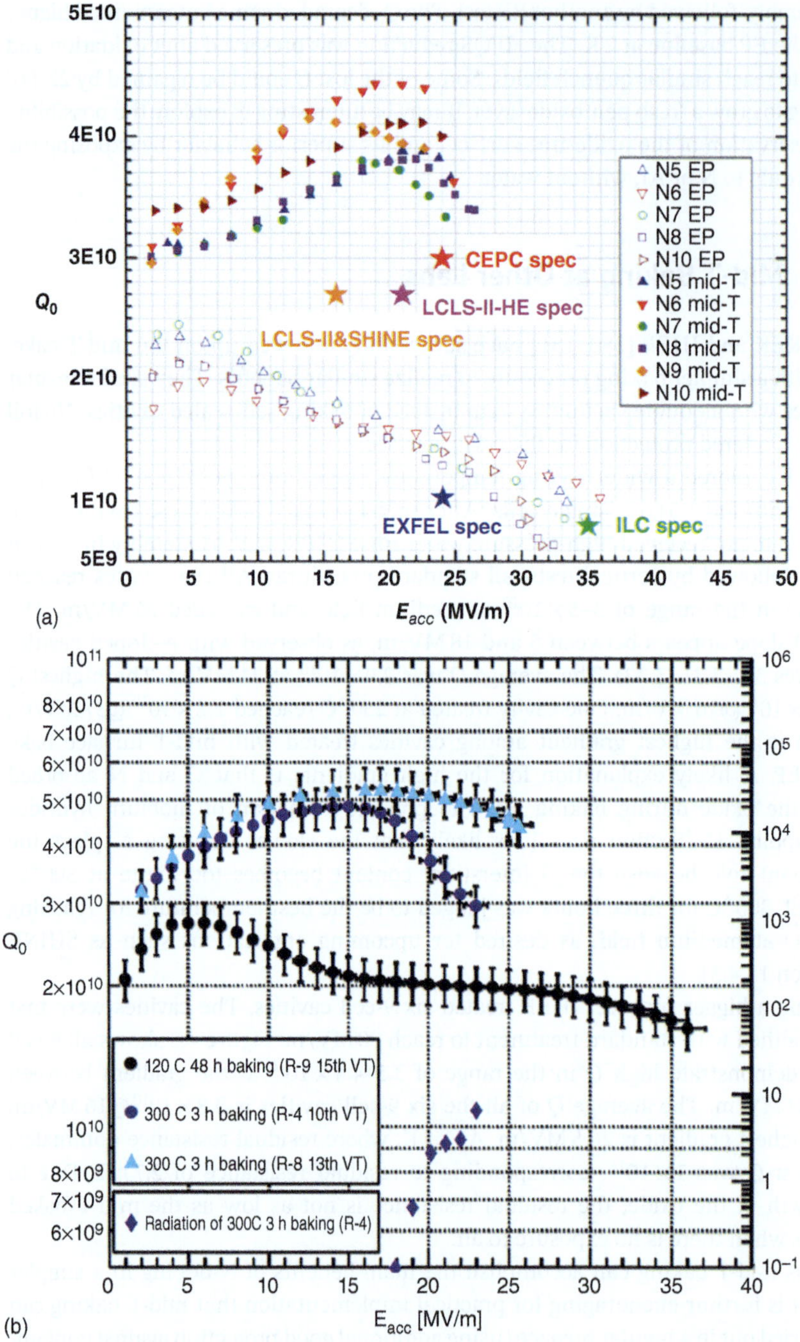

Figure 4.3 (a) IHEP (China) results on mid-T baking for 9-cell cavities compared to results on the same cavities with the standard ILC treatment. Source: He et al. [180]/CC0 1.0. (b) mid-T baking results from KEK. Source: [181]/H. Ito et al./Oxford University Press/CC BY 4.0.

KEK studied mid-T furnace baking with 1-cells at various temperatures from 200 to 800 °C for three hours with and without HPR. 200 °C gave the same *Q versus E* result as for standard 120 °C 48-hour baking with HFQS absent, confirming the baking studies discussed in the Section 6.6 on HFQS mitigation. The highest *Q* value of 5×10^{10} at 26 MV/m (Figure 4.3b was obtained at 300 °C, and did not deteriorate with HPR. This result showed an anti-*Q*-slope, as with N-doping. Higher sensitivity of residual resistance to dc magnetic was confirmed as for N-doped cavities. Other baking temperatures were not as successful. For 500–800 °C HFQS was observed once more.

Overall, there are some distinct advantages of mid-T Baking over N-doping, as carried out at IHEP and KEK in a standard 800 °C furnace used for H-degassing, and proven with 9-cells at IHEP. Mid-T baking appears to accomplish N-doping with results superior to 2–6 doping and comparable to 2/0 and 3/60 doping recipes developed for LCLS-HE. The mfp from the Fermilab study is about 50 nm. Systematic studies of the mfp from the other labs are missing. The high gradients (close to or exceeding H_{c1}) suggest that the mid-T baking method is successful in avoiding defects that cause flux entry. Due to the absence of N gas exposure, mid-T bake likely also prevents the formation of NbN precipitates, which become defects. Another advantage is the absence of the need for EP after doping, again due to absence of NbN. The fact that the results remain of high quality after clean air exposure and HPR is very encouraging for practical applications. Clearly, a high-quality furnace is necessary, which IHEP achieves with the double vacuum chambers.

4.4 The Low-Field Q-Slope (LFQS) and 340 °C Baking Cures

In the GHz frequency range, niobium cavities generally show a decrease in the quality factor at fields lower than 5 MV/m. This familiar effect is called the LFQS. Typically at 2 K, the *Q* falls from a few $\times 10^{10}$ to about 10^{10}. At 1.5 K, for a cavity capable of higher *Q*'s, the *Q* falls from near 10^{11} to about $7\text{–}8 \times 10^{10}$. If the field is lowered to ~0.2 MV/m, the *Q* continues to fall [63, 188, 189] to about 3×10^{10} (Figure 4.4a). When *Q* measurements are extended to ultra-low temperatures ($T < 20$ mK) using a dilution refrigerator, the *Q* continues to drop down to about 2×10^9 at 10 mK (Figure 4.4b). We expect that ultra-low temperatures eliminate RF losses from the temperature-dependent part of the surface resistance, such as from thermally excited quasiparticle, as described by Mattis and Bardeen [60]. Therefore any surface resistance in this temperature region must be a part of the residual surface resistance.

It is interesting to explore the behavior of *Q* at even lower fields, down to levels that correspond to single photons [190]. The microwave setup for such a delicate measurement includes a series of attenuators on the cavity input line, as well as both cryogenic and room-temperature amplifiers on the RF pickup line. A low-noise cryogenic amplifier (HEMT- high electron mobility transistor) makes it possible to reliably measure the photon lifetimes down to an average cavity population of ~10

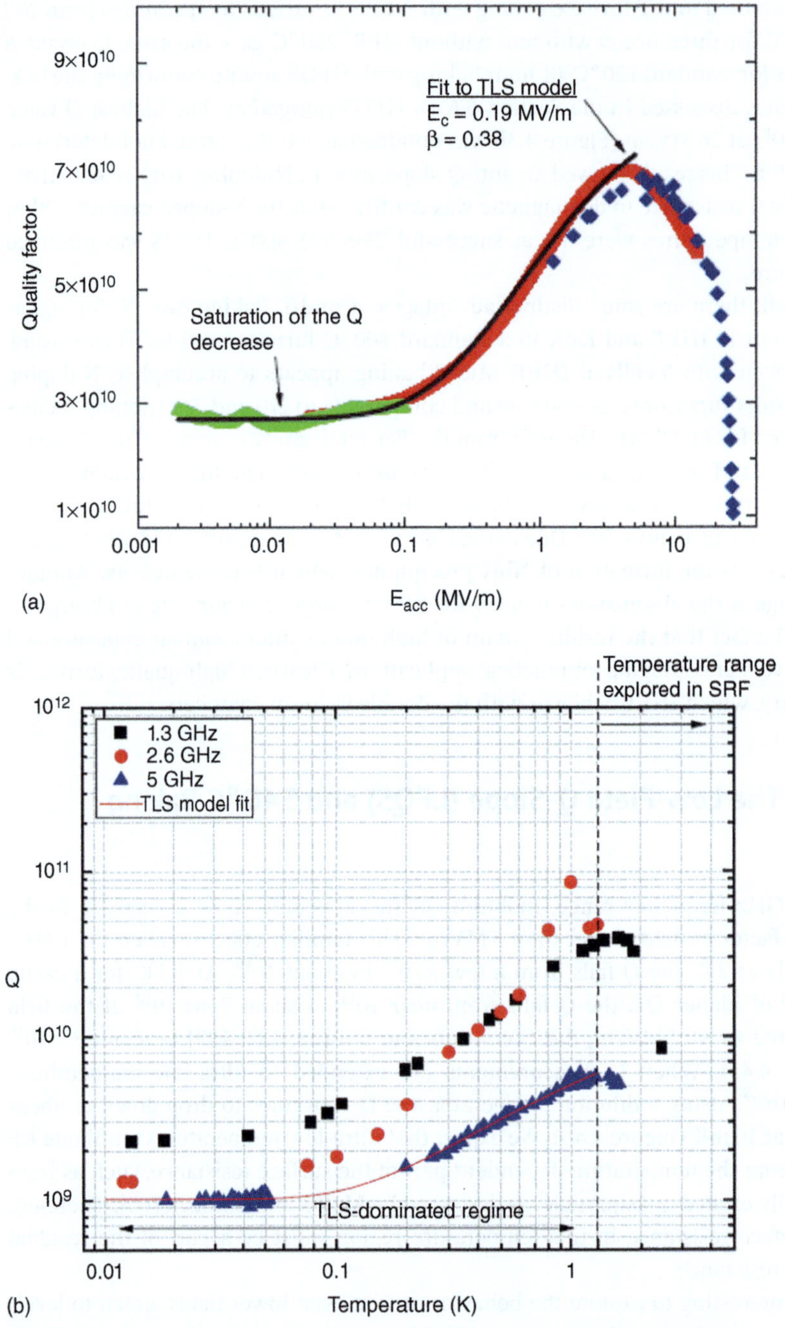

Figure 4.4 (a) $Q(E)$ at 1.3 GHz and 1.5 K measured over a wide field range for an electropolished 1-cell cavity. The fit to the TLS model (Section 4.6) is also shown. Source: [190]/A. Romanenko and D. Schuster/American Physical Society/CC BY 4.0. (b) $Q(E)$ extended to ultra-low temperatures for cavities at several frequencies [191]. (c) Additional surface resistance from thicker oxide by anodizing, compared with surface resistance at 5 MV/m as a function of the RF field amplitude, measured at T = 1.5–1.6 K. Dashed lines are fits to TLS model. Source: [190]/A. Romanenko and D. Schuster/American Physical Society/CC BY 4.0.

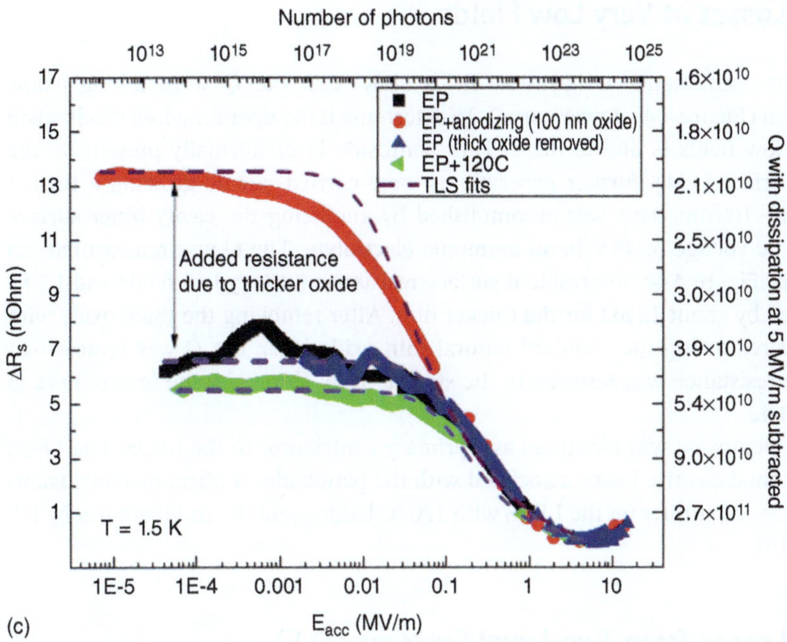

Figure 4.4 *(Continued)*

photons, while a Josephson parametric amplifier (JPA) in the output line extends the sensitivity further to single-photon levels! The average photon number is calculated from the cavity stored energy.

The JPA [192] is a superconducting amplifier that operates near the quantum limit (one photon). The JPA is formed by a superconducting quantum interference device (SQUID) combined with a superconducting coplanar waveguide resonator. (For a more complete discussion of the Josephson effect and the SQUID, refer to Section 13.6.) The combined system acts as a tunable nonlinear microwave resonator, whose frequency can be varied in situ via an external magnetic field. As a mechanical analogy, the eigenfrequency of a pendulum of variable length, can be tuned by its length. Tunability of the nonlinear microwave resonator can be exploited to parametrically pump the JPA via application of a strong microwave signal at twice the resonant frequency. This results in a strong parametric amplification of weak signals at the JPA. Thus the operation of the JPA requires a flux bias to bring the amplifier nearly on resonance with the cavity and a charge or flux pump to enable the amplification. The JPA can reach >20 dB low-noise amplification of quantum signals.

Usual power measurements are typically not bandpass filtered and are therefore limited by various sources of the RF noise present in the broad frequency range. Special methods are needed since vector network analyzers do not have sufficient frequency stability to measure high *Q*s. Figure 4.4c shows *Q* values at ultra-low field levels obtained by single decay measurements [193] of the transmitted power with a narrow 10–10 000 Hz bandpass filter around the resonance.

4.5 Losses at Very Low Fields

The Q measurements at ultra-low fields show that the Q stops falling below 0.1 MV/m (Figure 4.4c–lower curve). To determine if the operating loss mechanism at very low fields is due to the 3–5 nm pentoxide layer normally present on the cavity surface [194], further experiments were carried out by growing a thicker oxide of ∼100 nm. This was accomplished by anodizing the cavity inner surface using a dc voltage of 48 V in an ammonia electrolyte. The Q was remeasured. As shown in Figure 4.4c, the residual surface resistance (at very low fields and 1.5 K) increased by about 12 nΩ for the thicker film. After removing the thick oxide with HF, and regrowing the standard natural thin oxide layer, the Q was remeasured and the resistance was restored to the starting level shown by the lower curve of Figure 4.4c.

Thus, pentoxide was identified as a primary contributor to the LFQS. The LFQS arises from dielectric losses associated with the pentoxide, confirming conclusions from previous studies on the LFQS with 120 °C baking and by oxide removal by HF rinsing [5].

4.6 Losses from Two-Level Systems (TLS)

Martinis et al. proposed two-level systems (TLS) as the main cause of the increased low RF field losses in planar superconducting resonators [195]. Two established signatures for this type of dissipation are similar to those found for Nb cavities (i) saturation of losses below a certain threshold, and (ii) dependence on the amount of the amorphous material exposed to the electric fields.

These findings suggest that TLS lossy mechanisms operating within the Nb_2O_5 layer are the root cause of LFQS phenomena. Similar losses found in planar resonators [196] have been attributed to TLS losses present in the native niobium oxide. Superconducting circuits made from aluminum or niobium Josephson junctions are also limited by losses attributed to two-level defects within the amorphous oxide layers (Figure 4.5a).

One source of TLS is due to individual atoms tunneling between two local energy minima within the amorphous part of the resonator. For cavities, the TLS losses originate from the native oxide layer on the surface. For Josephson junctions, the dielectric oxide (tunneling barrier) between the superconducting electrodes is responsible for TLS losses.

Two-level defects dominate the low-temperature behavior of glassy systems, such as amorphous Nb_2O_5. These defects are not due to impurities, but rather emerge from the deviations away from crystalline order that characterize the amorphous state. The low-temperature behavior of these materials is generally described by the standard two-state tunneling model (Figure 4.5b), hence the name TLS.

An attractive picture is that the two-level system is formed by the movement of an atom or small group of atoms between two potential minima (Figure 4.5b). This still leaves open the question of which atom or group is actually "moving." In one model,

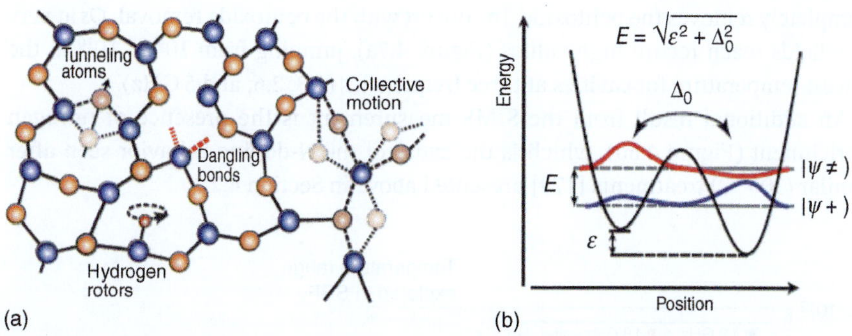

Figure 4.5 (a) Example mechanisms of TLS formation in an amorphous material: tunneling of single atoms and collective motion of small atomic groups, dangling bonds, and Hydrogen defects. (b) Double-well potential model for TLS. The energy difference E between the TLS eigenstates is determined by the asymmetry energy E and the inter-well tunneling rate Δ_0. Source: [196]/C. Müller et al. with permission of IOP Publishing.

TLS is perceived as atomic-sized electric dipoles, which couple to the oscillating electric fields. According to the prevalent theory [197], TLS-induced losses emerge from the dipole moments of "loose" atoms coupling to the electric field at the surface of resonators. This mechanism provides a model to fit the Q data for LFQS in Figure 4.4. In the TLS-dominated regime, the model provides an excellent fit with a characteristic temperature dependence: $1/\tanh\left[\alpha\left(\frac{\hbar\omega}{2kT}\right)\right]$. For the decreasing quality factors at very low temperatures, the $Q(T)$ for all the cavities clearly follows the TLS model, as shown by the fits in Figures 4.4. Due to the tunneling mechanism, the RF losses increase exponentially with local (electric) field.

4.7 Eliminating TLS Losses

Removing the pentoxide by vacuum baking at 340 °C for several hours provides a definitive test that TLS losses come from the pentoxide. TOF-SIMS [198] measurements on Nb cavity cutouts presented in Figure 4.6a show that the 340 °C baking

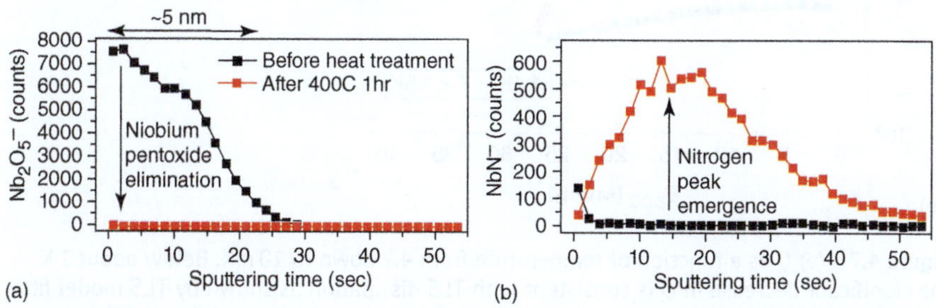

Figure 4.6 TOF-SIMS depth profiles from a cavity cutout. (a) The Nb_2O_5 pentoxide layer is completely dissolved after the 400 °C treatment. (b) Increase of the nitrogen level within 10 nm from the surface, similar to Figure 4.2b. [198]/A. Romanenko et al/American Physical Society/CC BY 4.0.

completely removes the pentoxide. In concert with the pentoxide removal, *Q*s at very low fields reach record high values (Figure 4.7a), jumping from 10^9 to 10^{10} at the lowest temperature for cavities at three frequencies (1.3, 2.6, and 5 GHz).

An additional result from the SIMS measurement is the presence of nitrogen enrichment (Figure 4.6b), which is the cause of the N-doping behavior seen after similar (300 °C) treatments [179] presented above in Section 4.2.

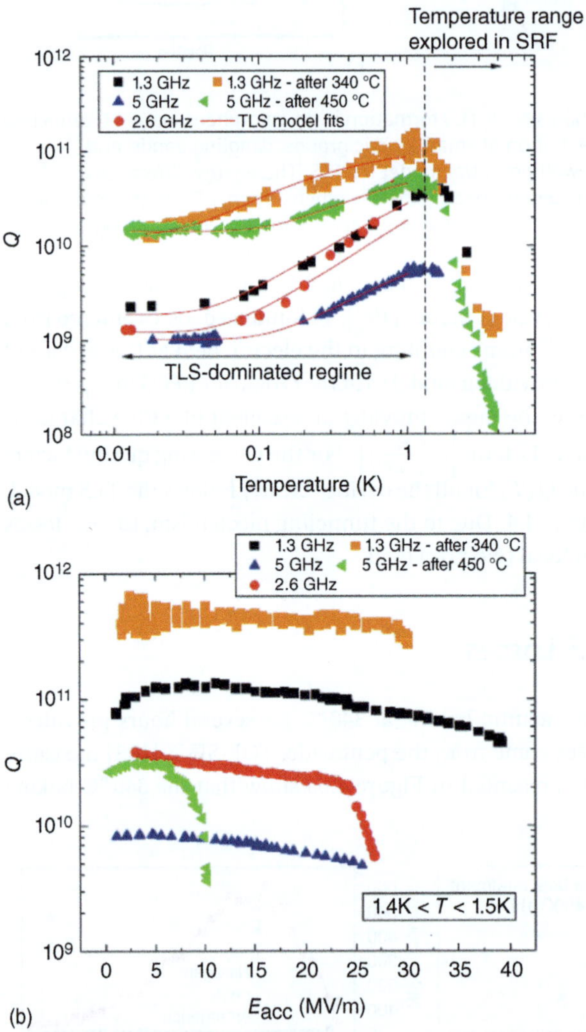

Figure 4.7 (a) Q as a function of temperature from 4.2 down to 10 mK. Below about 1 K the significant decrease in Q is consistent with TLS dissipation as shown by TLS model fits. The oxide-eliminating heat treatment increases the Q dramatically for 1.3 and 5 GHz cavities [3D] [198]. (b) Quality factors at high fields for 1.3, 2.6, and 5 GHz cavities as a function of the accelerating field at about 1.5 K, before and after bake. Note the record high Q values well above 10^{11} after 340 °C bake. Source: [198]/A. Romanenko et al./American Physical Society/CC BY 4.0.

As with mid-T baking discussed in Section 4.2, after the heat treatments, the 1.3 GHz cavity had an extremely high-quality factor $Q = 4\times10^{11}$ at 1.5 K (Figure 4.7b) over a broad range of fields showing that the heat treatment suppressed the residual resistance at all fields. This is consistent with the removal of TLS loss mechanisms along with the pentoxide layer [190]. At temperatures lower than 1 K (Figure 4.7a) Q degradation (from $>10^{11}$ to 10^{10}) could arise from remaining sources of TLS. Examples could be other niobium oxides (e.g. NbO) and their interfaces, or surface adsorbates. Heat treatment at higher temperatures may help dissociate these suboxides to further raise the Q values for Quantum Computing and Sensing applications in the future.

With Q values in the 10^{10} range at 10 mK and low fields, SRF cavities become attractive for future application to quantum computing (Chapter 13) and quantum information systems. 1.3 GHz Nb cavities offer a potential for a thousand-fold increase in the prevailing photon lifetimes, and similarly for cavity-stored quantum state coherence times. Best Q's of 10^8 are reported in various quantum regime studies [195, 196] with corresponding best coherence times of the order of 1 msec. After heat treatment, the achieved photon lifetimes of 0.5–2 seconds in 1.3 GHz cavities correspond to vast improvements over previously reported lifetimes [201]. Even for 5 GHz, Nb cavities without heat treatment photon lifetimes are about 30 msec, which is several times higher than the previous record near 10 msec [201].

5

High Q's from DC Magnetic Flux Expulsion

5.1 Trapped Flux Losses, Sensitivity

When a superconducting cavity is cooled below T_c, there is the potential to trap local ambient dc magnetic flux at defects, preventing the Meissner effect from expelling the flux. As a result, cryostat and cryomodule designers make strong efforts to lower ambient dc magnetic fields by adding magnetic shields in cryomodules and test dewars to achieve ambient dc field values typically in the range of 5–10 mGauss. The achieved values become occasionally higher due to residual magnetization of assembly components, some left inside accidentally.

Trapped flux losses are especially relevant to high-duty cycle accelerators based on SRF, such as LCLS-II and SHINE, which rely on high Q values for affordable operation. With better understanding of flux trapping and flux expulsion discussed in this chapter, it is now possible to achieve higher Q values, especially those made possible with N-doping (Chapter 3) by implementing new ways to minimize the trapped residual dc magnetic field during the cooldown. As we will discuss in Section 5.6.1, the amount of trapped flux depends on the cooldown details, which affects the magnetic flux trapping efficiency.

The trapped flux *sensitivity*, S describes the amount of cavity losses per unit of trapped flux, and is defined as:

$$\text{Sensitivity, S} = \frac{R_{fl}}{B_{trap}} \tag{5.1}$$

S depends on materials properties. The flux is trapped in vortices that oscillate in the RF field of the cavity, giving rise to an important component of the residual resistance, which also increases with the RF field. Since residual resistance is roughly linear in the RF field, the power dissipation becomes cubic in the RF field [202–204]. The linear dependence of residual resistance on the density of trapped flux [205] suggests that vortices do not interact strongly with one another, and overall flux dissipation can be modeled with losses from single flux lines. S also depends on the orientation and geometry of the cavity with respect to the trapped field, which becomes important for complex-shaped cavities, especially the low-beta resonators.

Superconducting Radiofrequency Technology for Accelerators: State of the Art and Emerging Trends, First Edition. Hasan Padamsee.

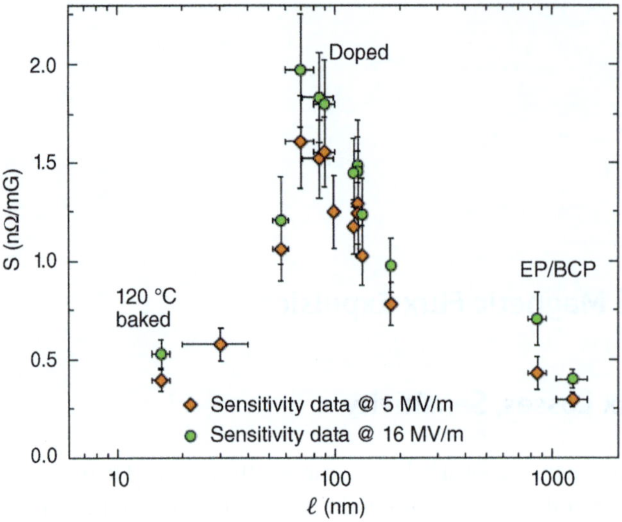

Figure 5.1 Trapped flux sensitivity at 5 MV/m (orange diamonds) and 16 MV/m (green dots) as a function of the mean free path. Source: M. Martinello et al., Fermilab [182]/AIP Publishing LLC/CC BY 4.0, & Courtesy M. Martinello.

The sensitivity for EP and BCP cavities [134] is about 0.35 nΩ/mG, measured as for example on a 1.5 GHz cavity made out of high-purity niobium sheet. Reference [206] shows somewhat larger values. Impurity-doped niobium cavities show stronger sensitivity to trapped flux losses, depending on their electron mean free path, which depends on the interstitial N or Ti content.

Figure 5.1 gives measurements from Fermilab which show that the sensitivity *versus* electron mean free path follows a bell-shaped curve. At one end of the curve, the sensitivity is minimized for very small mean free paths (e.g. 120 °C baked cavities) and for the other end, at very large mean free paths, i.e. EP and BCP cavities without doping or baking. The maximum sensitivity of residual resistance to trapped flux is found to be at $\ell \simeq 70$ nm. The 2/6 N-doping recipe, developed for the LCLS-II cavities, gives sensitivity values for 1.3 GHz doped cavities with ℓ around 120–180 nm. This value falls in the mean free path region where the vortex dissipation is not at its maximum but climbing up.

5.2 Trapped Flux Sensitivity Models

We discuss two models (local and nonlocal) to explain the experimentally established bell-shaped curve, including estimates for the calculated sensitivity. As discussed in Chapter 2, the magnetic flux carried by a vortex is quantized and equal to Φ_0. In the nonlocal description of the vortex, a quantized flux line is represented as a modulation of the order parameter of the superconductor that tends to zero at the center of the vortex, and approaches its finite value far from it [207]. In the local description [72], the vortex is described as a normal conducting core with radius of the order of the coherence length ξ_0, with superconducting currents spinning outside the core, and screening the magnetic flux confined inside. A sketch of the two different descriptions of the vortex is shown in Figure 2.14.

A simple approach to calculate losses due to trapped flux is that the normal-conducting core of a vortex has RF dissipation as any material in the normal state. Ignoring any losses due to flux motion, the trapped flux surface resistance is then calculated as the normal conducting surface resistance times the fraction of the normal-conducting area [4].

Once the external field B_{ext} is trapped over a surface of the cavity, it generates N vortices, each containing the fluxoid Φ_0. In a local description of the coherence length, a vortex may be described as a normal conducting core with radius of the order of the coherence length ξ. The vortex contains a single flux quanta, Φ_0, and the magnetic field decays over a distance λ_L from the center of the core [72, 85]. The trapped flux surface resistance is then the normal conducting surface resistance times the fraction of the normal-conducting area:

$$R_{fl} = N\frac{\pi a^2}{\Sigma}R_{\mathrm{n}} = \frac{\pi a^2 R_{\mathrm{n}}}{\Phi_0}B_{\mathrm{e}} \tag{5.2}$$

Considering that all the external flux is trapped, the resulting sensitivity is given by:

$$S = \frac{\pi a^2 R_{\mathrm{n}}}{\Phi_0} \tag{5.3}$$

From the Bardeen and Stephen analysis [85], both the radius a of the normal conducting core, and its resistance R_n depend on the mfp. In the clean and dirty limits the radius is:

$$a_{\mathrm{clean}} = 1.16\xi_0, a_{\mathrm{dirty}} = \sqrt{\frac{\pi}{3}\ell\xi_0} \tag{5.4}$$

The normal-conducting resistance R_n of niobium at 2 K and 1.3 GHz is defined by the normal skin effect when $\ell < 500$ nm, and by the anomalous skin effect when $\ell > 500$ nm [4]. The surface resistance in these two limits is:

$$R_n^{\mathrm{normal}} = \sqrt{\frac{\mu_0\omega}{2\sigma_n}} = \sqrt{\frac{\mu_0\omega v_{\mathrm{F}}}{2ne^2\ell}} \tag{5.5}$$

$$R_n^{\mathrm{anomalous}} = \left[\sqrt{3}\pi\left(\frac{\mu_0}{4\pi}\right)^2\right]^{1/3}\omega^{2/3}\left(\frac{\ell}{\sigma_n}\right)^{1/3} \tag{5.6}$$

The majority of N-doped cavities show $\ell < 400$ nm, so the case of normal skin effect seems appropriate. Here the sensitivity can be calculated as:

$$S_{\mathrm{clean}} = \frac{\pi}{\Phi_0}(1.16\xi_0)^2\sqrt{\frac{\mu_0\omega v_{\mathrm{F}}}{2ne^2\ell}} \tag{5.7}$$

$$S_{\mathrm{dirty}} = \frac{\pi^2}{3\Phi_0}\ell\xi_0\sqrt{\frac{\mu_0\omega v_{\mathrm{F}}}{2ne^2\ell}} \tag{5.8}$$

Note that, in dirty limit:

$$S_{\mathrm{dirty}} \propto \sqrt{\ell} \tag{5.9}$$

This simple model of static core losses is capable of predicting the experimental bell shape (Figure 5.1). In the clean limit, the sensitivity decreases as the mean free path increases because the resistance of the normal-conducting core decreases for cleaner materials, while the area of the core remains constant. In the dirty limit, the sensitivity increases as the mean free path increases because the area of the normal-conducting core increases, and this effect prevails over the decreasing of R_n. The boundary between clean and dirty limit is around $\ell/\xi_0 \sim 1$.

The local model provides a simple picture. However, the estimated values of sensitivity are overestimated, as shown by the comparison in Figure 5.2 with the measured results.

The local description of the vortex is a crude approximation; the nonlocal picture provides a proper portrayal of the superconductivity theory for the vortex [85]. In the nonlocal case, the normal-conducting core is more accurately a zero-point singularity of the density of Cooper pairs (Section 2.4.9). In the nonlocal picture, there is no dissipation from a static "core." Dissipation is caused only by vortex oscillations driven by the RF field. But the mathematics of the non-local description becomes very difficult with the "singularity". To simplify the mathematics Bardeen and Stephen [85] replace the non-local superconductor by an ideal local model, and proceed to calculate the dissipation due to vortex motion. Checchin [214] carries out a new calculation based on this model, as discussed below.

5.3 Vortex Physics

Vortices are subject to various forces: *the Lorentz force* due to flowing currents (dc or rf), the *thermal force* due to temperature differentials, the *pinning force* which acts as a restoring force to keep vortices pinned to defects, and the *viscous force* that acts as a drag on vortex motion.

Assuming the vortex magnetic field is in the z direction when a current density j_x is applied along the x direction, the vortex experiences a *Lorentz force* F_L per unit length along the y direction:

$$\boldsymbol{F_L} = \boldsymbol{j_x} \times \phi\hat{z} \tag{5.10}$$

The *thermal force* acts to displace vortices to increase the entropy of those areas [73, 205, 208, 209].

The generation of a vortex comes with a loss of condensation energy. When a superconductor contains defects or normal-conducting inclusions, i.e. areas where the condensation energy is smaller than the surrounding, the vortex prefers to sit in those positions, since the loss in condensation energy is smaller there. Random pinning centers in the material define a pinning landscape for vortex movement. There is an energy cost to move the vortex away from the pinning site. The pinning force is therefore associated with a pinning potential. A simple analytic form for the pinning potential for small displacements is a parabola, so that the *pinning force* may be described by a simple elastic restoring force. Possible pinning sites are material imperfections or defects, normal-conducting and dielectric inclusions, dislocations, and dislocation clusters (Section 5.11). A dislocation occurs when atoms are out of place from the perfect crystal structure. An efficient pinning

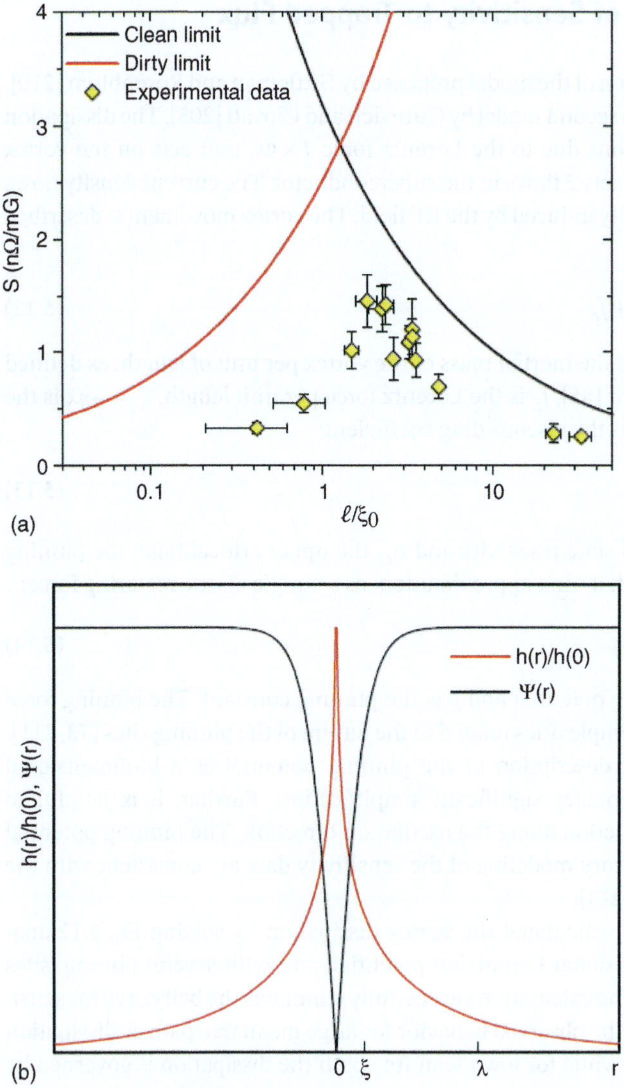

Figure 5.2 (a) Sensitivity as a function of mean free path for the clean (black curve) and dirty (red curve) limits calculated considering dissipation from the normal-conducting core of the vortex. The yellow diamonds correspond to the experimental data at zero field [49]. (b) Reduced magnetic filed $h(r)/h(0)$ and order parameter $\Psi(r)$ as a function of the radial distance r from the center of vortex, in the Abrikosov nonlocal vortex description. Source: M. Martinello, Illinois Institute of Technology, Thesis [49] Courtesy of M. Martinello, Fermilab.

center has dimensions comparable to the coherence length ξ near T_c when flux trapping is in play, and where $\xi \cong \mathbf{150\text{–}300}\ \boldsymbol{nm}$.

The *viscous drag force* is generated by Joule dissipation due to the normal-conducting currents induced in the vortex core by the flux motion [85]. It acts against the motion of vortices and is proportional to the velocity:

$$\boldsymbol{F}_v = -\eta \dot{\boldsymbol{x}} \tag{5.11}$$

5.4 Calculation of Sensitivity to Trapped Flux

We start with a discussion of the model proposed by Gittleman and Rosenblum [210]. Later we will consider a second model by Gurevich and Ciovati [205]. The dissipation is from vortex oscillations due to the Lorentz force $J \times \Phi_0$ that acts on the vortex whenever a current density J flows in the superconductor. The current density flows at the surface of the cavity induced by the RF field. The vortex movement is described as a damped oscillator:

$$M\frac{\partial^2 x}{\partial t^2} = f_L + f_v + f_p \tag{5.12}$$

M being analogous to the inertial mass of the vortex per unit of length, as defined by Bardeen and Stephen [85], f_L is the Lorentz force per unit length, $f_v = -\eta\dot{x}$ is the viscous force, where η is the viscous drag coefficient:

$$\eta = \frac{\Phi_0 B_{c2}}{\rho_n} \tag{5.13}$$

where ρ_n is the normal state resistivity and B_{c2} the upper critical field; the pinning force f_p can be assumed, in first approximation, as a simple elastic restoring force:

$$f_p = -\frac{\partial}{\partial x}U_p = -px \tag{5.14}$$

where U_p is the pinning potential and p is the pinning constant. The pinning force description can have complexities related to the nature of the pinning sites [73, 211]. However, an idealistic description of the pinning potential as a bi-dimensional Lorentzian potential enables significant simplification. Further, it is helpful to limit the pinning interaction along the oscillation direction. The pinning potential values used for satisfactory modeling of the sensitivity data are consistent with the data for niobium [212, 213].

Checchin et al. [214] calculated the vortex dissipation by solving Eq. 5.12 analytically, for a bi-dimensional Lorentzian potential, and with several pinning sites along the z-direction. The calculation successfully simulates the bell curve for sensitivity vs mfp. It models the observed behavior for large mean free path as dissipation described by flux flow, while for low mean free path the dissipation is governed by flux pinning.

In the flux flow regime, the pinning forces can be neglected. Therefore, the main force acting on the vortex during the flow is the viscous drag force. The dissipation depends only on the mean free path through the viscous coefficient η that increases as the resistivity ρ_n, decreasing with increasing mean free path.

At small values of mean free path, the viscous drag force becomes negligible, while the pinning force is large. This is the pinning regime. Even when pinned, the vortex flux line dissipates power because of the oscillation induced by the RF currents due to the Lorentz force. The higher the pinning center strength, the higher the pinning force and the more constrained the oscillation. The larger the pinning strength, the smaller the sensitivity. Dirtier or more defective materials have larger pinning strength. The model predicts the sensitivity vs mfp quite well as shown in Figure 5.3a.

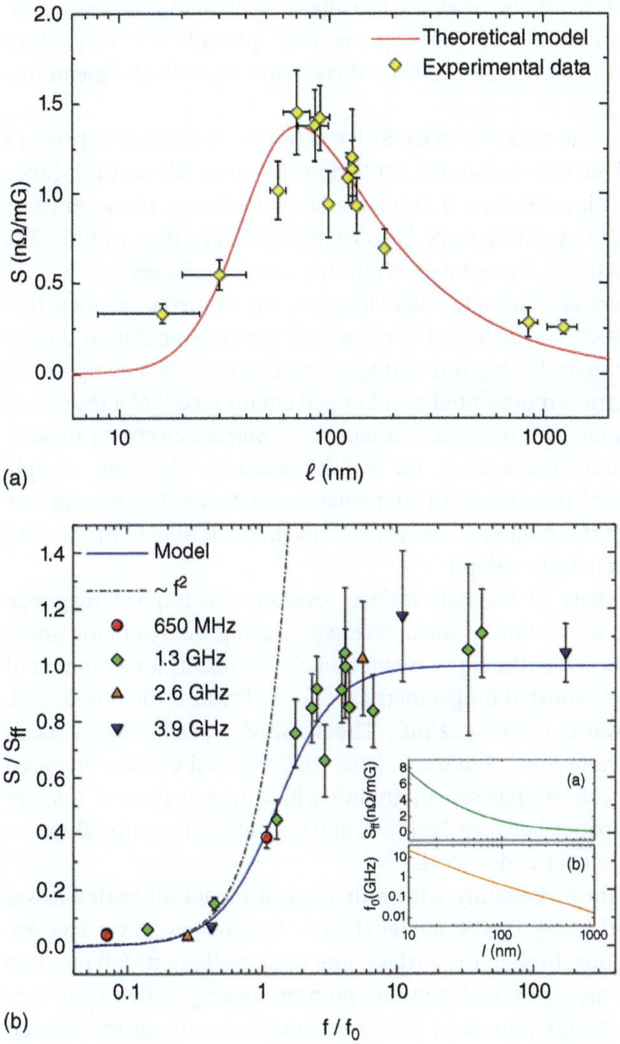

Figure 5.3 (a) Calculated sensitivity as a function of the mean free path due to vortex oscillation in an RF field. Source: Checchin et al. [214]/IOP Publishing/CC BY 3.0 (red curve) and comparison with experimental data at zero field. (b) Frequency dependence of trapped flux sensitivities compared to data. The flux-flow sensitivity as a function of the mean-free-path is shown in inset a. The depinning frequency as a function of the mean free-path is reported in the inset b. The frequency is normalized with respect to the depinning frequency. Source: M. Checchin et al., Fermilab [204]/AIP Publishing LLC/CC BY 4.0.

Checchin extends his model [204] to predict that the trapped flux surface resistance is proportional to ω^2, in the pinning regime with small mfp. This means that for low values of mean free path, the sensitivity increases with increasing frequency and then saturates at higher frequency. It is therefore expected that nitrogen-doped cavities with low resonance frequency, as 650 MHz elliptical cavities, show *lower* values of trapped flux sensitivity, as confirmed by the data (Section 5.5).

In summary, the motion of the vortices is subject to viscous drag and pinning, which introduce dissipation. The viscous drag prevails for very clean superconductors, while the pinning prevails for dirty materials with low mean free path values.

The model can also explain why Nb on Cu SRF cavities show substantially lower values of vortex-related surface resistance [214] than for pure Nb cavities. Since sputtered niobium thin films (Section 9.1) have more defects (e.g. more porosity, columnar growth, etc.), the pinning force is substantially larger than in bulk Nb. Both the pinning potential and the number of pinning points are larger.

Weingarten [217] offers an alternate calculation for the observed trapped flux losses in Nb on Cu cavities. He treats the vortices as freely moving due to the RF frequency being greater than the depinning frequency. There are two components to the losses. The first term is represented by a lumped circuit model of a resistance and inductance in parallel. The resistance comes from normal electrons present in the core, and the inductance from the superconducting electrons. Simple circuit analysis gives a $\omega^{2/3}$ power law for frequency dependence. The second loss contribution comes from the motion of the vortex. His model is able to fit the data of fluxon losses versus penetration depth.

At Cornell, the sensitivity of residual surface resistance to trapped magnetic flux was also studied as a function of mean free path during the N-doping study [95, 148, 206]. Figure 5.4a shows the experimental sensitivity results for a number of 1.3 GHz N-doped cavities, adjusted for geometrical effects. Towards the non-doped, clean limit, S tends towards 0.2–0.5 nΩ/mG. The sensitivity peaks near a mean free path of 5–10 nm, where $S = 2.5$ nΩ/mG. Note that N-doped cavities prepared at FNAL show qualitatively similar sensitivity but with a higher peak at a lower value of mfp. This discrepancy between Fermilab and Cornell for the mfp values for maximum sensitivity is not yet understood.

Cornell results compare successfully with a theoretical model for vortex losses by Gurevich and Ciovati [205]. In the model (Figure 5.5), magnetic vortices are strongly pinned at locations through the surface, and their oscillations fall into two regimes sensitive to the mean distance between pinning sites ℓ_p. When ℓ_p is very short, the vortices are strongly pinned in place, with only a small portion of each trapped vortex contributing to the dissipative losses. As ℓ_p increases, vortex oscillations between pinning sites increase in magnitude, increasing losses. On the other hand, when ℓ_p is long, vortices are largely unpinned and free to oscillate with large amplitude and low line tension; dissipative losses are limited, decreasing further with increasing ℓ_p. There is a crossover between these two regimes where the losses peak at an intermediate value of ℓ_p. A bell-shaped behavior is predicted. Viscosity of flux line movement is not used, as Checchin does. The model also predicts that higher frequencies shift the peak to small *mfp* values, and increase the maximum.

5.5 Dependence of Sensitivity on RF Field Amplitude

Figure 5.6 shows the sensitivity as a function of accelerating field and frequency for cavities prepared by various treatments [204] with short, medium, and long

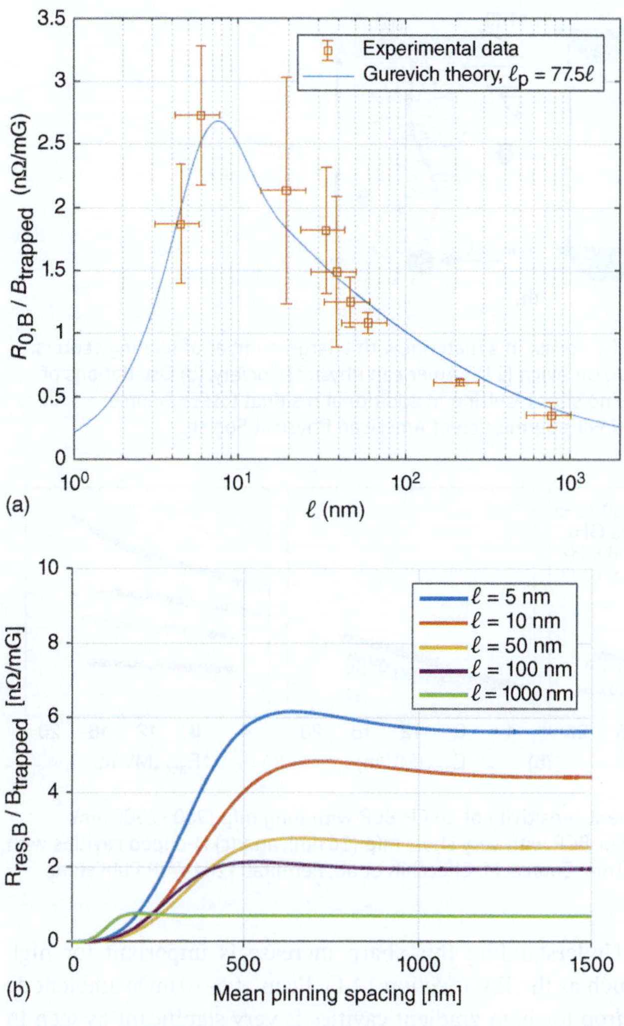

Figure 5.4 (a) Sensitivity to trapped magnetic flux of the residual resistance as a function of electron mean free path for Cornell nitrogen-doped cavities. Source: J. Maniscalco, Cornell University, Thesis [148] Courtesy of J. Maniscalco, Cornell University. Line indicates theoretical prediction from [205], using $\ell_p = 77:5\ \ell$ as the simple relationship between pinning center spacings and the mean free path. (b) Trapped flux sensitivity versus spacing between pinning sites for various mean free paths. Source: [95] Courtesy of D. Gonnella, Cornell University/ based on Gurevich's theory of losses from vortex oscillations [205].

mean free paths. There is a linear dependence below about 70 mT (16–20 MV/m). The linear increase can be understood in terms of a nonlinear pinning force [219], or by hysteretic losses in the framework of a mean-field description of pinning [220].

At higher RF fields, the sensitivity becomes nonlinear with RF field amplitude. Figure 5.7a shows the nonlinear behavior for several treatments, suggesting an

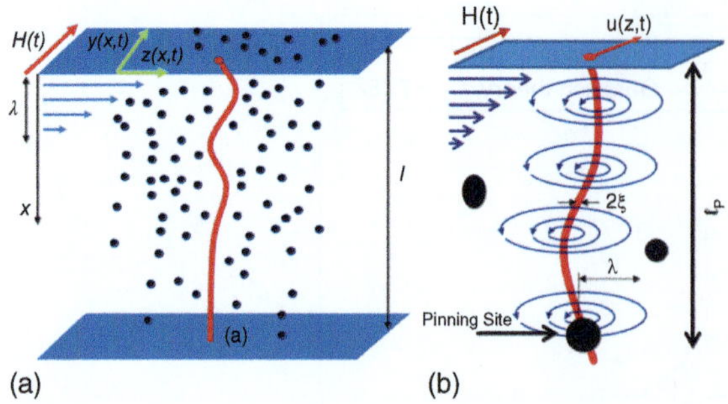

Figure 5.5 (a) Location of a vortex in a material with a large number of pinning centers. Source: W. Pathirana and A. Gurevich [218] American Physical Society. (b) Oscillations of vortex lines between pinning sites resulting in additional residual losses. Source: [205]/Gurevich and Ciovati with permission of American Physical Society.

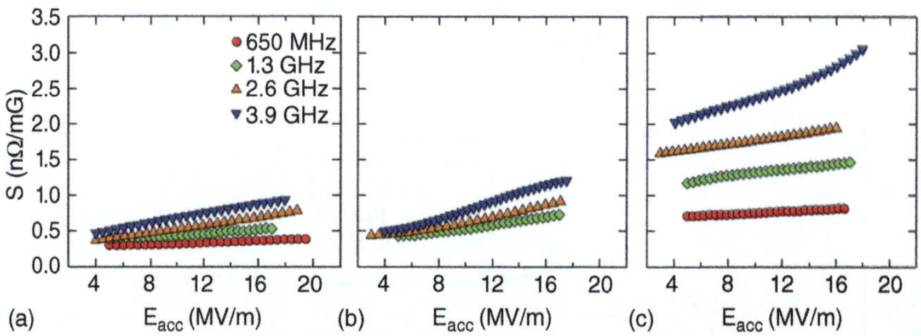

Figure 5.6 Field-dependent sensitivity of (a) EP/BCP with long mfp (900–2500 nm), (b) 120 °C baking after EP or BCP with very short mfp (16 nm), and (c) N-doped cavities with intermediate mfp 80–120 nm. Source: M. Checchin et al., Fermilab [204]/AIP Publishing LLC/CC BY 4.0.

exponential increase. Understanding this sharp increase is important for high gradient applications such as the ILC (Section 12.1). Even at 5–10 mOe ambient dc magnetic field, the *Q*-*d*rop for high gradient cavities is very significant as seen in Figure 5.7b.

Further vortex dynamics simulations and further measurements with higher levels of trapped flux up to 100 mOe [221] suggest a rough picture shown in Figure 5.8. At low RF amplitudes, the linear dependence arises from vortex oscillations within the pinning potential, as discussed. The slope change from linear occurs due to depinning of vortices from the pinning sites. At the highest RF field levels, there is a saturation when the vortex is fully depinned.

5.6 DC Magnetic Flux Expulsion

5.6.1 Fast *versus* Slow-Cooling Discovery

For a long time in the development and application of SRF, better shielding cavities from the ambient dc magnetic field was the best way to reduce trapped vortices, and

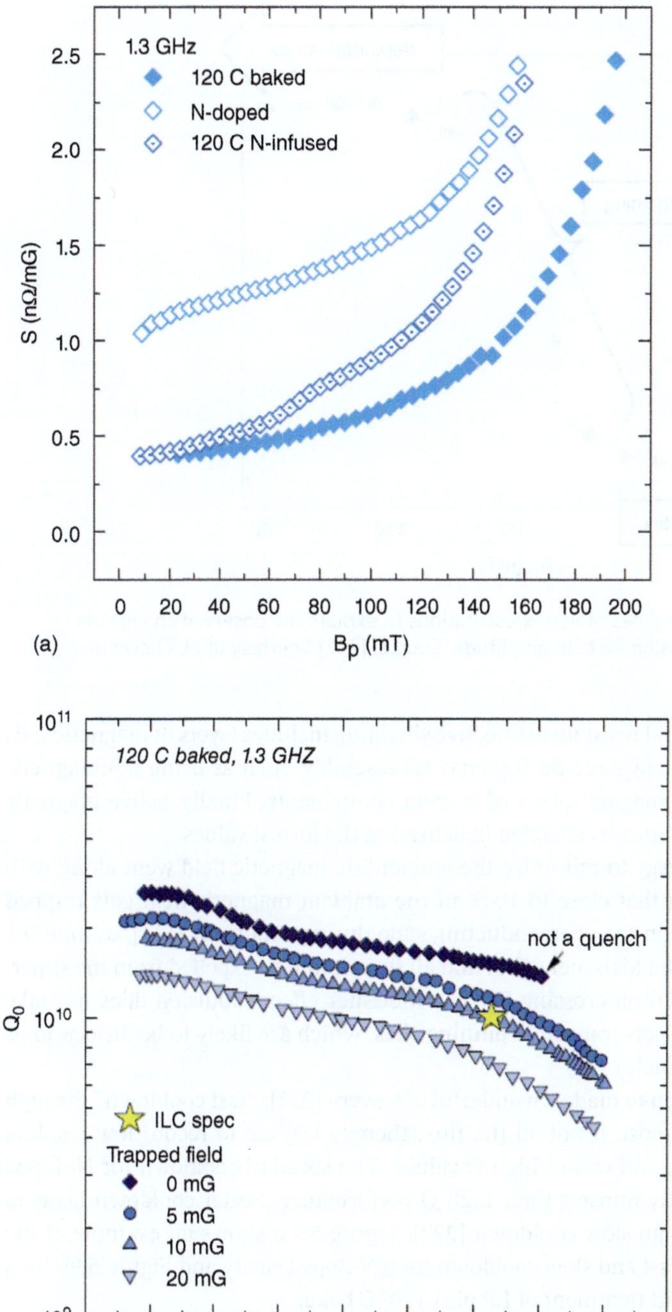

Figure 5.7 (a) Nonlinear growth of sensitivity to trapped flux at high fields. Source: [221]/with permission of Mattia Checchin. (b) Impact on Q for low ambient dc magnetic fields (0–20) mG, at high gradients. Source: [221]/with permission of Mattia Checchin.

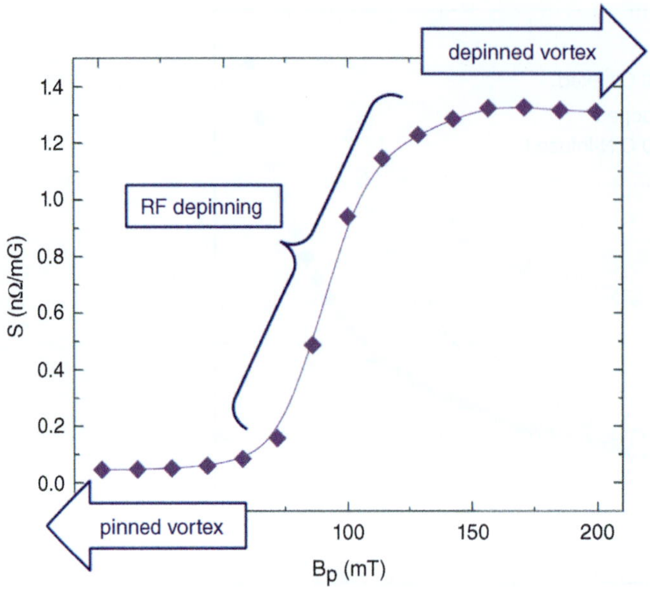

Figure 5.8 Three regimes of vortex oscillations to explain the observed changes in behavior with increasing RF field amplitude. Source: [221] Courtesy of M. Checchin, Fermilab.

the resulting residual resistance. Effective shielding includes layers of magnetic foils and good magnetic hygiene during cryostat assembly, such as using nonmagnetic components, or demagnetization of suspect components. Finally, active magnetic field compensation proves effective in achieving the lowest values.

The overall strategy to minimize the ambient dc magnetic field went along with the understanding that close to 100% of the ambient magnetic field gets trapped during the transition to superconducting state during cooldown [134], despite the expectation from the Meissner effect that all flux should be expelled from the superconducting material on crossing T_c. The Meissner effect expulsion does not take place because flux gets trapped at pinning sites, which are likely to be dislocations, impurities, or vacancies.

In 2015, Romanenko made a wonderful discovery [222]: "fast cooldown" through T_c helps to expel most, if not all the flux, thereby leading to record low residual resistance at 1.5 K, and record high Q values. The speed of cooldown for N-doped cavities significantly impacts their high Q performance. Faster cooldown leads to much better Q's than slow cooldown [222]. Figure 5.9 a shows an example of the contrast between fast and slow cooldown for a N-doped cavity and Figure 5.9b for a cavity with standard treatment of EP plus 120 °C bake.

The full high Q benefits of N-doping or mid-T bake can only be realized together with fast cooldowns to successfully exclude flux. This is especially important as N-doping recipes result in higher sensitivity of RF losses to trapped flux due to their shorter mfp that leads to higher viscosity, and eventually to higher pinning. The better flux expulsion effect with fast cooling was also observed on cavities prepared

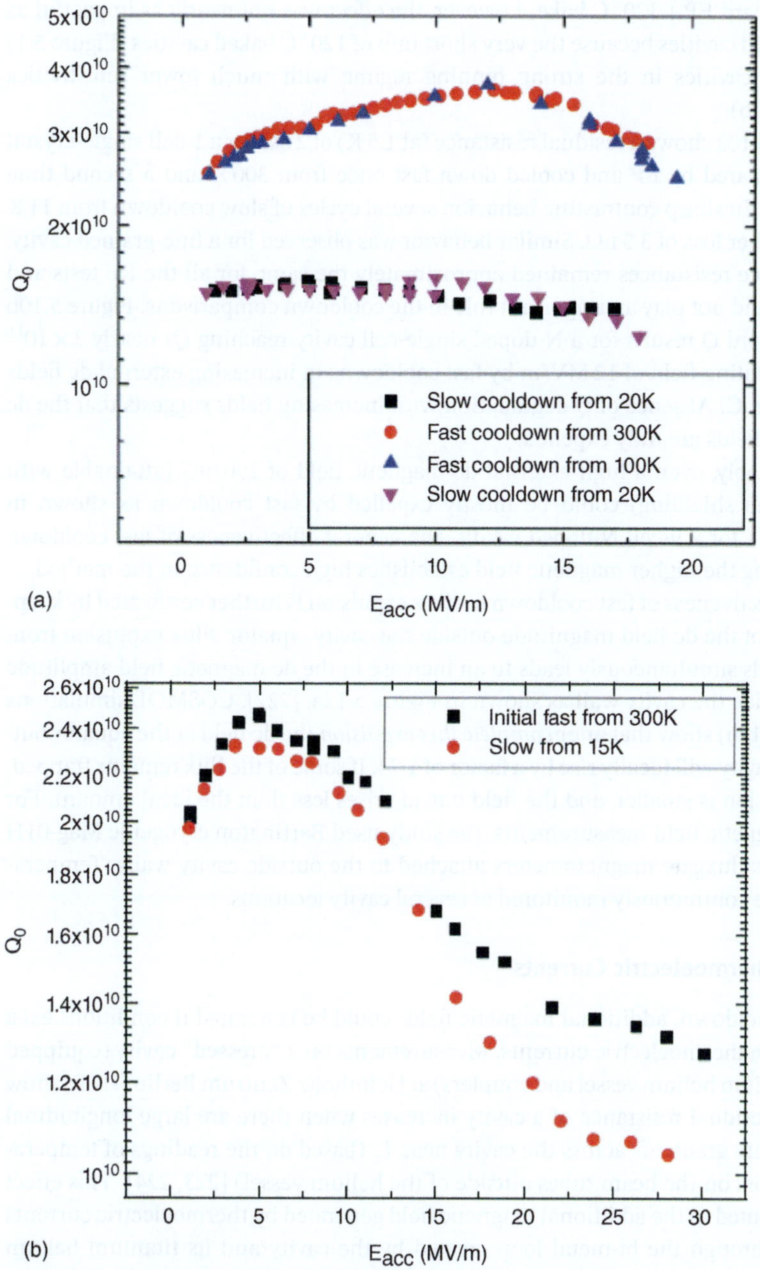

Figure 5.9 (a) Effect of cooldown speed (fast versus slow on Q_0 *versus* E_{acc} at 2.0 K of a N-doped 1.3 GHz, 9-cell cavity and (b) for a EP + 120 °C bake 9-cell cavity. Source: A. Romanenko et al., Fermilab [222]/AIP Publishing LLC/CC BY 3.0.

with standard EP + 120 °C bake; however, the effect was not nearly as impactful as for N-doped cavities because the very short mfp of 120 °C baked cavities (Figure 5.1) put these cavities in the strong pinning regime with much lower sensitivities (Figure 5.9b).

Figure 5.10a shows a residual resistance (at 1.5 K) of 2 nΩ for a 1-cell single-crystal cavity prepared by EP and cooled down fast once from 300 K and a second time from 20 K. In sharp contrasting behavior, several cycles of slow cooldown from 11 K gave a higher loss of 3.3 nΩ. Similar behavior was observed for a fine-grained cavity. BCS surface resistances remained approximately the same for all the RF tests and therefore did not play an important role in the cooldown comparisons. Figure 5.10b shows record Q results for a N-doped single-cell cavity reaching Qs nearly 3×10^{11} for accelerating field of 12 MV/m by fast cooldowns in increasing external dc fields of 2 to 23 mG. Absence of Q degradation with increasing fields suggests that the dc magnetic fields are fully expelled.

Remarkably, even a high external dc magnetic field of 190 mG (attainable with little or no shielding) could be mostly expelled by fast cooldown as shown in Figure 5.11 for a 9-cell N-doped cavity. The general effectiveness of fast cooldown in expelling the higher magnetic field establishes high confidence in the method.

The effectiveness of fast cooldown for flux expulsion is further confirmed by keeping track of the dc field magnitude outside the cavity equator. Flux expulsion from cavity walls simultaneously leads to an increase in the dc magnetic field amplitude right outside the cavity wall as shown in Figure 5.12a. [222]. COSMOL simulations (Figure 5.12b) show that after *complete flux expulsion* the dc field at the equator outside the cavity will ideally rise by a factor of 1.74. If some of the flux remains trapped, the expulsion is smaller, and the field outside rises less than the ideal amount. For these magnetic field measurements, the study used Bartington cryogenic Mag-01H single-axis fluxgate magnetometers attached to the outside cavity walls. Temperatures were continuously monitored at several cavity locations.

5.6.2 Thermoelectric Currents

During cooldown, additional magnetic fields could be generated if conditions exist to produce thermoelectric currents. Measurements on a "dressed" cavity (equipped with titanium helium vessel and couplers) at Helmholtz Zentrum Berlin (HZB) show that the residual resistance of a cavity increases when there are large longitudinal temperature gradients across the cavity near T_c (based on the readings of temperature sensors on the beam tubes outside of the helium vessel) [223, 224]. This effect was attributed to the additional magnetic field generated by thermoelectric currents flowing through the bi-metal loop created by the cavity and its titanium helium vessel. The magnetic field generated by the Seebeck effect gets trapped in the cavity during the cooldown through T_c. The lack of magnetic field probes inside the helium tank during the HZB test precluded direct measurements of any induced magnetic field that would lend definitive support to the explanation by the thermoelectric currents.

But the HZB findings appear contradictory to the FNAL experience with fast cooling necessary to expel flux. The difference could possibly arise from the

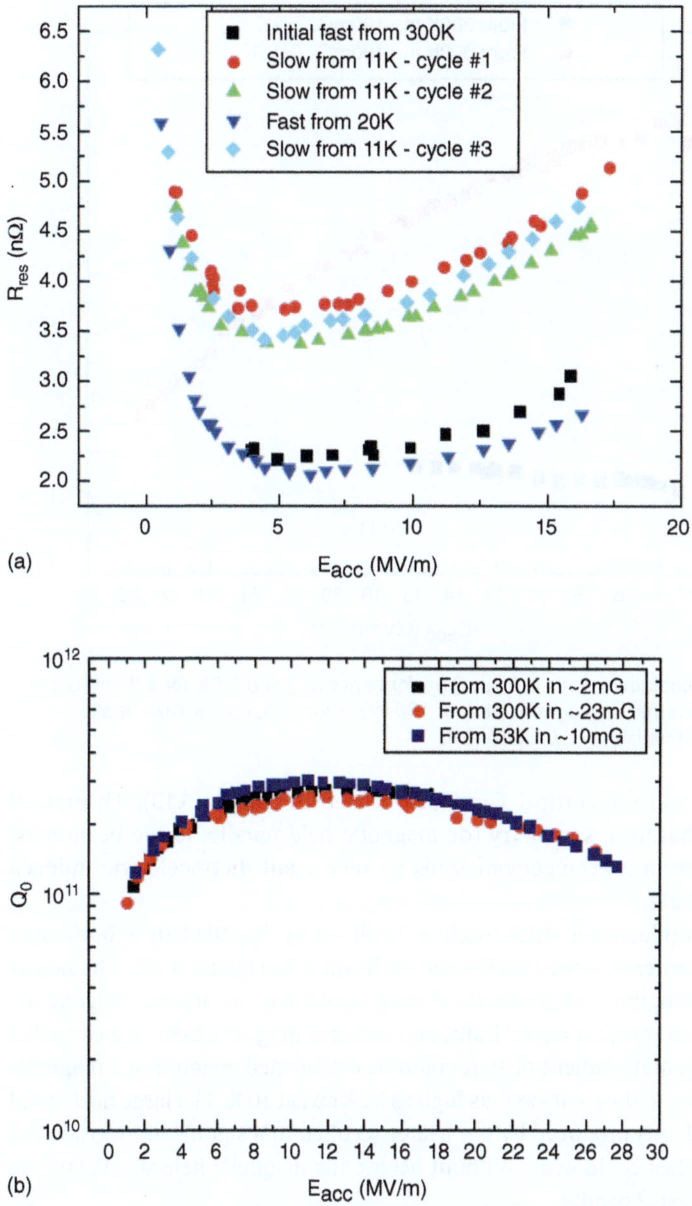

Figure 5.10 (a) Residual resistance of the single crystal 1-cell electropolished cavity after different fast and slow cooling cycles; fast cooling results in 2 nΩ whereas slow cooling gives 3.5 nΩ. Residual resistance was measured at 1.5 K. Source: A. Romanenko et al., Fermilab [222]/AIP Publishing LLC/CC BY 3.0. (b) Record high Q values of 3×10^{11} at 1.5 K at external dc fields up to 23 mG, much higher than usually present in cryostats. Source: Romanenko et al., Fermilab [171]/AIP Publishing LLC/CC BY 3.0.

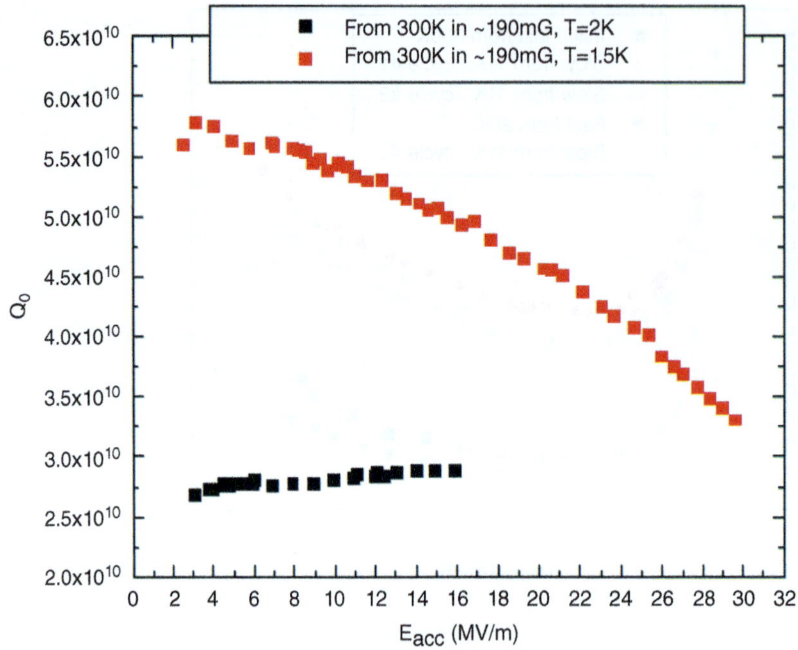

Figure 5.11 Fast cooldown leads to high Q performance at 2 and 1.5 K for a 9-cell cavity in a chosen high external dc magnetic field of 190 mG. Source: A. Romanenko et al. Fermilab [171]/AIP Publishing LLC/CC BY 3.0.

horizontal HZB and the vertical Fermilab geometries (Figure 5.13). Theoretical analysis shows that axial symmetry (dc magnetic field parallel to the beam axis) in a vertical cavity test arrangement leads to very small thermoelectric induced magnetic fields [225].

Cornell [95] conducted a study with a 9-cell cavity installed in a horizontal cryostat, with a heater mounted on the cell #9 beam tube (Figure 5.13). The heater produced large longitudinal gradients during cooldown, to artificially generate thermoelectric currents that would induce an ambient magnetic field as well as flux trapping. A horizontal gradient of 30 K could be established generating a magnetic field perpendicular to the cavity axis as high as 62.8 mG at 10 K. The large horizontal temperature gradients produced by the heater resulted in a significant degradation of the Q even for fast cooldowns. Without heater, the magnetic field was 0.3 mG to produce the highest Q results.

Therefore, to achieve low residual resistance and highest Q, horizontal cavities in cryomodules should be cooled with as small a horizontal temperature gradient as possible to reduce thermoelectric currents, while keeping vertical temperature gradient as large as possible to reduce trapping of any ambient magnetic fields.

If good shielding and good magnetic practices achieve low ambient magnetic fields, flux expulsion by fast cooldowns may not be necessary. Cooldowns with small horizontal spatial temperature gradients would avoid thermoelectric effects. However, if the ambient magnetic field is large (>5 mG), it becomes necessary to

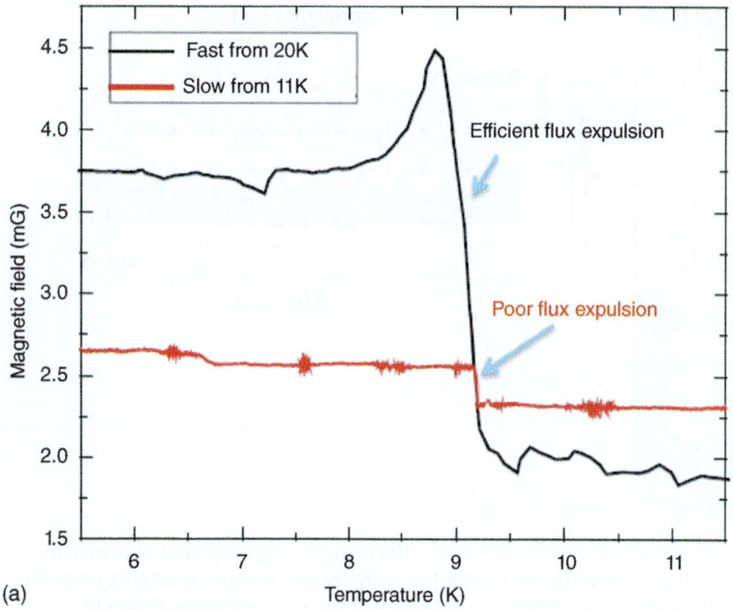

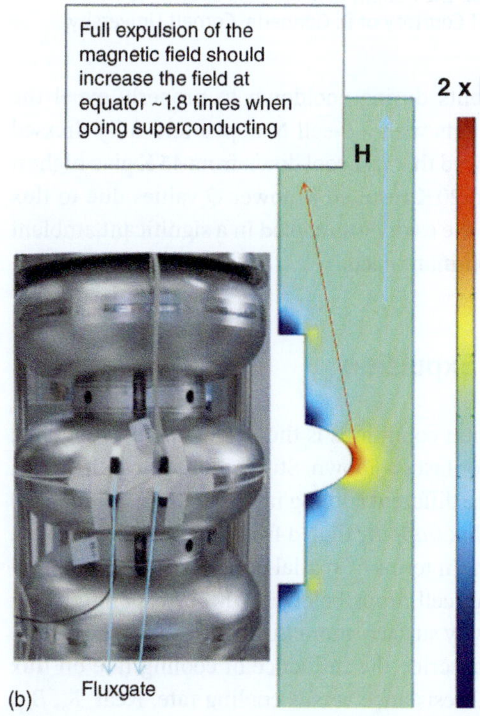

Figure 5.12 (a) Fast cooling (from high temperatures on the right to low temperatures on left) shows how the magnetic field at the equator increases through transition. But slow cooling (lower red curve) shows very little expulsion. Slow cooling leads to a much smaller change in the magnetic field outside of the cavity walls. (b) COSMOL simulations show that full flux expulsion should increase the magnetic field on the outside wall at the equator by a factor close to 1.74. Source: A. Romanenko et al., Fermilab [222]/AIP Publishing LLC/CC BY 3.0.

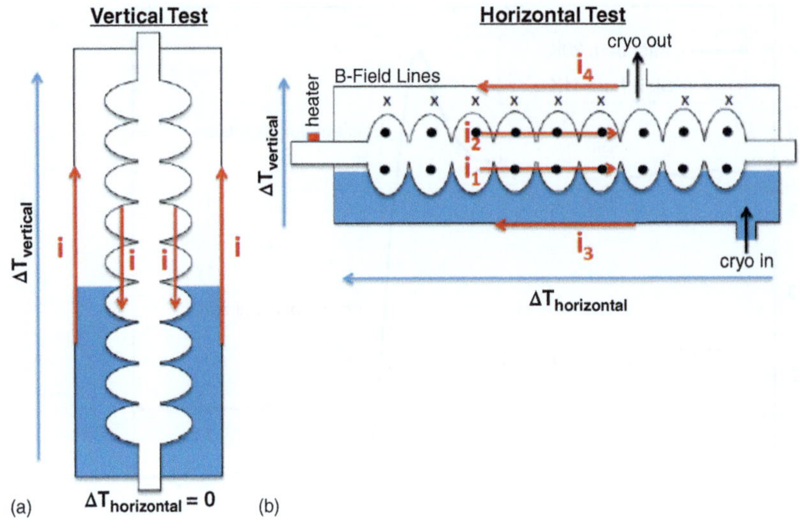

Figure 5.13 (a) Cylindrical symmetry reduces the induced magnetic field at the cavity surface in vertical test. (b) In the horizontal test, vertical temperature gradients break the symmetry resulting in nonzero magnetic field at the cavity inner surface. A simple schematic of the magnetic field lines inside the helium tank are also shown. Source: D. Gonnella, Cornell University, Thesis [95] Courtesy of D. Gonnella, Cornell University.

use large vertical temperature gradients during cooldown to properly expel the magnetic flux. For example, experiments with a 9-cell N-doped and fully dressed cavity in a horizontal cryomodule showed that fast cool down from 45 K give highest Q values [227], but slow cooling from 20 K resulted in lower Q values due to flux trapping from ambient fields. Clearly, the cavity was cooled in a significant ambient field when good expulsion was the dominant need.

5.7 Cooling Rates for Flux Expulsion

Most relevant for the best flux expulsion conditions is the cooling rate through T_c, rather than the starting temperature for fast cooldown. Studies for fast cooling from 300, 100, and 20 K, 15 K or 11 K explore different cooling rates. A slow (2–5 mK/sec) cooling through T_c leads to stronger flux *trapping* than a fast (>30 mK/sec) cooling, which leads to efficient flux expulsion. In terms of spatial gradients, when the temperature difference across the top half-cell drops below 1–2 K corresponding to a gradient of 0.1–0.2 K/cm along the cavity surface, more flux gets trapped.

Kubo [228] developed a model to describe the influence of cooling rate on flux trapping and to calculate the residual resistance versus cooling rate. Near T_c, B_{c1}, and B_{c2} become smaller than the ambient magnetic field, B_a. Two-phase transition fronts form: one at $B_{c2} = B_a$ and the other at $B_{c1} = B_a$. The region in between falls in the vortex state domain. As the material cools down, the phase transition fronts sweep the material and so does the vortex state domain. Vortices are trapped by

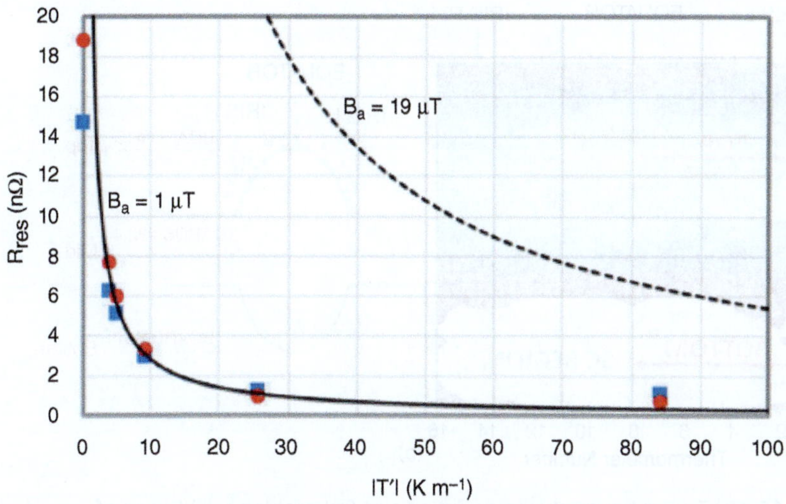

Figure 5.14 Comparison between model calculations [229] and measurements [171] for the residual resistance due to trapped flux as a function of temperature differentials on cooldown. Source: T. Kubo, KEK [229]/Oxford University Press.

pinning centers randomly distributed in the material. When poor thermal gradients are present the mixed state covers a wider spatial range along the cavity during the cooldown, compared to when thermal gradients are large. The presence of the wider range increases the probability of flux trapping.

Kubo [229] estimates the number of trapped vortices by considering where the phase fronts are at rest and the collisions between moving vortices and pinning centers. The results of the model are in good agreement with the data as shown in Figure 5.14. The residual resistance due to trapped flux drops from 20 nΩ to about 1 nΩ when the temperature differentials increase from less than 10–30 K/m during cooling in an ambient dc field of 10 mG (1 μT). The calculation also predicts the residual resistance versus cooling rate for a large ambient field of 190 mG chosen for the experiment in Figure 5.11. At the fastest cooling rate, the residual resistance can be lower than 5 nΩ, corresponding a residual $Q > 5 \times 10^{10}$ to be compared with the experimental result shown in Figure 5.14.

5.8 Flux Expulsion Patterns

It is interesting to examine the interesting thermal stages for flux expulsion during fast cooldown, and trapping during a slow cooldown. During a fast cooldown, liquid helium is poured into the bottom of the warm dewar. The superconducting phase emerges at the cavity bottom, and progresses to the top. A fast-moving propagating phase boundary sweeps out the magnetic flux. During slow cooldown, the superconducting phase nucleates essentially simultaneously at multiple locations over the cavity due to isothermal conditions over the cavity volume. As cooling continues,

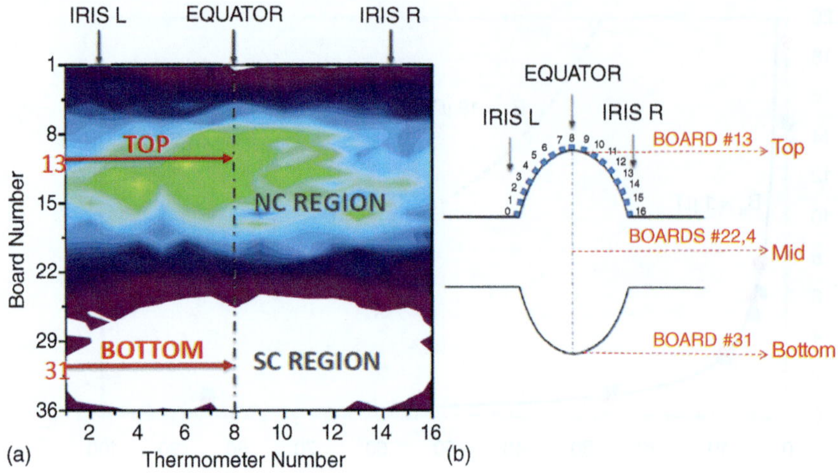

Figure 5.15 (a) Typical temperature map during cool fast cooldown. (b) Layout of thermometers. Source: M. Martinello, Illinois Institute of Technology, Thesis [49] Courtesy of M. Martinello, Fermilab.

these regions encircle some areas of normal phase where flux gets trapped. For magnetic flux contained in these normal "islands" to get expelled, the vortices need to pass through adjacent superconducting regions, which is energetically unfavorable. Hence the flux gets trapped inside the superconductor [222].

A sequence of temperature maps taken for a horizontally placed cavity confirms this simple picture contrasting fast with slow cooling. The horizontal orientation is representative of cavities cooled in an accelerator. The layout of the thermometry system and a typical temperature map is shown in Figure 5.15a,b. The geometry is more prone to flux trapping situations when the external magnetic field is perpendicular to the cavity beam axis (Section 5.9).

For fast cooldown, the sequence of T-maps from upper to lower panels in Figure 5.16 [230] show how the superconducting-normal conducting interface moves cleanly across the cavity, promoting efficient magnetic field expulsion. The large thermal gradients present during the fast cooldown exert a depinning force on trapped vortices, which pushes the flux out of the superconductor [171]. Vortices move in the same direction as the thermal gradient, out from the Meissner state region to the mixed state region, minimizing the Gibbs free energy.

During slow cooldown, the sequence of T-maps from upper to lower panels in Figure 5.17 shows how the superconducting phase nucleates at several locations over the cavity. On further cooling, those interfaces encircle areas of normal phase which contain flux. Now temperature gradients are small, so there is not enough thermal depinning force for flux expulsion. Moreover, the flux would need to pass through the encircling superconducting areas, which is not thermodynamically allowed. The flux gets trapped.

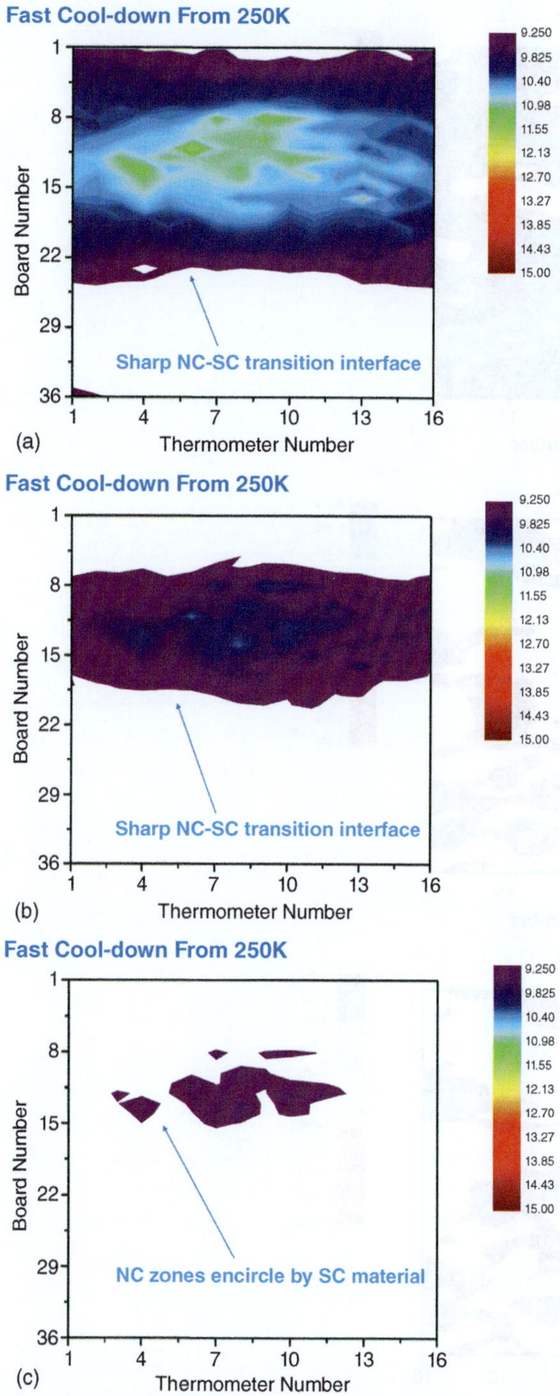

Figure 5.16 Fast cooldown from 250 K: Propagation of the normal conducting-superconducting phase boundary from upper to lower panels. The white areas have become superconducting. The colored areas (normal conducting) range in temperature from 9.25 to 12 K. The colored areas grow smaller with time as the boundary sweeps the cavity. Source: M. Martinello, Illinois Institute of Technology, Thesis [49, 230] Courtesy of M. Martinello, Fermilab.

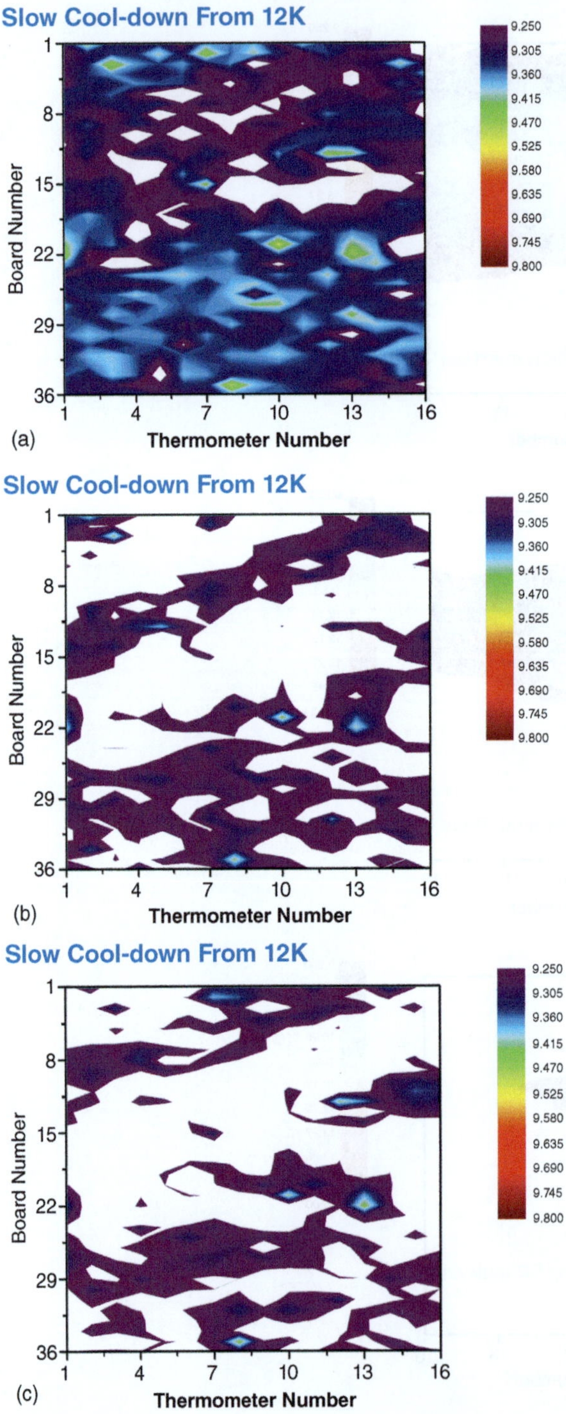

Figure 5.17 During slow cooldown, isothermal conditions allow several separate phase transitions to nucleate over many regions, which creates normal conducting islands surrounded by the super-conducting sea. Magnetic flux expulsion from the normal islands becomes thermodynamically unfavorable. Flux gets trapped in the islands. Source: [49, 230] Courtesy of M. Martinello, Fermilab.

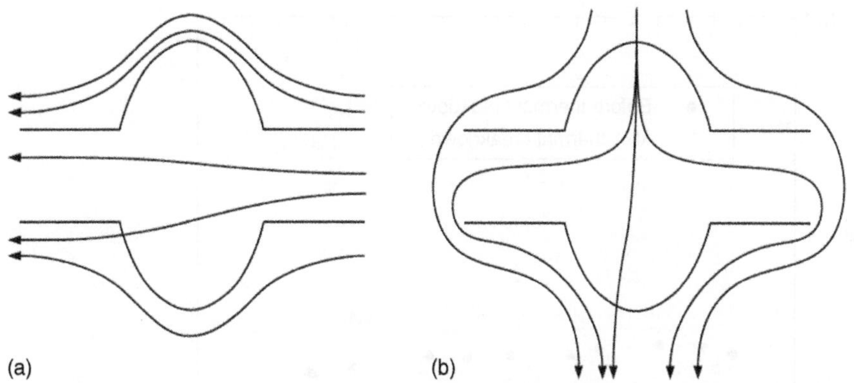

Figure 5.18 (a) When the dc magnetic field is parallel to the beam axis, the flux lines exit smoothly with the moving transition boundary. (b) When the magnetic field is perpendicular to the beam axis superconducting regions encircle vertical field lines to concentrate flux at the very top of the cavity. Source: M. Martinello et al., Fermilab [231]/AIP Publishing LLC/CC BY 3.0.

5.9 Geometric Effects – Flux Hole

When the magnetic field is parallel to the cavity beam axis (Figure 5.18a), large thermal gradients provide efficient conditions for flux expulsion because flux lines that cross the cavity walls have an easy path to follow along with the NC-to-SC transition boundary. When the external magnetic field is perpendicular to the cavity axis as shown in Figure 5.18b, and the transition boundary sweeps upward from the bottom, flux eventually gets trapped in the final region that crosses T_c, which is at the top equator, forming a "flux-hole" [231]. Magnetic flux lines concentrate in the normal-conducting region at the top of the cavity. This region becomes smaller as the transition boundary advances until the magnetic field is completely squeezed in at the very top of the cavity equator. The top point of the cell is the last to be cooled below the transition temperature, T_c, becoming a flux hole in the superconductor, from which it is not energetically favorable for flux to escape. It cannot cross the already superconducting regions. Such a flux hole can be harmful for cavity performance leading to both Q degradations and lower quench fields. As the RF field increases, heating at the top of the cavity is observed, compatible with the flux hole scenario.

Flux trapping measurements at IPNO [232] on several types of low-beta resonators (QWR, Spoke) reveal a geometrical dependence of the sensitivity to dc magnetic flux trapping. The normal component of the residual magnetic field is trapped during the superconducting transition.

5.10 Flux Trapping With Quench

Q degradations and local surface resistance increases are often observed after quenching, suspected to be due to trapped ambient magnetic field during the quench

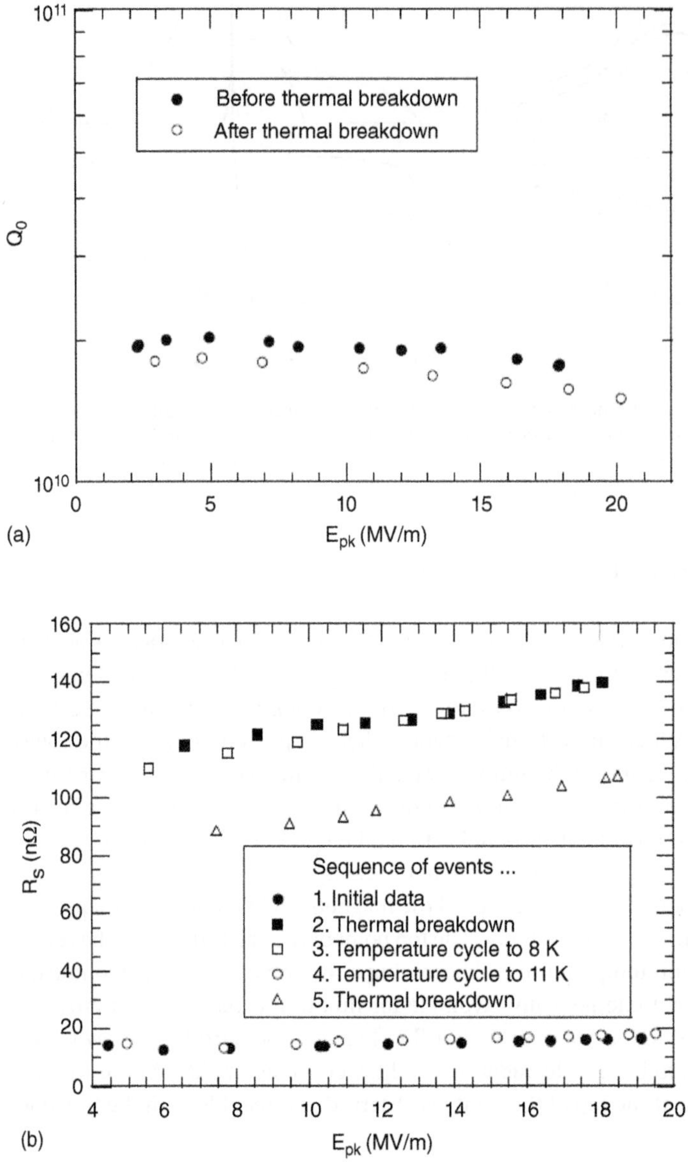

Figure 5.19 (a) Q_0 versus E_{pk} on a 1-cell cavity, prior to and after thermal breakdown events. Source: [10]/with permission of Jens Knobloch. (b) Surface resistance recorded by thermometry for the cavity with Q-drop measured in (a). The data sequence is shown. Note how thermal cycling above the critical temperature restores the surface resistance to the value before quenching. Source: J. Knobloch, Cornell University, Thesis [10] Courtesy of J. Knobloch, Cornell University.

events. There is a full recovery of Q and surface resistance when the cavity is warmed up above T_c and re-cooled (Figure 5.19), confirming trapped flux as the cause. Previously the origin of the trapped magnetic flux was thought to be due to thermal currents driven by the local thermal gradients driven by the quench zone [10, 233].

However, recent studies [68] show that the trapped flux enters the quench region from the ambient dc field. It is not created by thermoelectric currents. During a quench, a normal conducting hole opens on the cavity wall, and the magnetic field that was previously expelled from the superconducting cavity volume can now penetrate. Orthogonal ambient fields cause a larger drop in Q, as for trapped flux during cooldown. Confirming the recent discovery, the study found no drop in Q due to a quench, if the ambient magnetic field is very low, although thermal currents from the quench are still present. Cavities prepared by several different methods shown in Figure 5.20 were quenched in a near-zero ambient field achieved by compensation using external Helmholtz coils. Figure 5.20 shows the clear absence of Q drops due to quenching.

Interestingly, when the ambient magnetic flux is trapped during a quench, it can also be removed by re-quenching, instead of warming up above T_c, provided the second quench takes place in zero ambient fields.

Previous studies [235–238] of the quality factor degradation in high and medium-β superconducting resonators studied the amount of flux trapped during the quench to find a dependence of the quench-related degradation on the locally applied external magnetic field. Recovery of the cavity quality factor by "quench annealing" in the absence of an external field was also reported in these studies.

In further quench studies, a dc magnetic field from 10 to 200 mOe was applied outside the cavity before quenching. The degradation of the Q-factor recorded after every quench was seen to grow with the strength of the outside field. These studies clearly show that the magnitude of the surface resistance increase (and corresponding Q-drop) is closely related to that of the external dc field present.

5.11 Material Quality Variations

Efficient flux expulsion not only depends on thermal gradients obtained by fast cooldown protocols, but also by material properties, especially bulk properties, and by material treatment, as for example by annealing.

There is significant variation in flux trapping behavior of Nb materials from different production batches. Figure 5.21a shows differing flux expulsion properties of cavities from two batches of fine-grain Nb procured for LCLS-HE cavities [239]. Elevated temperature heat treatments prior to N-doping greatly improve flux expulsion. This is especially important for LCLS-HE requirements, because sensitivity of residual resistance to trapped flux increases with higher gradient needed (Section 5.5).

Some Nb production runs can have very poor expulsion. There can be large variability between different vendors, or even for different batches from the same vendor [241]. As discussed in Section 5.12, these differences arise from various distributions of pinning forces for flux pinning defects. For cavities that expel flux well during cooldown, temperature differences as low as 2 K over the cavity cell are sufficient to expel nearly all the external field. For cavities made from material that expel flux poorly, most of the flux is trapped even for temperature differentials as large as 10 K.

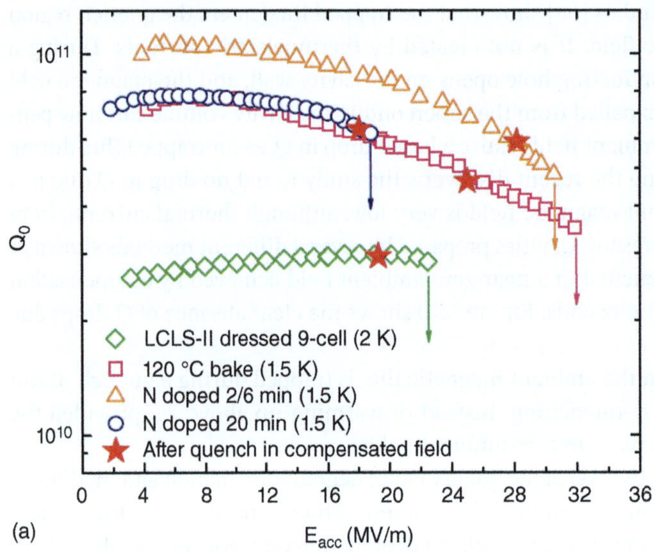

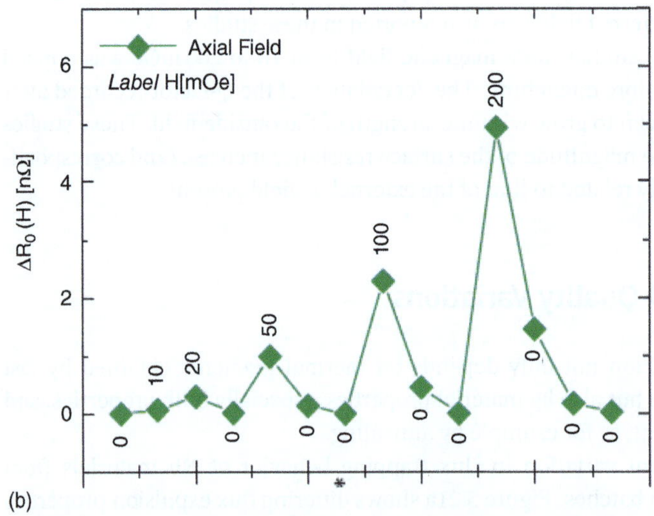

Figure 5.20 (a) Q_0 versus accelerating field after cooldown in near-zero ambient field before any quench. After quenching >10 times [234] red stars correspond to the Q_0 points acquired. There is no Q_0 drop after the quenches. (b) Variation of the residual resistance due to quenches in presence of external magnetic fields of increasing values from 10 to 200 mOe [68, 234]. Source: M. Checchin et al., Fermilab [234]/American Physical Society/CC BY 3.0.

Flux expulsion behavior depends primarily on bulk Nb properties. Surface properties do not play a big role, as demonstrated by similar flux expulsion behavior for different surface treatments, such as EP or BCP, or N-doping, or 120 °C baking, provided the bulk Nb properties are similar [239]. The expulsion efficiency also does not depend on surface pinning properties determined by the mfp. For example, 120 °C baked cavities (Figure 5.21b), which have a very low mfp (e.g. 2 nm) at the surface (and therefore different pinning strength) [171] exhibit qualitatively similar

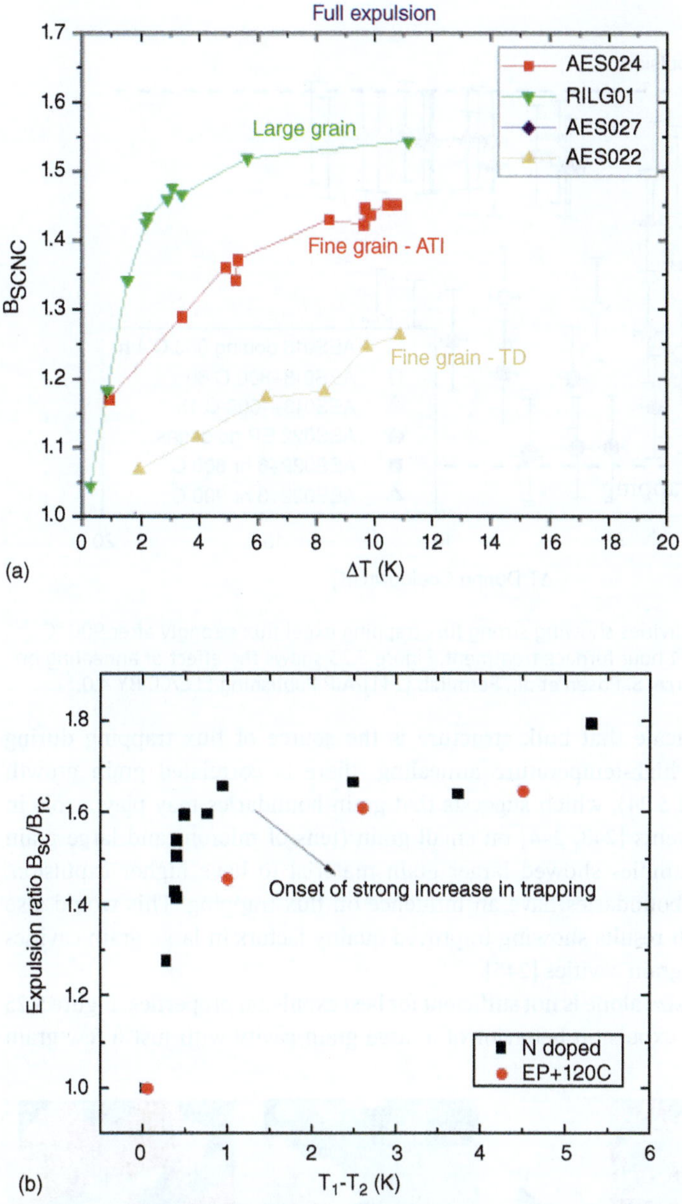

Figure 5.21 (a) Flux expulsion ratio for materials from different vendors showing different flux expulsion properties. Source: M. Martinello, Fermilab [240] Courtesy of M. Martinello. (b) Similar flux expulsion ratios for cavities with different surface mfps (achieved by different treatments), showing that mfp is not the key parameter for good flux expulsion properties. Source: A. Romanenko et al., Fermilab [171]/AIP Publishing LLC/CC BY 3.0.

flux expulsion behavior as N-doped cavities with longer mfp. Strong mechanical work such as heavy deformation can degrade expulsion, suggesting the role of stress-induced dislocations [242].

Annealing at 800 °C improves flux expulsion, as shown in Figure 5.22, but some materials need to be treated to 900 °C, or even 1000 °C [241]. The benefits

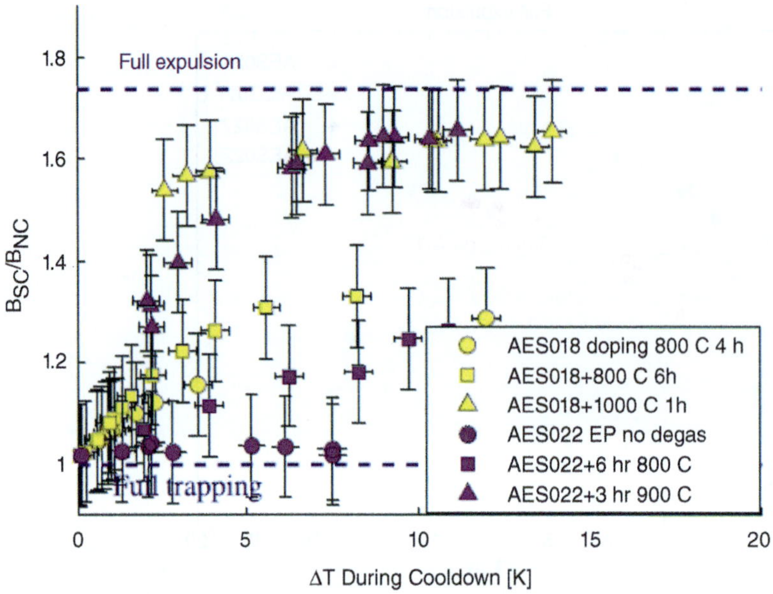

Figure 5.22 Two cavities showing strong flux trapping expel flux strongly after 900 °C 3 hour and 1000 °C 1 hour furnace treatment. Figure 5.23 shows the effect of annealing on the grain sizes. Source: S. Posen et al., Fermilab [241]/AIP Publishing LLC/CC BY 4.0.

of annealing indicate that bulk structure is the source of flux trapping during cooldown. With high-temperature annealing, there is correlated grain growth (Figures 5.23 and 5.24), which suggests that grain boundaries may play a role in pinning. Experiments [243, 244] on small grain (tens of micron) and large grain niobium (mm) samples showed larger grain material to have higher expulsion, suggesting grain boundaries have an influence on flux trapping. This would also be consistent with results showing improved quality factors in large grain cavities compared to fine grain cavities [245].

But large grain size alone is not sufficient for best expulsion properties. Figure 5.25 includes the flux expulsion behavior of a large grain cavity with just a few grain

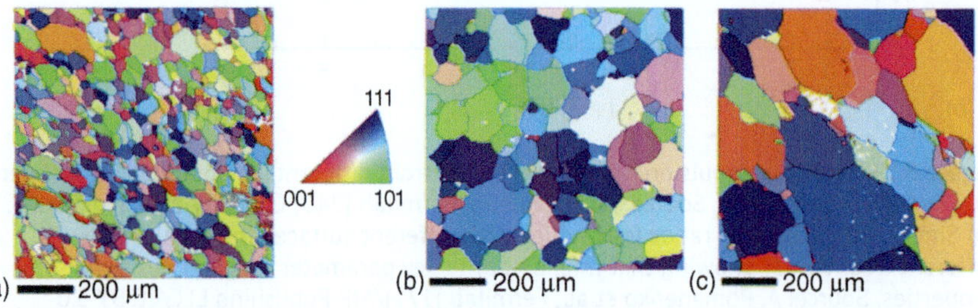

Figure 5.23 Crystal orientation maps of niobium samples with three different treatments: (a) as received, (b) treatment at 900 °C for 3 hours, and (c) treatment at 1000 °C for 1 hour. Significant grain growth is observed in cases (b) and (c). Source: S. Posen et al., Fermilab [241]/AIP Publishing LLC/CC BY 4.0.

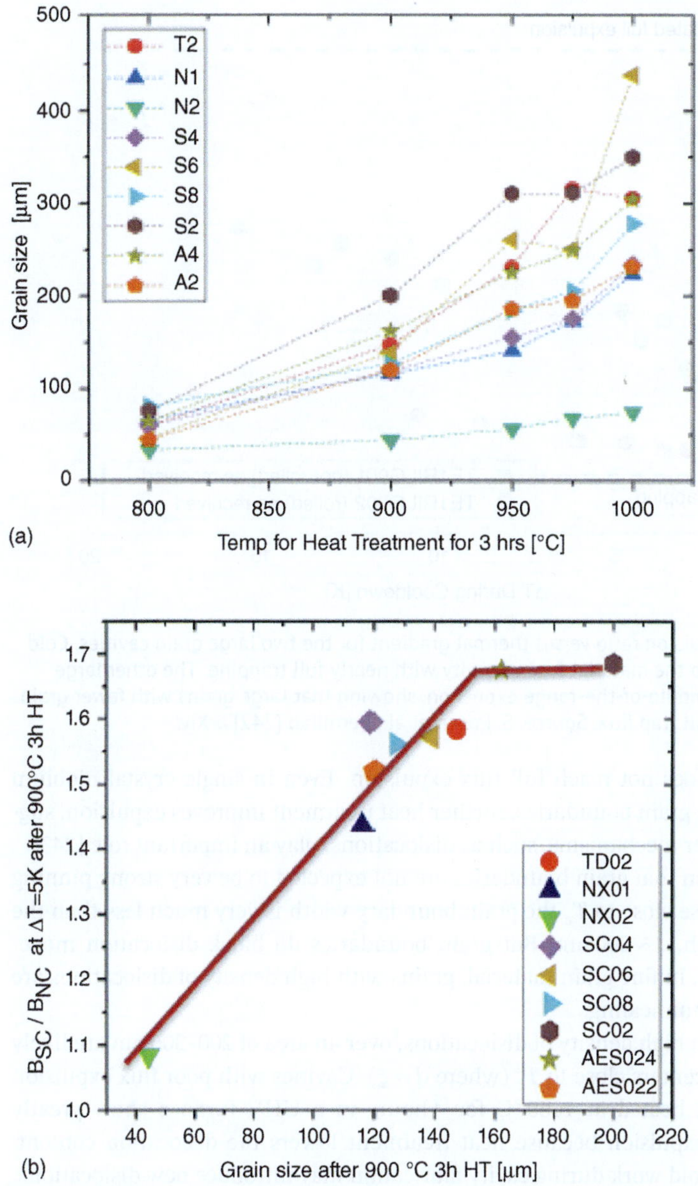

Figure 5.24 (a) Correlation between grain size and heat treatment temperature. Source: [168]/with permission of M. Martinello et al., Fermilab. (b) Flux expulsion data is weakly correlated with average grain size of different batches of material annealed at 900 °C. Note how various materials with similar (120 mm) grain size show different expulsion ratios from 1.4 to 1.6. Source: M. Martinello et al., Fermilab [168] Figure Courtesy of Z. Sung, Fermilab.

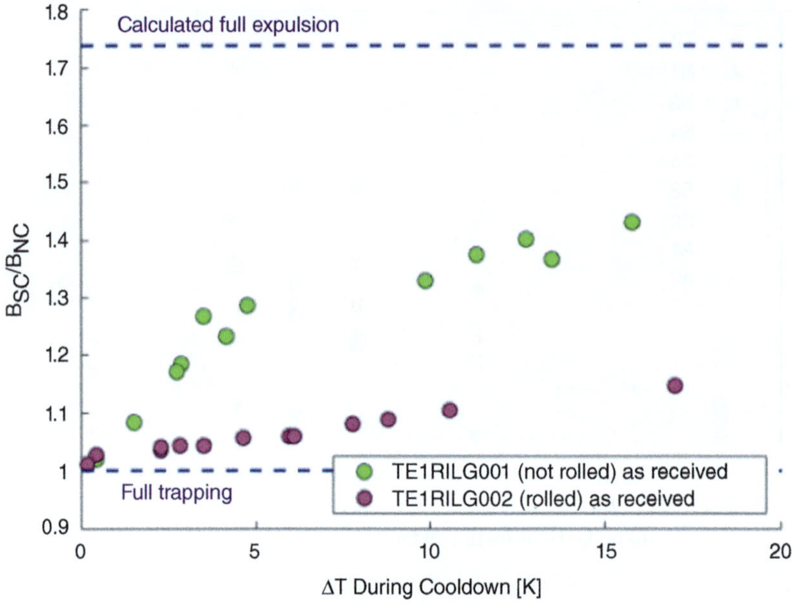

Figure 5.25 Expulsion ratio *versus* thermal gradient for the two large grain cavities. Cold work was added to the material for the cavity with nearly full trapping. The other large grain cavity has middle-of-the-range expulsion, showing that large grains with fewer grain boundaries can still trap flux. Source: S. Posen et al., Fermilab [242] arXiv.

boundaries. It does not reach full flux expulsion. Even in single crystal niobium samples with no grain boundaries, further heat treatment improves expulsion, suggesting that other mechanisms, such as dislocations, play an important role [243].

Another reason that grain boundaries are not expected to be very strong pinning centers is because close to T_c the grain boundary width is very much less than the coherence length, $\xi \sim 200$ nm. But grain boundaries do block dislocation movement. Therefore, in fine grain material, grains with high density of dislocations are found even after annealing.

Regions with a high density of dislocations, over an area of 200–300 nm are likely strong pinning centers close to T_c (where $d \sim \xi$). Cavities with poor flux expulsion efficiency when heated at 1000 °C for 4 hours in a UHV furnace show greatly improved flux expulsion because heat treatment lowers the dislocation content. But additional cold work during cavity fabrication may introduce new dislocations.

Scanning electron microscopy (SEM) imaging techniques such as electron back-scatter diffraction (EBSD) and electron channeling contrast imaging (ECCI) help to identify bulk material properties responsible for flux trapping. EBSD studies on fine grain material show many grains with high levels of local misorientation which is an indicator of dislocation density. ECCI can image crystal defects such as dislocations. ECCI makes use of the *electron channeling* mechanism. When electrons channel down through the crystal, local distortions in the crystal lattice yield a modulation in the backscatter electron signal.

Figure 5.26 a and b show examples of EBSD on materials with good and poor expulsion properties. Using a cavity with material that shows poor flux expulsion,

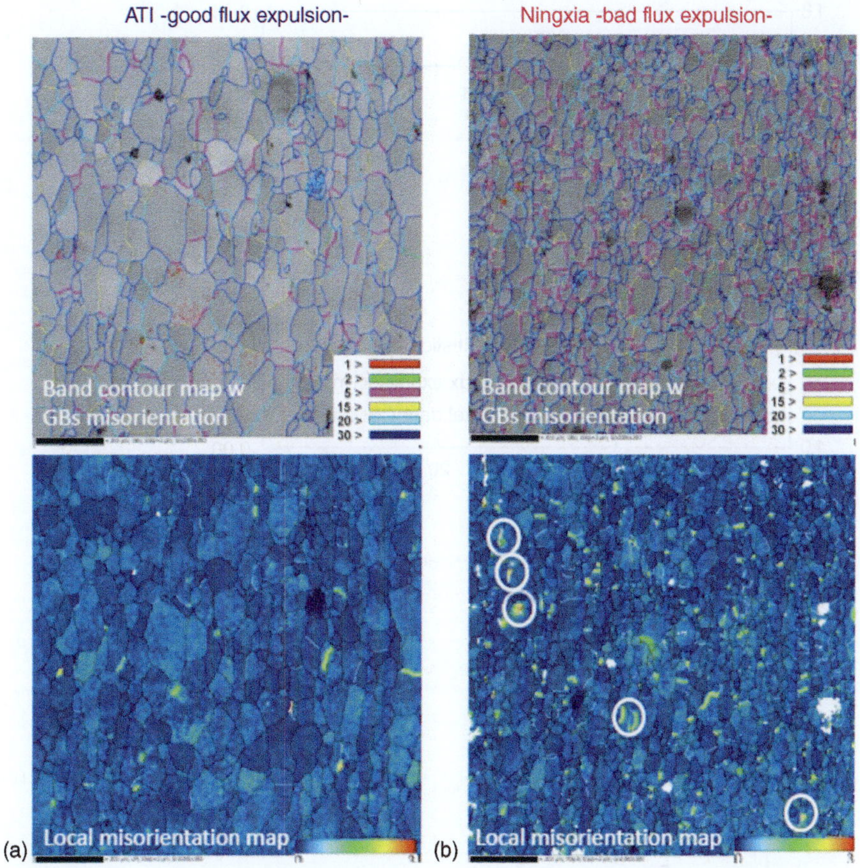

Figure 5.26 (a) EBSD image for material with good flux expulsion showing few crystal misorientations (left panels). (b) EBSD image for material with poor flux expulsion showing substantial disorientations (right panels). Source: [240]/Courtesy of M. Martinello, U.S. Department of Energy Office of Science.

surface studies with ECCI and EBSD have been carried out on hot spots identified by T-maps, then cut out from regions that show strong trapping despite fast cooldown [157, 240]. These hot spots show that small grains and areas with large local misorientation have a larger density of dislocations and lattice defects. Areas in which dislocations are closely spaced to each other are stronger pinning centers compared to isolated dislocations, especially when their region spreads over 200–300 nm, close to the coherence length near T_c

5.12 Modeling Flux Trapping From Pinning Variations

During a superconducting transition, the thermodynamic force from temperature gradients moves free vortices. As discussed in Section 5.8, during a fast cooldown, the superconducting transition starts from the bottom (where the liquid helium drops) and propagates to the top of the cavity. Vortices follow the same

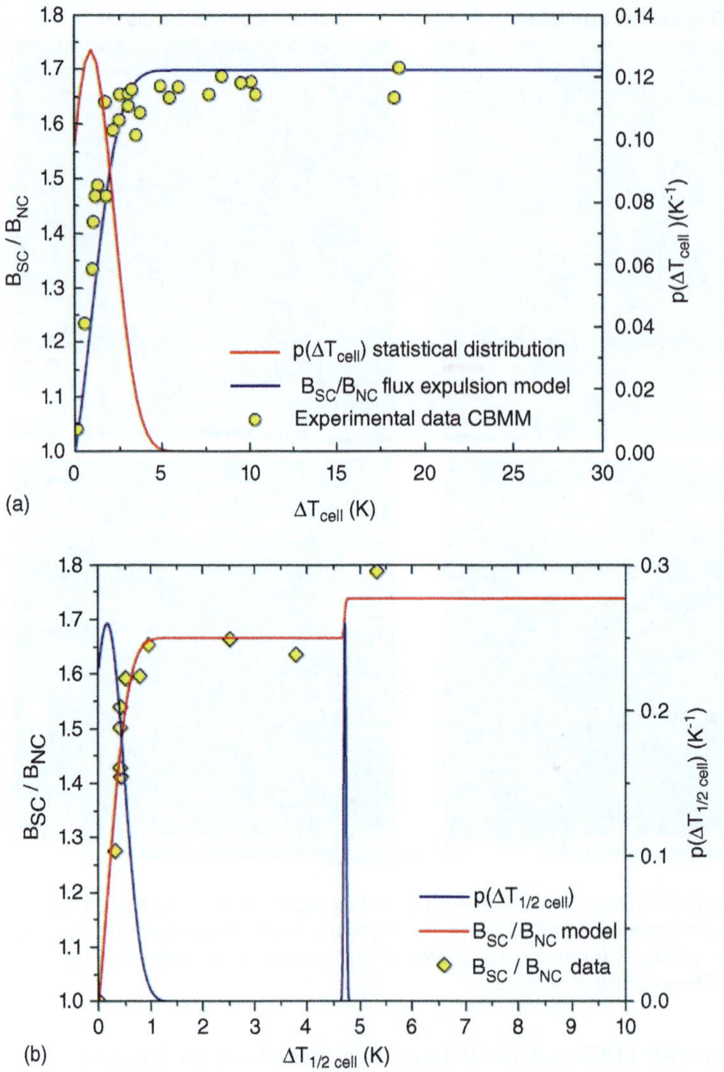

Figure 5.27 (a) Modeling flux expulsion and pinning. Comparison between data (yellow dots) and simulations (blue curve) of flux expulsion for cavity CBMM. The red curve shows a single-peaked probability density function for pinning. (b) Comparison for cavity AES011 using a double-peaked probability density function (blue curve). Source: [49] Courtesy of M. Martinello Fermilab.

path, continuously pushed up toward the warm areas of the cavity. However, since defects in the material pin down vortices, the thermodynamic force from the temperature gradient needs to be greater than the pinning force. Thus, the microstructure of the cavity material plays an important role in the ease of expulsion [241, 246]. A simple statistical simulation [49] models measured flux expulsion properties that result when the thermodynamic force overpowers the pinning force.

The pinning force arises from inhomogeneities or defects in the material. One definition for this force is in terms of a critical current density needed to depin vortices. Here the Lorentz force acting on a vortex overwhelms the pinning force to move vortices. Since fast cooldown generates not a Lorentz force but a thermodynamic force for flux expulsion, one can define a minimum thermal gradient ΔT which overcomes the pinning force. The critical thermal gradient depends on the temperature as well as on the nature and concentration of pinning sites. The minimum thermal gradient is related to the temperature difference (measured during the superconducting transition) along the cavity cell, from bottom iris to top iris, or along half cavity cell, from equator to upper iris.

A statistical simulation of the flux expulsion behavior [49] assumes a distribution of defects characterized by a Gaussian probability density function for expelling a pinned vortex. The function is centered around a most probable value of pinning force, proportional to the critical thermal gradient ∇T_{crit}. Depending on the material, more than one distribution of pinning centers may be present. Each group of defects as dislocations, grain boundaries, precipitates, etc. defines a certain probability to expel pinned vortices.

Experimental data of flux expulsion are usually represented as the flux expulsion ratios (B_{SC}/B_{NC}) as a function of the thermal gradient along the cavity, from equator to upper iris, or from bottom iris to upper iris. The model takes into account the distributions of pinning centers through either a single or a double-peaked probability density function. Figure 5.27 shows the flux expulsion data of the cavity CBMM [241] along with a single peak statistical model (red curve) optimized for a good fit (blue curve) to the experimental points (yellow data points). The single peak simulation fits data from a cavity that shows good flux expulsion. The CBMM cavity has the flux expulsion ratio that saturates around $\simeq 1.7$, close to ideal.

However, a single peak distribution, cannot properly model a cavity capable of only partial expulsion. Figure 5.28 shows data for a cavity [241] for which flux expulsion does not improve by increasing the temperature difference along the cavity cell to increase the thermal gradient force. This limitation is due to the presence of a different distribution of pinning centers with larger average pinning forces, and therefore larger needed ΔTs. A double-peaked distribution function is found to be more successful in modeling the data. In Figure 5.28, the simulations are compared with the flux expulsion data of the cavities CBMM and ACC002 shown in Reference [241].

It is clear that in both cases, for good and poor flux expulsion, the model successfully interpolates the experimental data with the appropriately chosen distributions.

Most cavity flux expulsion data can be successfully described by a double Gaussian distribution of defects. Figure 5.28 shows how the magnetic flux expulsion data of two cavities made from material with vastly different flux expulsion behavior have been successfully described with the model. The model also shows that to achieve complete flux expulsion, it is necessary to reach the maximum critical thermal gradient of the second peak of the probability density function. This explains why it is rare to achieve ideal flux expulsion ratios equal to the ideal value of 1.74.

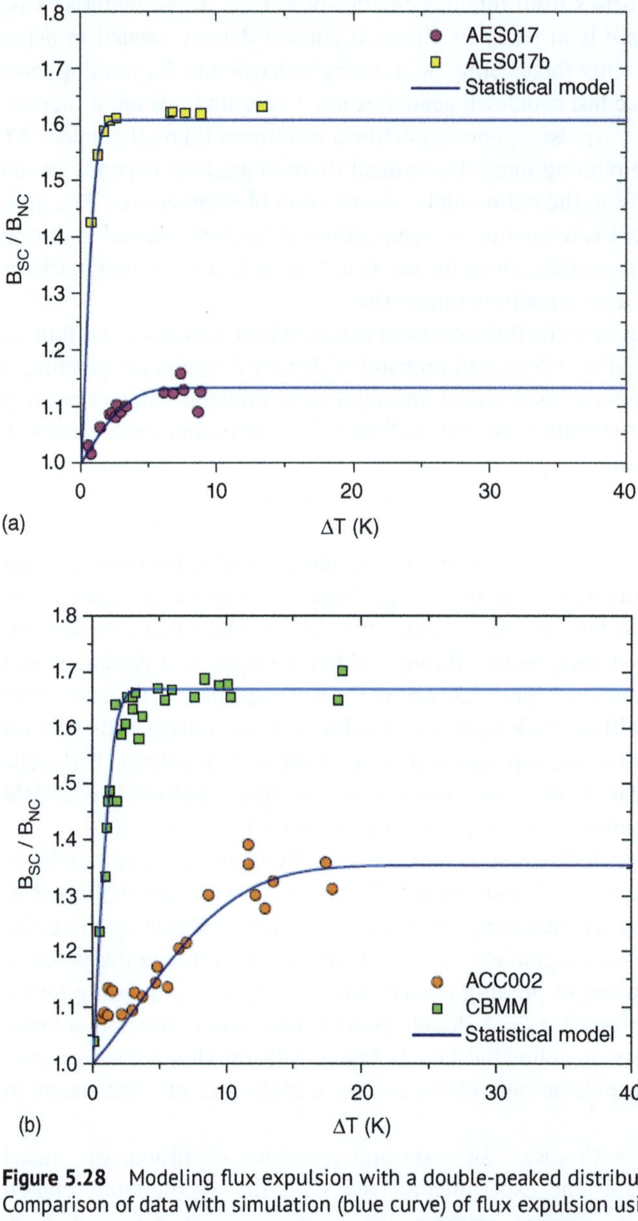

Figure 5.28 Modeling flux expulsion with a double-peaked distribution of pinning. Comparison of data with simulation (blue curve) of flux expulsion using double Gaussian distributions of pinning centers. For cavities (AES017b, CBMM) made of material that expels flux well, a temperature difference of a few K across the cell is sufficient to achieve nearly complete flux expulsion. On the other hand, cavities ACC002 and AES017 show incomplete expulsion even for temperature gradients near 20 K. Source: [49] Courtesy of M. Martinello Fermilab.

Part III

High Gradient Frontier: Performance Advances and Understanding

6

High-Field Q Slope (HFQS) – Understanding and Cures

6.1 HFQS Summary

As mentioned in Introductory Chapter 2, Q versus E curves for niobium typically show three distinct regions of Q changes: low, medium, and high-field Q slopes [5]. As discussed in Chapter 4, the low field Q-slope (LFQS) arises from two-level-system (TLS) losses in the amorphous pentoxide. The losses increase for thicker oxide layers and substantially reduce when the layer is dissolved into the bulk with 300–400 °C baking. The medium field Q-slope (MFQS) shows falling Q with increasing field for most surface treatments, except for the interesting treatment of doping with N (or Ti), which shows rising Q, as discussed in Chapter 3. In this chapter, we address mainly the causes and solutions for the high-field Q slope (HFQS). Understanding and controlling HFQS is of great importance for particle accelerators demanding highest gradients to lower the real estate costs. At appropriate places, there will also be opportunities to discuss features of the MFQS.

We start with a summary of the salient features of the HFQS summarized in [4]. HFQS is a strong Q degradation effect that significantly limits performance of niobium superconducting RF (SRF) cavities. The characteristic signature of HFQS is a rapid decrease in Q starting from the accelerating gradients of ≈ 20–25 MV/m for high-β cavities, corresponding to about 80–100 mT surface magnetic field for 1.3 GHz Tev energy superconducting linear accelerator (TESLA) shape cavities. The HFQS is present in cavities made from fine-grain, large-grain, and even single-grain niobium, prepared by either buffered chemical polishing (BCP) or electropolishing (EP), or after annealing at 800 °C, followed by exposure to air and high-pressure rinsing (HPR). Temperature maps show that the RF heating from the falling Q takes place in high magnetic field regions only [5]. The most obvious difference between the BCP and EP cavities is the higher surface roughness for a BCP surface due to sharp steps at grain boundaries. When roughness increases with increasing amounts of BCP, it lowers the onset field of the HFQS.

The important property of the HFQS is that mild baking at 120 °C for 48 hours removes the Q-drop, allowing gradients to increase above H_{c1} and toward H_{sh}. Mild baking is most effective for cavities prepared by EP, and gives some improvement for cavities prepared by BCP. Even though the Q-slope improves after baking, most small-grain BCP cavities quench at field values slightly higher than the onset field

Superconducting Radiofrequency Technology for Accelerators: State of the Art and Emerging Trends, First Edition. Hasan Padamsee.

for the HFQS. This is most likely due to local field enhancements from the surface roughness due to BCP, but needs to be better supported by temperature maps that should show no pre-heating as expected for magnetic induced quench. Large-grain cavities prepared by BCP reach much higher fields after baking compared to small-grain cavities, due to fewer grain boundaries. Baking times for large-grain cavities can also be reduced from 48 hours to 12 hours [247]. EP-baked large-grain cavities reach higher fields than BCP-baked cavities [245], again showing that roughness plays a role, since large-grain cavities have very few grains where sharp steps are present.

Smoothness plays an important role in the baking benefit. Baked EP cavities with a smoother surface (Figure 2.5) reach much higher fields than baked BCP cavities. Large-grain and single-crystal cavities [248] prepared by BCP also benefit greatly by the baking effect [247], due to overall higher smoothness from the absence of grain boundary steps from BCP. Baking times of 6 hours and 12 hours have given good results for single-crystal and large-grain cavities.

But roughness is not solely responsible for the HFQS since EP-treated large-grain and single-grain cavities, with no roughness from grain boundary steps, also show HFQS before 120 °C baking. Also, 120 °C baking has no effect on surface roughness, but greatly mitigates HFQS, which confirms that roughness alone is not responsible for HFQS.

The baking benefit is preserved even after the baked cavity is exposed to air and water, showing that adsorbed surface layers are not responsible for the HFQS, or the baking benefit. The baking benefit does not substantially change when the 5 nm pentoxide layer is removed, and a new one is grown. Similarly, the baking benefit does not change when the oxide layer is made thicker by baking in air, or by anodizing to about 30 V (~ 60 nm additional oxide thickness).

HFQS has been an intense area of R&D. A decade of HFQS research between 2000 and 2010 formed the basis for a few different hypotheses [5, 249–251], but they failed to provide a complete explanation of HFQS and its baking cure or to satisfy all the experimental findings. Understanding has progressed substantially in the following decade from several new measurements and new models, which we present in this chapter. Understanding the baking benefit is the key to the unravell the mechanism of the HFQS. Besides 120 °C baking, new treatment procedures such as N-doping (Chapter 3) and nitrogen infusion (to be discussed in Chapter 7) also impact the HFQS.

6.2 HFQS in Low-β Cavities

Low-β quarter wave resonator (QWR) and half wave resonator (HWR) generally prepared by the more convenient BCP process all show the HFQS, as seen from the performance of a large number of cavities for Facility for Rare Isotopes Beams (FRIB) at Michigan State University (Section 11.2.7) [252]. Since the FRIB cavities are built from fine-grain niobium and treated with BCP (for ease of chemistry), the 120 °C bake for 48 hours does not help much with the HFQS, and can also lead to a quench, likely due to local field enhancements from BCP surface roughness. However, 120 °C bake reduces R_{BCS} significantly, due to shortening of the mfp, while slightly increasing R_{res} due to conversion of some of the pentoxide to suboxide [187, 253, 254]. The Q benefit from baking is especially significant for the low-β, low-frequency cavities operating at 4.2 K. At this temperature, R_{BCS} is the dominant contribution to the surface resistance for the low-frequency QWR and HWR cavities. As seen for 1.3 GHz cavities, if low-β cavities were fabricated from large-grain materials, 120 °C bake would likely help to cure the HFQS and lead to higher gradients. But the low-β accelerator laboratories have not taken advantage of the potential HFQS benefits of large-grain niobium.

6.3 Deconvolution of R_{BCS} and R_{res}

As for the N-doped cavities in Section 3.6, deconvolution studies [133] of the surface resistance show which component (temperature-dependent R_{BCS} or R_{res}) is responsible for the HFQS. The studies are carried out for various treatments: BCP, EP, each before bake and after bake (Figure 6.1), and they reveal several interesting aspects of both HFQS and the MFQS. Deconvolution shows that HFQS primarily arises from R_{res}. The baking effect drastically lowers the residual resistance component above the onset field of the HFQS.

Electropolished cavities before baking exhibit a mild field dependence of R_{res} up to 80 mT, above which R_{res} increases sharply in the HFQS regime, corresponding closely to the onset field for the HFQS. Baking at 120 °C eliminates the sharp rise in R_{res}, leaving a "residual" slope at $B > 80$ mT (Figure 6.1a). EP + 120 °C is most effective in mitigating the HFQS out to the highest field. These results clearly show that the onset of HFQS arises primarily from R_{res}.

Section 6.13 discusses a model for the HFQS based on nano-hydrides turning normal at the onset field. 120 °C baking suppresses the formation of the hydrides due to hydrogen trapping at interstitials and/or vacancies released during the bake. TEM studies presented in Section 6.8 show that the suppression of nano-hydrides is incomplete, which may leave a large number to contribute to the continuation of the MFQS.

If the cavity is heat treated at 800 °C to remove H dissolved in Nb during chemistry, the HFQS is still present, but the onset field is a bit higher; again, the 120 °C bake cures the HFQS and helps reduce the MFQS slope. These changes with 800 °C bake suggest that bulk H removal has a beneficial effect on both MFQS and HFQS.

Coming primarily from R_{res}, HFQS cannot be an intrinsic property of the superconducting condensate. This finding puts constraints on models to explain the HFQS. Proposed models must be based on extrinsic mechanisms for RF losses.

The field dependence for the temperature-dependent component ([BCS] resistance) is similar for EP and BCP (Figure 6.1 b). For both treatments, baking at 120 °C has a dramatic effect. At low fields, the absolute value of the R_{BCS} component decreases by a factor of 2, due primarily to the low mfp from baking, as expected from the BCS theory. The expected low field result for decrease in BCS resistance

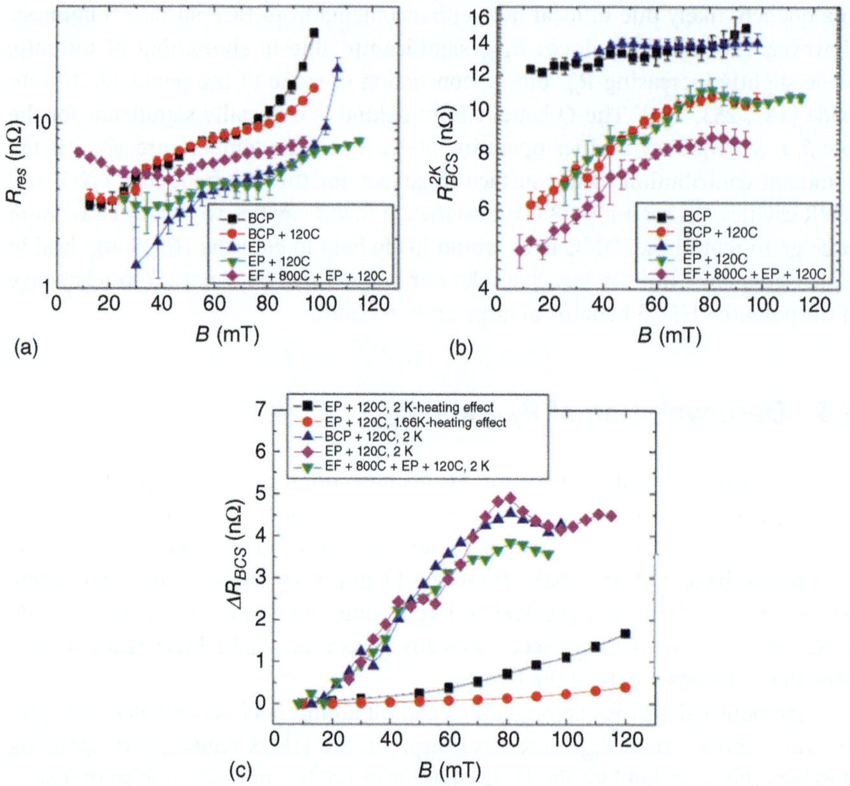

Figure 6.1 Average (a) R_{res} at low T, and (b) R_{BCS} at $T = 2$ K of a niobium cavity as a function of the peak surface magnetic field. The error bars refer to the fit errors. (c) Comparison of the BCS component of the MFQS with the expected increase in R_{BCS} from heating due to the BCS resistance. Source: [133]/A. Romanenko et al. with permission of AIP Publishing LLC. Various cavity treatments give similar MFQS.

gives confidence that the deconvolution method successfully picks out the BCS component of the surface resistance.

Many aspects about the MFQS were discussed in Padamsee and coworkers [5, 255]. The decomposition study here gives further clues about the MFQS (Figure 6.1 b). Before bake, R_{BCS} is high (12–14 nΩ) but relatively flat. After 120 C bake, R_{BCS} falls to a maximum of 5 nΩ, and develops a strong field dependence in the medium field region. It is possible that when the density of nano-hydrides decreases after 120 °C bake, the associated losses will also come down as seen for the drop in the magnitude of the MFQS. The slope that becomes apparent in the BCS component after the bake may be due to the field-dependent behavior of the remaining nano-hydrides.

Figure 6.1c compares the BCS component from the decomposition analysis with the predictions from a thermal feedback effect [42]. As discussed in Refs. [133, 255], the MFQS is too large to be due to a thermal feedback effect from the RF surface temperature increase via the exponential temperature dependence of R_{BCS}. Rather the MFQS must emerge from some change in the properties of niobium.

It would be interesting to carry out deconvolution studies for large-grain or single-grain Nb cavities since these cavities need short baking times compared to fine-grain cavities.

6.4 Depth of Baking Effect

6.4.1 From Anodization

Determining the depth of the 120 °C baking effect provides clues to how it works to cure the HFQS. One approach to determine the depth of the layer affected by the bake is to convert the modified Nb layer into oxide by anodization in several steps. Studies show [254, 256] that with 40 V anodization after 120 °C bake, the HFQS forms again. Re-baking at 120 °C removes HFQS again. During anodization, the oxide layer grows at a rate of about 2 nm/V, consuming about 0.75 nm/V of Nb. Hence the baking-modified layer estimate from this method is about 30 nm deep.

6.4.2 From HF Rinsing

Another procedure is stepwise material by successively removing and regrowing the oxide layer, yielding a removal of about 1.5–2 nm/step of Nb material [253]. Removal is by hydrofluoric acid (HF) rinse, which dissolves the pentoxide. Reoxidation is performed in water. Water rinse regrows a new oxide layer of 4–5 nm thick, consuming about 1.5–2 nm of niobium as estimated from the corresponding densities.

After 120 C baking, the first HF rinse usually results in an increase in Q at low fields due to the removal of lossy Nb suboxides introduced by the bake [254]. By repeating these cycles, removal of about 10 nm of material starts to bring back the HFQS. As Figure 6.2 shows, after 24 HF rinse cycles (36–48 nm of niobium consumed), Q versus H curve is essentially back to the prebaking shape. Therefore, the depth of the layer affected by mild baking is between 40 and 50 nm, in line with diffusion of O results, discussed in Section 6.6. Re-baking at 120 °C restores the benefits of the baking effect.

The HF rinsing study also shows that LFQS is governed by the properties of the surface within the first nanometers (Section 4.4 further discusses the LFQS). Both MFQS and HFQS are determined by the material thickness of about 20–50 nm, suggesting related causes.

6.4.3 Depth of Magnetic Field Penetration by LE-μSR

Besides the depth of the baking benefit layer, it is important to know the RF penetration depths before and after 120 °C bake. Low-energy muon spin rotation, relaxation and resonance (LE-μSR) is a relatively new technique to measure penetration depths. In the μSR technique [257], low-energy, spin-polarized muons are implanted into niobium. The time evolution of the polarization is followed as a precise diagnostic of the local magnetic field. Muons spin rotate in magnetic

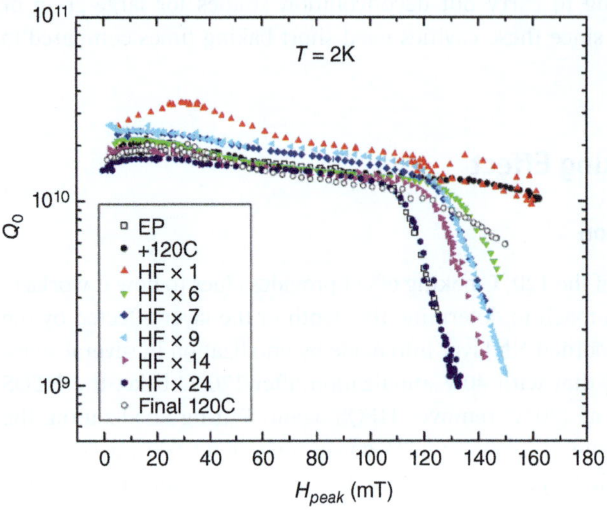

Figure 6.2 RF test results after multiple HF rinse cycles for an electropolished cavity, after bake and after up to 24 HF rinsing cycles. No field emission was present except for the final 120 °C bake test. Source: A. Romanenko et al., Fermilab [253]/American Physical Society/CC BY 3.0.

fields and emit a positron preferentially in the spin direction after decay (lifetime τ_μ = 2.2 μs). The precession frequency, proportional to the magnetic field, is measured by detecting the anisotropic muon decay (Figures 6.3, [42]). The decay positrons are registered in detectors surrounding the sample to determine the time evolution of the muon spin and thereby the local magnetic field strength. The low-energy technique [131, 258] allows precise microscopic measurements of the magnetic field profile $B(z)$ inside superconductors to a depth range of 130 nm and over a spot size of about 1 cm, making LE-μSR an ideal probe to study HFQS and the impact of 120 °C baking. It can also be used to measure the penetration depth for other techniques, such as N-doping (Section 3.5.1).

The first μSR measurements were carried out by Grassellino on the surface muon beamline at TRIUMF using the Los Alamos Meson Physics Facility (LAMPF) spectrometer [257]. As the kinetic energies of these beams were near 4.1 MeV, the

Figure 6.3 (a) 100% spin-polarized muons are deposited in a sample and spin rotate in the local magnetic field. The muons decay with emitted positrons correlated with the spin direction. The initial muon spin polarization of the muons is perpendicular to the applied static magnetic field H. The silver mask upstream ensures that the positive muons stop only in the central region of the sample. Right and left detectors record positrons correlated with the time of arrival. The time evolution of the asymmetry in the two signals gives a measure of the local field in the sample. Source: A. Grassellino [257]/APS CC BY 3.0 and T. Junginger and Laxdal, TRIUMF [258]/Talk (b) The precession frequency is proportional to the magnetic field, which allows direct measurement of field attenuation in the Meissner state, and hence the London Penetration Depth. Source: T. Junginger and Laxdal, TRIUMF [258]/Talk.

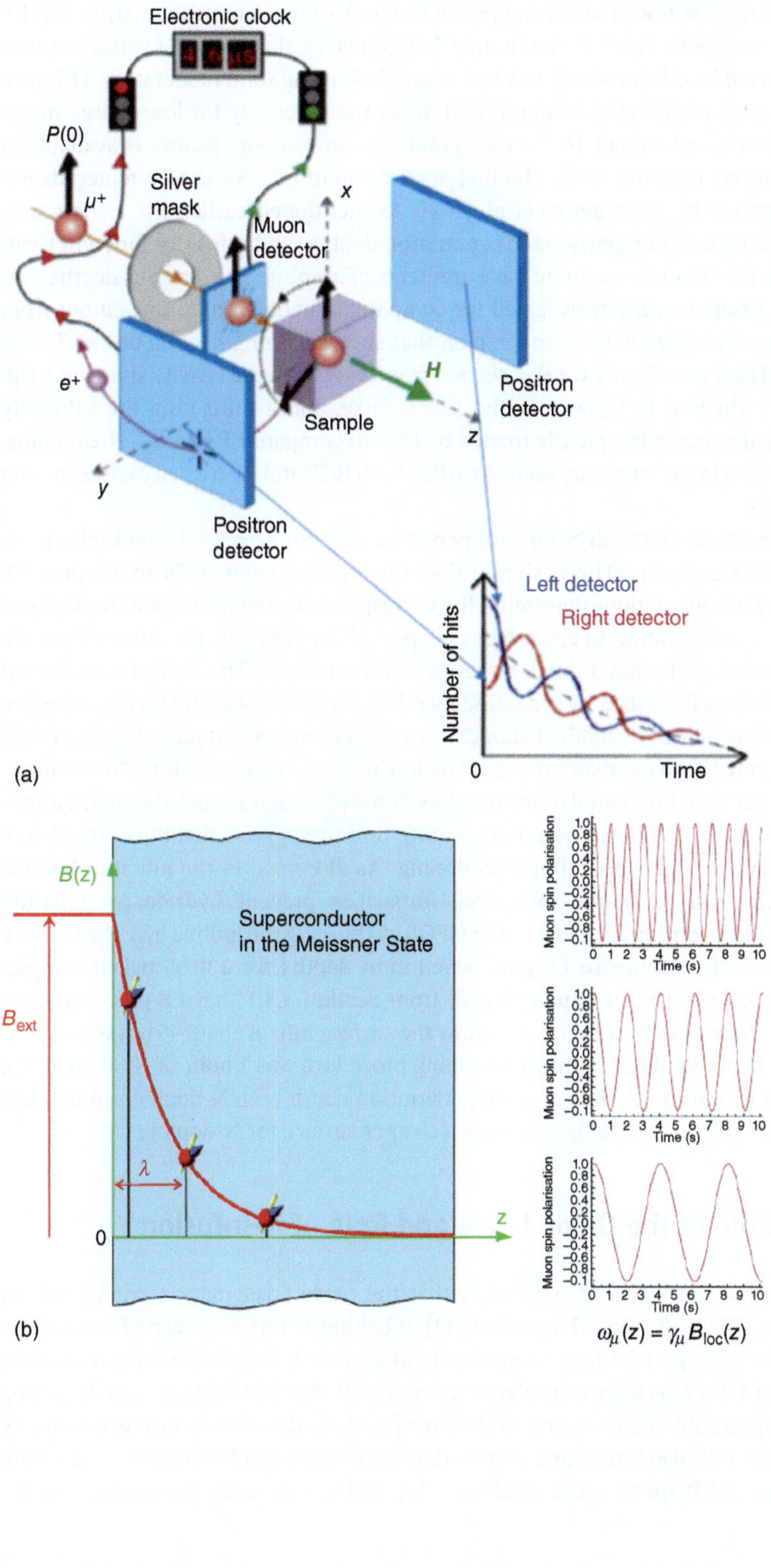
Electronic clock
4.6 μs
P(0)
μ+
Silver mask
Muon detector
x
Positron detector
H
e+
Sample
z
y
Positron detector
Left detector
Right detector
Number of hits
0
Time
(a)
B(z)
Superconductor in the Meissner State
B_ext
λ
0
z
Muon spin polarisation
Time (s)
(b)
$\omega_\mu(z) = \gamma_\mu B_{loc}(z)$

spin-polarized surface muons stopped in the bulk of the sample (at 300 μm depth). To probe magnetic fields in the thinner London layer, the energy of surface muons has to be reduced from 4 MeV to a few tens of keV using solid moderators. This process requires a high surface muon current as the efficiency for low-energy muon production is only about $10^{-4} - 10^{-5}$ [258]. The low-energy facility is available at Paul Scherrer Institute (PSI). The first application of LE-μSR to SRF materials has been reported by Romanenko et al. [131]. As mentioned earlier, the technique is useful for relative comparisons of penetration depths (and mfps) for different treatments, rather than for accurate measurements of absolute penetrations depths.

Three samples were investigated for comparison with LE-μSR, one cutout from the EP cavity before bake from a region that showed strong heating on the T-map during HFQS (see Figure 6.4a), a second from a 120 °C baked cavity showing a significant reduction in heating by the 120 °C bake, and a third from the EP cavity (before bake) but subsequently treated by BCP to compare EP and BCP treatments. As discussed in the summary section earlier, both BCP and EP treated cavities exhibit the HFQS.

The results show the BCS/Pippard penetration depth λ_L = 24–25 nm for both the EP and BCP samples. (This is shorter than the reported value of 38 nm for pure Nb [96].) Model calculations determine the corresponding mean free path to be about 400 nm, corresponding to clean Nb, as expected. In contrast, the cutout from the 120 °C baked cavity has a much larger penetration depth. The decay for the baked sample is best described by a depth-dependent electron mean free path; a surface layer with penetration depth of about 50 nm shows a very short mfp of 2 nm, changing to about 16 nm at greater depths. The low mfp is consistent with a "dirty limit", as expected from interstitial impurities (Section 6.6), which are mostly oxygen atoms dissolved during baking. The very short mfp finding suggests that the mitigation of HFQS via baking is due to impurity doping. As discussed in the niobium-hydride model (Section 6.13), the interstitial impurities prevent hydride precipitation responsible for strong RF losses of the HFQS by trapping the mobile hydrogen atoms.

It is useful to compare LE-μSR penetration depths for 120 °C baked samples with those for nitrogen doping. Recall from Section 3.5.1 that LE-μSR studies of heavy N-doped sample (Figure 3.9) show the surface mfp of about 40 nm with a penetration depth of about 28 nm. The doping procedure was 1 hour of N (at 1000 °C), followed by 60 μm EP. Note that the penetration depth with N-doping is much less than that with a 120 °C bake, showing a cleaner surface for N-doping.

6.5 Role of the Oxide Layer and Role of N-Infusion

An important experiment establishes that the oxide layer plays a crucial role in removing HFQS via the baking effect [42]. If baking at 120 °C is carried out *without* the oxide layer present (e.g. immediately after 800 °C bake), the bake does NOT get rid of HFQS, as shown in Figure 6.5 a. Recall that 800 °C hydrogen degassing in a furnace fully dissolves the oxide layer [260], so that the surface is essentially oxide-free. It is also interesting to note that the onset field of HFQS moves up from the usual 25 MV/m to about 30 MV/m. A possible reason for the increase in the

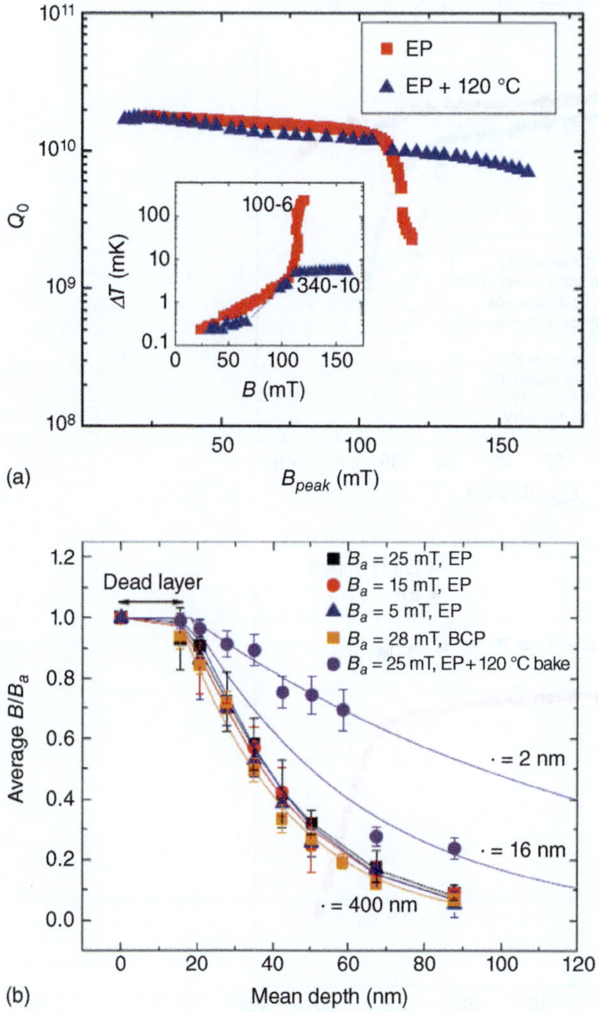

Figure 6.4 (a) Performance of unbaked and 120 °C baked EP cavities from which samples were cut out; the inset shows the RF heating ΔT (B) for cutout samples. Sample 100-6 is from a region that shows strong heating during HFQS. Sample 340-10 is from a region that shows low heating after 120 °C baking. Another sample (30-6) was cut out from the EP unbaked cavity and treated with buffered chemical polishing (BCP) for 20 μm material removal to represent unbaked BCP-treated cavities. (b) Magnetic field penetration depths measured by LE-μSR. Solid lines for unbaked EP (100-6) and BCP (30-6) samples show global fits based on the Pippard/BCS model to extract mean free path values from the penetration depth. Solid lines for the baked cavity cutout (340-10) show Pippard/BCS model calculations for two different values of the electron mean free path ℓ, one for the near-surface region (<50 nm) and one for deeper (>50 nm). Source: A. Romanenko et al., Fermilab [131]/AIP Publishing LLC/CC BY 3.0.

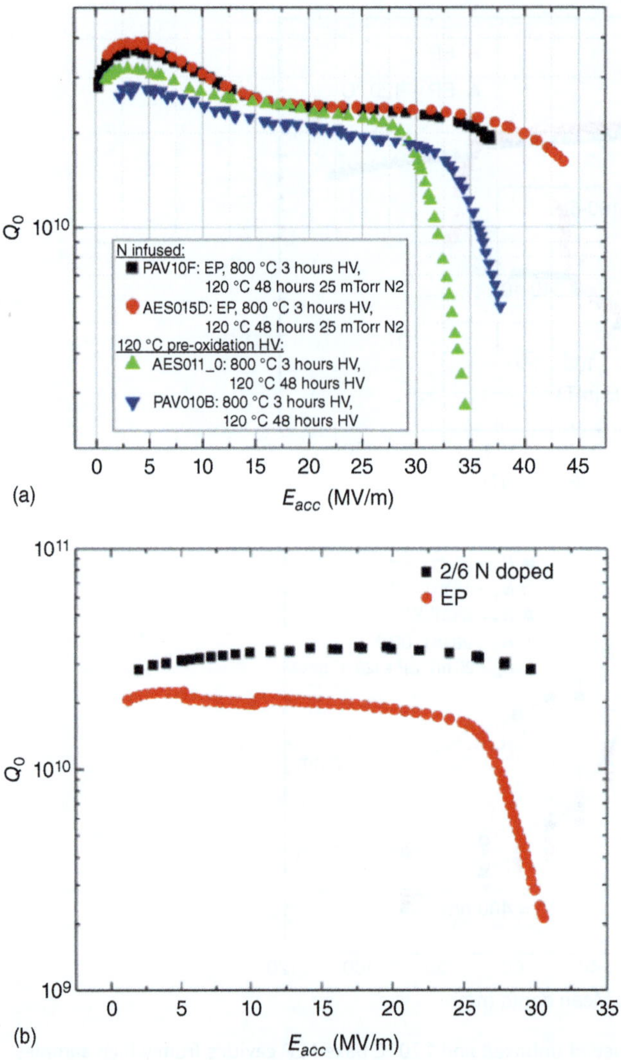

Figure 6.5 (a) Failure of the baking effect in the absence of the oxide layer. HFQS is still present after the 120 °C bake when carried out in the same furnace after 800 °C baking, which renders the surface *oxide-free*. The result is confirmed by the same experiment on two cavities. However, when the 120 °C bake is carried out on two oxide-free cavities but with the presence of 25 mTorr of N, the HFQS is cured. (b) N-doping cures the HFQS. An N-doped cavity prepared with the 2/6 recipe reaches above 30 MV/m with no HFQS compared to the standard HFQS for an EP-treated cavity with an onset field of 25 MV/m. Source: [44] Courtesy of A. Grassellino Fermilab.

onset field could be that less H is present in the cavity due to 800 °C bake, which fits within the framework of the hydride model for the HFQS (to be discussed in Section 6.13). The experiment was repeated to compare the behaviors of large-grain and fine-grain cavities, with the same result that without the presence of the oxide layer, 120 °C baking does *not* cure the HFQS [261].

Figure 6.5 a shows another effect important for understanding HFQS. Baking at 120 °C in the presence of 25 mTorr N mitigates the HFQS, moving the onset out to 40 MV/m and beyond. This treatment has been dubbed "N-infusion," which is covered in depth in Section 7.3.

These experiments allow us to postulate that either O or N diffusion into Nb as interstitials is playing a role in curing the HFQS. Again, within the framework of the hydride model (Section 6.13), both O and N interstitials inhibit the formation or growth of hydrides. We also discussed similar effects of N interstitials in reducing the residual resistance in Section 3.7.2 for N-doping. The N-infusion treatment also has a strong effect on the maximum field that can be reached, which we discuss in Section 7.3.

N-doping also helps to mitigate the HFQS, as shown in Figure 6.5b. This treatment was discussed primarily as a high Q treatment in Chapter 3. N was introduced directly after 800 °C heat treatment, so the oxide is absent. Once again, the presence of N as interstitials (in place of O) is likely responsible for the HFQS mitigation, which is consistent with the hydride model to be discussed (Section 6.13).

6.6 SIMS Studies of O, H, and OH Profiles

The preceding section established that the presence of an oxide layer is necessary for the 120 °C baking effect to cure the HFQS, providing an important clue that O plays a role in the 120 °C baking effect. Time of flight - secondary ion mass spectrometry (TOF-SIMS) studies on cavity cutouts from an EP-baked cavity show that the mild bake creates an extended ~60–70 nm layer of oxygen-rich region underneath the oxide (Figure 6.6) [262]. The secondary ion mass spectroscopy (SIMS) result clearly confirms the involvement of oxygen diffusion in the baking effect.

The observed depth of the oxygen enrichment by baking is roughly consistent with results from the previous studies on the depth of the baking benefit layer obtained by anodizing, or by HF rinse nano-profiling [253, 254, 256, 263], as well as by LE-uSR [131]. In Section 6.13, we discuss how these results fit well within the framework of the hydride model for HFQS because oxygen is an effective trapping agent for the interstitial hydrogen, thereby suppressing hydride formation, and precipitation after the 120 °C bake.

An important related finding from the SIMS studies above is that, prior to 120 °C baking, EP cavity cutouts show no oxygen impurities below the native oxide layer. This finding contradicts one of the older models proposed to explain the HFQS behavior. The "oxygen pollution model" proposes [183, 264] that an O-rich layer of many at% underneath the oxide is responsible for the HFQS. Baking at 120 °C dilutes this layer to suppress the HFQS. The latest TOF-SIMS studies do not find such a "pollution" layer. It is possible that past studies using, for example, dynamic SIMS did not have the benefit of spatial information, which could lead to possible false profiles. The modern TOF-SIMS systems used in this study utilize full 3D and imaging information which helps ensure that the depth profiles are not affected by any artifacts, such as the particles on the surface.

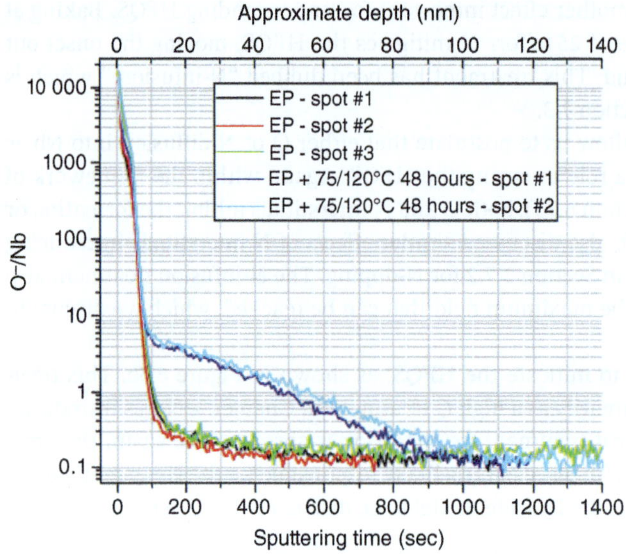

Figure 6.6 Oxygen depth profile comparison between cavity cutouts showing the effects of 120 °C baking. Before bake there is no enhancement of the oxygen concentration under the oxide layer. After bake there is a significant increase in O levels out to 70 nm. Source: A. Romanenko et al., Fermilab [262]/JACoW/CC BY 3.0, Courtesy of A. Romanenko.

Further studies on O diffusion were conducted on a cavity cut out from an EP cavity [265] that was subjected to successive *in situ* baking treatments at 120 °C for increasing durations, without compromising the vacuum. The SIMS results are shown in Figure 6.7 [262, 265].

The oxygen concentration profiles in Figure 6.7 a can be modeled by Fick's laws of diffusion [266]. Between 100 and 150 °C, the diffusivity of oxygen in niobium (in cm^2/s) is

$$D = 1.38 \times 10^{-2} e^{111{,}530/RT} \tag{6.1}$$

where R and T are the universal gas constant and temperature, respectively.

The model assumes that the oxide layer reduces upon heating [187], and that the concentration of oxygen at the interface between the oxide layer and the metal achieves the solubility limit. Thus, the oxide layer serves as an inexhaustible source of oxygen and maintains this value at the interface. The good agreement between SIMS data and diffusion calculations supports these assumptions.

To explore how the depth of O diffusion from different baking durations affects the HFQS, and to compare with the SIMS results above, a single-cell 1.3 GHz cavity was subjected to 120 °C baking treatments for durations of 3, 6, 12, and 15 hours [12, 267]. The cavities received a 40-μm EP reset of the surface prior to each baking to restore the HFQS. As a precaution against the H-related *Q*-disease from any H absorbed during EP (Section 6.11), the cavities were fast cooled to 4 K at approximately 10 K/min. A fast cooldown is effective in preventing the *Q*-disease.

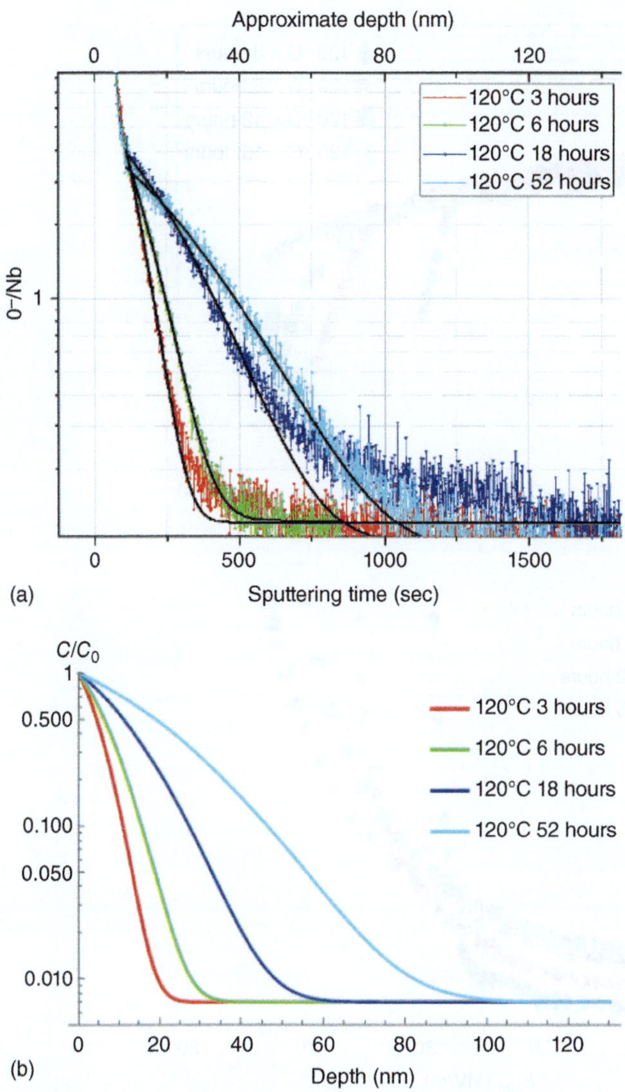

Figure 6.7 (a) TOF-SIMS profiles from an EP cavity cutout after successive *in situ* bakings at 120 °C for various durations [265]. (b) Calculated normalized oxygen concentration showing diffusion depth for the various bake durations in (a). Source: [265] Courtesy of A. Romanenko, Fermilab.

Figure 6.8a and b shows Q versus E measurements and extract R_{res} from accompanying decompositions. The increases in 120 °C baking times correspond to systematic increases in the onset field of the HFQS. The onset field for the rapid rise in the residual resistance determined from deconvolution confirms the increase in HFQS onset fields.

Figure 6.9a shows calculated oxygen diffusion profiles for each heat treatment, and Figure 6.9b shows the strong correlation between the HFQS onset field and

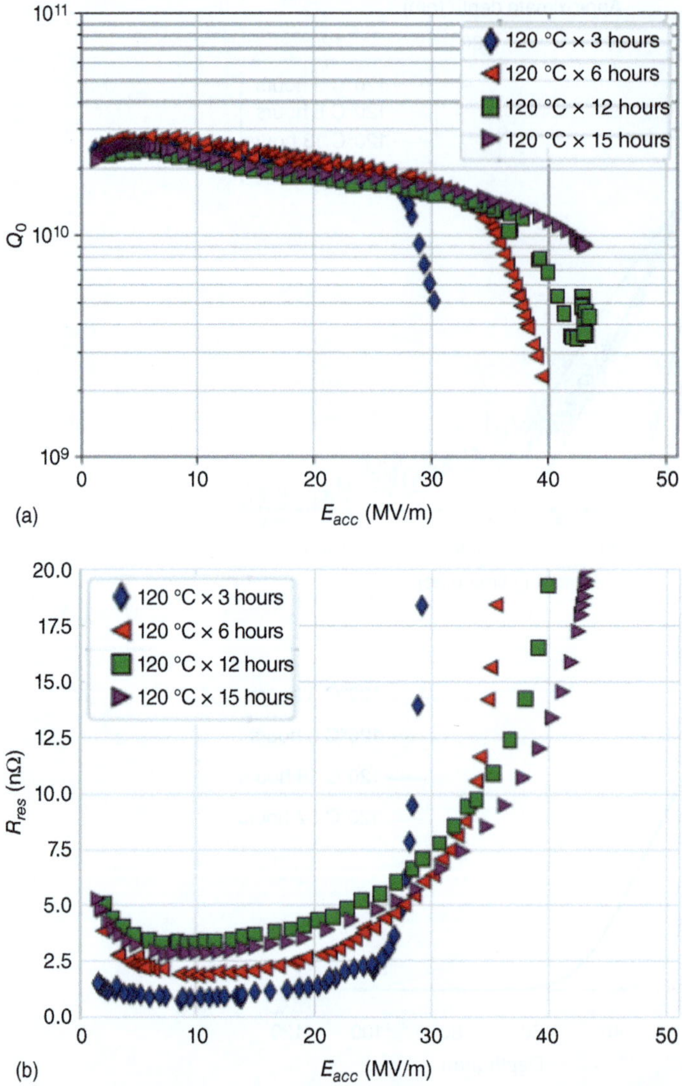

Figure 6.8 (a) Q versus E_{acc} curves for a cavity showing increasing onset fields of HFQS after increasing baking times of 3, 6, 12, and 15 hours. (b) Correlated behavior of R_{res} obtained from decomposition [12]. Source: [12] Courtesy of D. Bafia, Fermilab.

the overall diffusion depth. As oxygen diffuses further, it can capture free hydrogen from deeper in the RF layer, pushing the onset of HFQS to higher fields, as expected from the hydride model (Section 6.13). Extrapolation of the behavior in Figure 6.9(b) yields the encouraging prediction that any low-temperature bake that creates an oxygen profile extending to ~70 nm will be free from the HFQS up to a high quench field of 45 MV/m.

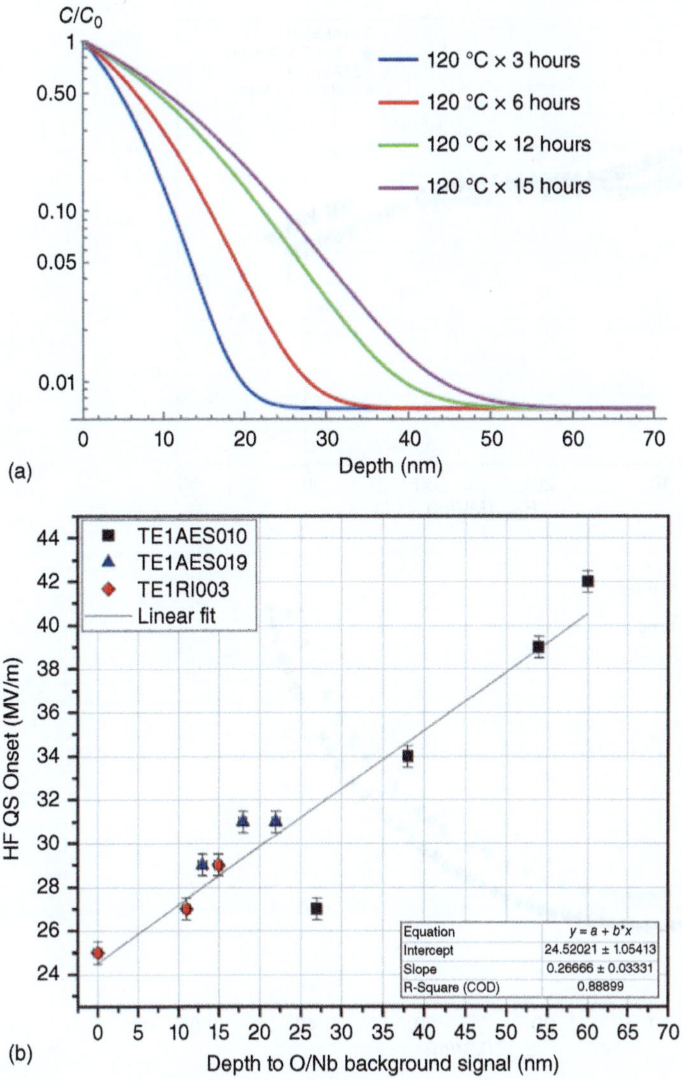

Figure 6.9 (a) Calculated diffusion curves for increasing baking times. (b) Correlation between HFQS onset field, and depth of O diffusion with increasing baking times [12]. Source: [12] Courtesy of D. Bafia, Fermilab.

As expected from the established correlation, two separate bake treatments (200 °C × 1 hour and 160 °C × 12 hours) with different baking time/temperature combinations, but with the same oxygen diffusion length, should yield a similar RF performance. Figure 6.10 a and the corresponding decomposition analysis in Figure 6.10b show how the RF performance and residual resistance of two such cavities are nearly identical. A practical consequence with economic benefits would be to optimize the standard baking treatment, replacing the usual long 48-hour bake at 120 °C with a short 1-hour bake at 160 °C.

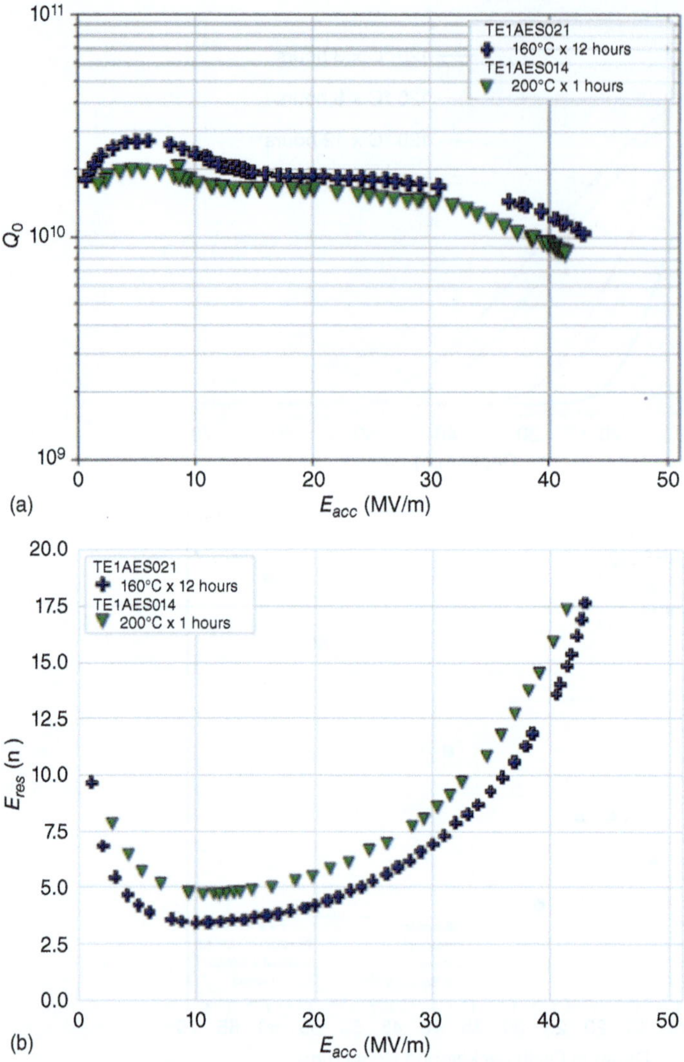

Figure 6.10 (a) Two different low-temperature baking treatments with the same oxygen diffusion length yield the same RF performance [12]. (b) Residual resistance for the same treatments via decomposition analysis [12]. Source: [12] Courtesy of D. Bafia, Fermilab.

6.7 Hydrogen Presence in HFQS

The TOF-SIMS studies [262] unlock additional clues. Accompanying the oxygen diffusion layer under the oxide, interesting hydrogen-related signals are revealed both before and after 120 °C bake. The NbH-/Nb- signal (Figure 6.11a) provides a measure of the hydrogen concentration in the lattice. Before baking, cutouts from

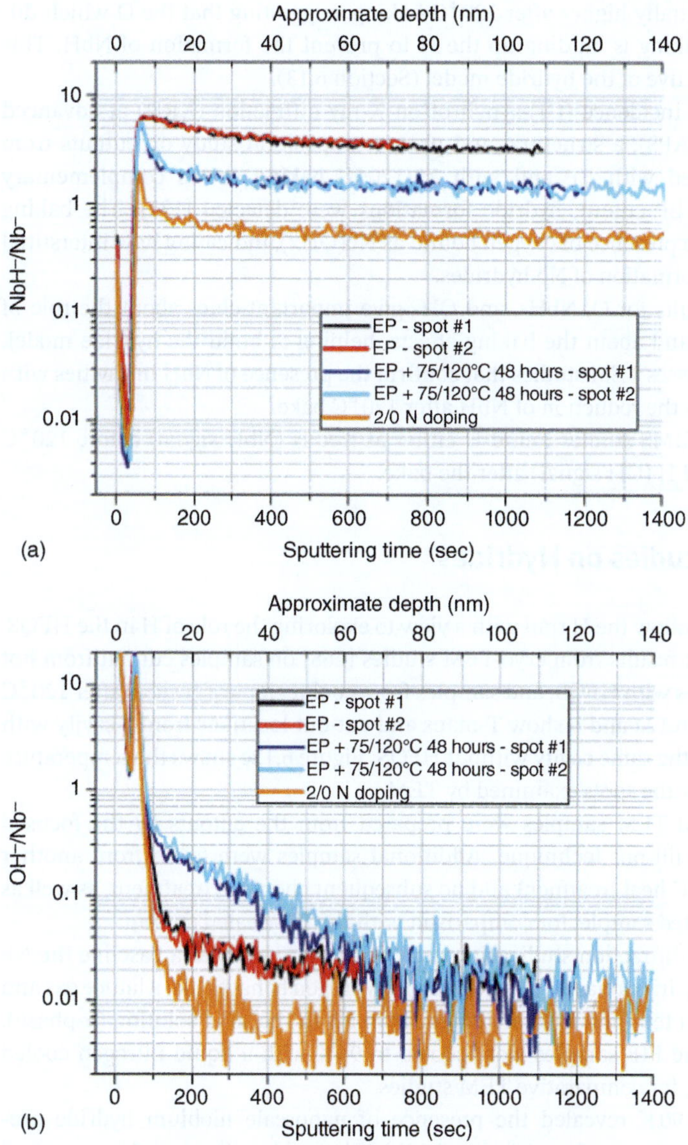

Figure 6.11 (a) NbH-/Nb- signal for cutouts from cavities with EP, cavities after bake, and an N-doped cavity. (b) OH-/Nb- signal from cutouts of the same cavities as in (a). Source: A. Romanenko et al., Fermilab [262]/JACoW/CC BY 3.0.

an EP-treated cavity show the highest hydride levels. After baking, there is a significant drop in the signal, indicating a reduction of hydride content. And after N-doping there is an even larger reduction.

Yet another clue comes from OH signals. Figure 6.11b shows OH-/Nb- signals, which provide a measure of the oxygen-bound hydrogen content. Note how this signal is substantially higher after 120 °C baking, suggesting that the O which diffused in after baking is binding up the H to prevent the formation of NbH. This finding is supportive of the hydride model (Section 6.13).

Using grazing incidence (GI)-synchrotron X-ray diffraction (XRD) at advanced photon source (APS), a structural and phase composition study on cutouts from cavities processed with EP and with 75/120 °C baking shows complementary results. For the EP cutout, hydride formations were detected [137]. The baking cutout shows the presence of oxygen bound as Nb(OHx), indicating how interstitial O prevents the formation of Nb hydrides.

The SIMS results for O, NbH-, and OH- give important clues about the role of H in the HFQS and about the baking effects, helping to build the hydride model. Section 6.8 discusses TEM studies that confirm the presence of NbH in cavities with HFQS, as well as the reduction of NbH after 120 °C bake.

In summary, SIMS studies found H either as free H (NbH signal) before 120 °C bake or as bound H (OH signal) after the bake.

6.8 TEM Studies on Hydrides

We continue to follow the H trail with a view to exploring the role of H in the HFQS. First, we present results from cryo-TEM studies [268] on samples cut out from hot regions of cavities with HFQS, and samples from cool regions of cavities after 120 °C bakeout. Figure 6.12a and b show T-maps and cut-out locations from a cavity with HFQS and from the same cavity without HFQS. Figure 6.12c shows the temperature rise with field for the spots examined by TEM.

Cross-sectional TEM samples were prepared from the cutouts by the focused ion beam (FIB) lift-out technique. Additional samples were taken from another cavity with 800 °C heat treatment and no subsequent chemical treatment, as well as from a BCP-treated sample, for comparison with the EP-treated cut-out.

TEM electron diffraction studies at room temperature show as a baseline the Nb lattice with no hydride phases (Figure 6.13a). Hydrogen behaves like a lattice gas and occupies random tetrahedral interstitial sites in Nb in the solid solution (α-phase). Subsequently, the FIB samples were cooled to 94 K with a liquid nitrogen cooled double-tilt stage, for comparative TEM studies.

Cooldown to 90 K revealed the presence of nanoscale niobium hydride precipitants in the near surface (Figure 6.13b). Phase identification demonstrated the presence of β- and ε-niobium hydride phases (see the Nb-H phase diagram in Figure 6.14). The sample extracted from the hot spot during HFQS heating (Figure 6.13b) showed a significant fraction (68%) of sites with hydrides of about 10 nm. The 120 °C bake cutout from the cold spot showed a substantially smaller fraction (27%) of niobium hydrides in the near-surface layer.

Note that the hydrides are not completely suppressed in the 120 °C baked sample, which may be correlated to the remaining reduced heating areas seen in the temperature map of Figure 6.12b as well as the continuing but diminished losses at higher fields seen in Figure 6.12c (red curve).

The BCP-treated sample showed a high concentration of nano-hydrides similar to the hot cut-out from the EP sample. Surprisingly, the sample with 800 °C heat

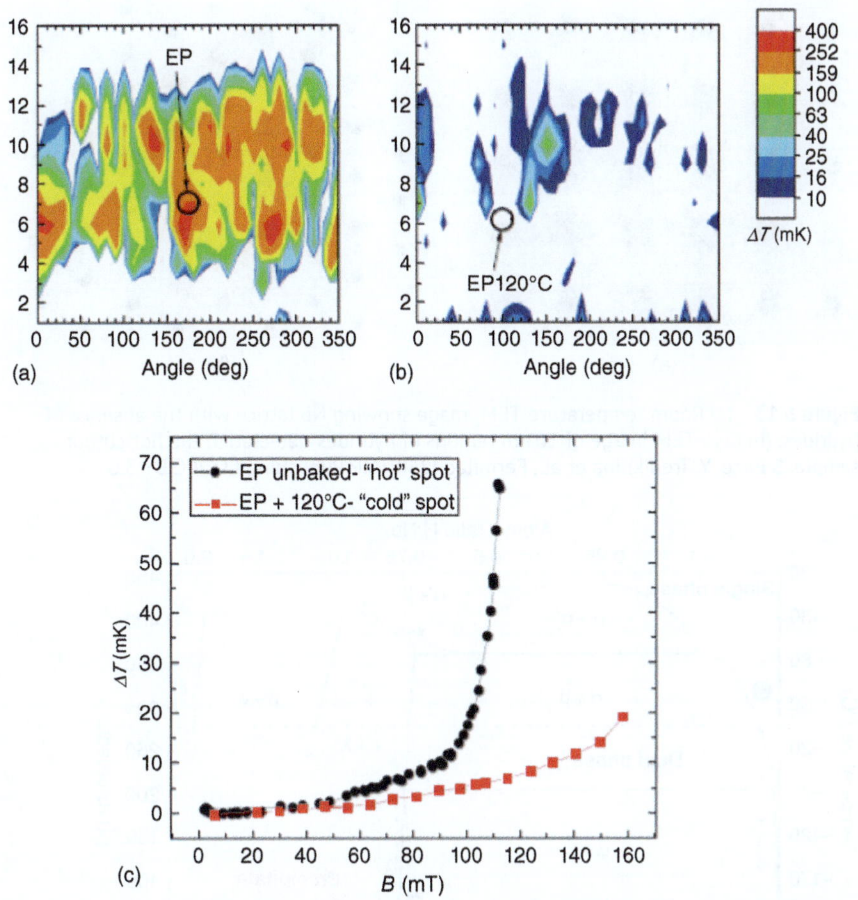

Figure 6.12 Temperature maps at E_{acc} = 28 MV/m of: (a) unbaked EP cavity, (b) EP + 120 °C baked cavity. Black circles pinpoint the cutout samples. (c) Temperature profiles for "hot" and "cold" spots. Source: Y. Trenikhina et al., Fermilab [268]/AIP Publishing LLC/CC BY 3.0.

treatment for H degassing (and no subsequent chemical treatment) also showed a high nano-hydride population, consistent with the presence of HFQS in such samples.

At this stage we may ask, why, in the first place, is H even present to form hydrides, especially if the sample or the cavities have been furnace treated at 800 °C specifically to remove H? Hydrogen could be reabsorbed in the furnace either during cooldown or during subsequent exposure to air. Section 6.10 shows how there is enrichment of H concentration at the surface.

Finally, an N-doped cutout shows minimal hydride population at 94 K, indicating that the mitigation of the HFQS with N-doping (Figure 6.5b) and with 120 °C baking are both connected to the suppression of hydride formation. Recall from Figure 6.11 that the SIMS data on NbH/Nb shows the N-doped sample to have the lowest H concentration. Calculations by the Center for Bright Beams at Cornell University find that interstitial nitrogen impurities near the surface of niobium suppress the formation of surface hydrides [141, 143].

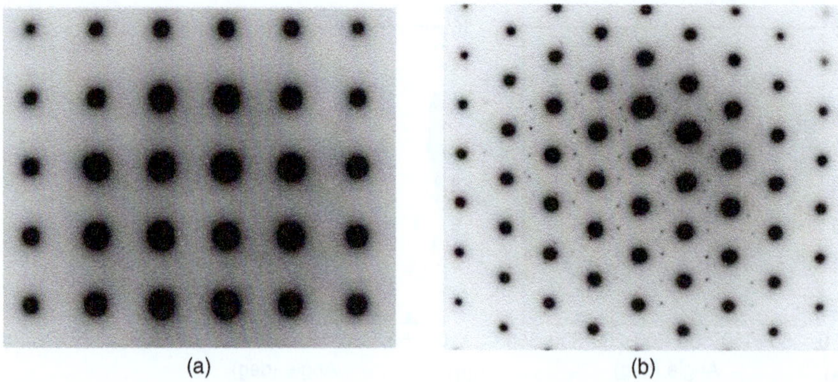

Figure 6.13 (a) Room temperature TEM image showing Nb lattice with the absence of hydrides. (b) Cryo-TEM image of 10 nm regions of hydrides detected in the hot cutout sample. Source: Y. Trenikhina et al., Fermilab [268]/AIP Publishing LLC/CC BY 3.0.

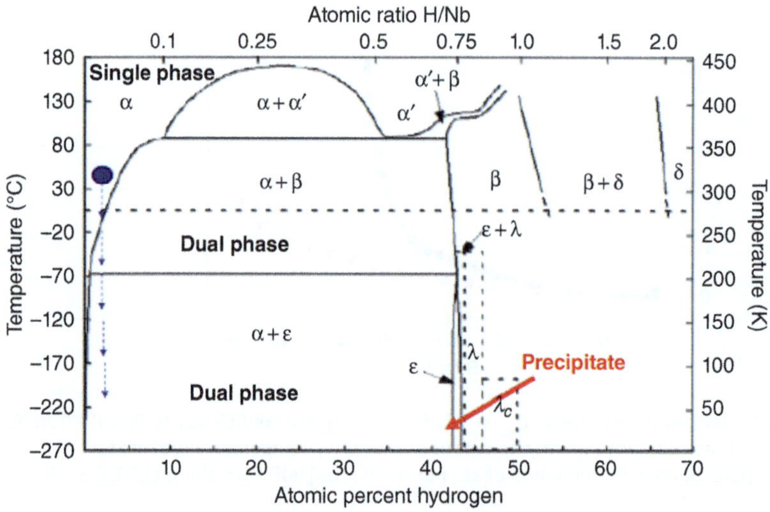

Figure 6.14 Phase diagram of the Nb-H system. Source: [269]/F. Manchester with permission of ASM International.

In summary, TEM studies show significant NbH presence in EP, BCP, and 800 °C treated samples. 120 °C baking shows a significant reduction in hydrides, which strongly points to hydrides as the root cause of the HFQS, as well as a reduction of hydrides as the reason behind the baking effect.

6.9 Niobium–hydrogen Phase Diagram

The Nb–H phase diagram (Figure 6.14) [269, 270] shows that at low atomic percentages, hydrogen exists in the α-phase at room temperature and behaves as a lattice gas with H distributed at lattice sites. Niobium can dissolve large amounts of hydrogen as interstitials. At room temperature, the solubility limit of this phase extends up to 4 at% of hydrogen. At low H concentration, niobium hydrides should not form at room temperature, except for the case of significant surface enrichment (Section 6.10).

As the temperature decreases, the H concentration to form other phases decreases. When hydrogen concentration is above the solubility limit, different stoichiometric hydrides can form. The relevant phases are β and ε, which correspond to NbH and Nb_4H_3 stoichiometries. The presence of both phases is confirmed by cryogenic TEM studies (above) as well as electron energy loss (EELS) studies [268].

The temperature range of 60–150 K is particularly dangerous, as the diffusion length of hydrogen remains large, allowing for diffusion toward hydride nucleation sites, such as niobium lattice vacancies, or dislocations, where niobium nano-hydrides grow [5, 271]. Below 60 K, the diffusion length of hydrogen becomes sufficiently low to arrest further hydride growth [5].

Hydrogen is extremely mobile in niobium and thus interaction with microstructural defects is very important for hydride precipitation. Hydrides can also decorate surface irregularities, such as scratches, because hydrogen gets segregated at the irregularities before the cooldown, or because such features provide nucleation centers for the hydrides to form.

Interstitial hydrogen interacts with defects including point defects (vacancies and interstitials), linear defects (e.g. dislocations), grain boundaries, surfaces, and interfaces [272]. Typically, surface and interface binding are the strongest, which leads to significant hydrogen segregation at the surface.

In particular, for cavity-grade niobium, hydrogen enrichment (estimated at several – 25 at%) at the niobium/oxide interface has been observed [273–275], to be discussed in Section 6.10. Further, the results to be discussed show that the near-surface hydrogen concentration is *not* strongly affected by vacuum heat treatments at 800–1000 °C typically applied to cavities to mitigate *Q*-disease. If the H in this region is responsible for HFQS, it would explain why 800 °C heat treatment does not remove of HFQS, even though it substantially reduces the bulk H content.

6.10 H Enrichment at Surface

The 800 °C furnace treatment serves to degas hydrogen absorbed during mechanical and chemical treatments. The bulk hydrogen content falls below 1 at%. Nevertheless, according to Section 6.9, TEM examinations of 800 °C annealed samples show niobium hydrides still form at 94 K, especially at the surface. Elastic recoil detection analysis (ERDA) measurements of the hydrogen concentration [273] show that hydrogen concentrates with a large peak (as high as 5 at% = 500 wt ppm) near the surface (Figure 6.15) even after 800 °C furnace treatment. In ERDA, a $^4He^+$ beam of a few MeV hits the sample with a grazing angle and ejects elastically H^+ ions of the niobium. A few 100 nm can be explored with a resolution of 30–40 nm. The sensitivity is ~0.1 at%. The width of the concentration peak found in Nb is about 40 nm, which is similar to the penetration depth and to the diffusion lengths of O during the 120 °C baking effect.

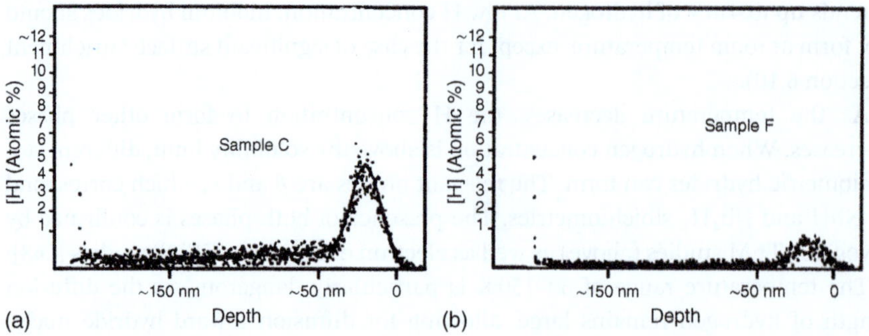

Figure 6.15 ERDA analysis for hydrogen concentration near the surface of a Nb with 200 RRR prepared by (a) BCP, (b) after heat treatment [273]. Source: C. Antoine et al. [273]/JACoW/CC BY 3.0.

Even a sample annealed at 1000 °C for several hours to degas most of the bulk H continues to show a H-rich peak at the surface (Figure 6.15b). H from the bulk naturally segregates in the first few tens of nanometers at the oxide–metal interface because of the local strain induced by the oxide layer, and due to the presence of impurities such as oxygen in the oxide layer.

We also expect some H to reenter the Nb from the residual H in the furnace. After 800 °C, the naturally formed oxide layer is no longer present on the surface of the niobium, so there is no barrier for hydrogen uptake into the material. Some H may also reenter the oxide-free Nb surface from the air when the cavity is removed from the furnace.

Further ERDA studies carried out later show higher concentration results [274, 275] on Nb samples treated by a variety of methods. Figure 6.16a shows higher surface concentrations (up to 25 at%) for chemically (BCP) treated samples, 110 °C baked samples, as well as for 800 °C annealed samples. In this case, the width of the hydrogen-rich region was found to be about 10 nm. One sample soaked in HF showed the H-rich surface layer to extend to 20 nm, caused by the removal of the natural oxide layer during HF rinsing, which normally serves as a barrier for additional hydrogen adsorption. The depth of the H-rich surface layer from this study is noticeably smaller than from the earlier ERDA study (Figure 6.15).

Samples extracted from hot and cold regions of a cavity show almost exactly the same surface enhancement profile (Figure 6.16b). We may expect such similar results because ERDA does not distinguish between hydrogen in the interstitials or H as NbH precipitates, which could be the main difference between the hot and cold cutout samples.

The shallow 10 nm depth of the surface H enrichment is somewhat contradictory to studies (Section 6.6), which show beneficial effects of O diffusion out to 60–70 nm. Possibly, the longer baking times or the higher baking temperatures are effective in raising the onset field of the HFQS by reducing the peak of the hydride concentration near the surface.

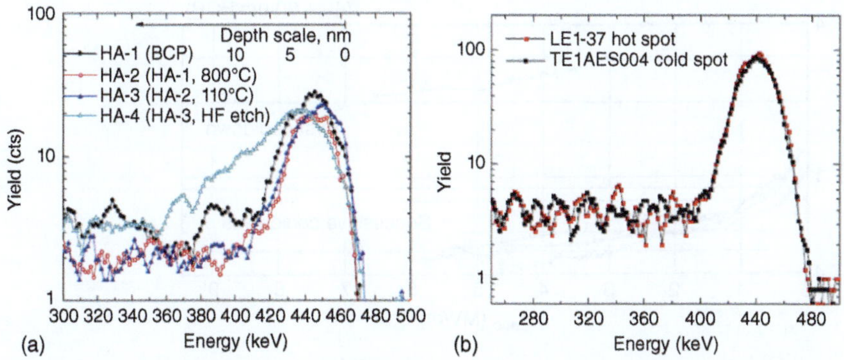

Figure 6.16 (a) Surface H enrichment and depth profile for niobium samples prepared by various cavity preparation techniques: BCP, 800, 110 °C mild bake, and HF etch. There is similar surface hydrogen enrichment (estimated at several to 25 at%) for all preparations. (b) H depth profiles from hot and cold cutouts show similar profiles and concentrations because the technique does not distinguish between H present as interstitials or as hydrides. Source: [275]/A. Romanenko and L. Goncharova, Fermilab with permission of IOP Publishing.

6.11 *Q*-disease Review

At this stage, it is appropriate to review the "*Q*-disease" [273, 276–279], which was recognized for the first time in the early 1990s. We also cover here the understanding of the disease and its cures.

In 1990, Deutsches Elektronen-Synchrotron (DESY) [276, 280, 281] and Wuppertal/Darmstadt [193, 282] discovered (Figure 6.17) that when they cooled their high (200) residual resistance ratio (RRR) cavities to operating temperatures for the second time, the *Q* values were significantly and permanently lower than when first cooled. Subsequent research over the next two years at several laboratories around the world came together to understand and cure the problem. This phenomenon depends on many factors, all related to the presence of large hydrogen concentrations in Nb. Here, it will be sufficient to give a summary of the phenomena, its understanding, and solution. The connection between the *Q*-disease and the HFQS will become clear in Section 6.13.

As delivered, commercial niobium typically has less than 1 wt ppm (0.01 at%) of dissolved hydrogen because the sheet material is annealed at the vendor facility near 800 °C for of recrystallization. At this temperature, most of the bulk H degasses from Nb. But H concentration in Nb cavities increases again during chemical processing, because H is absorbed freely when HF removes the protective pentoxide layer. It is well known that niobium has a high affinity for hydrogen [5, 135]. At room temperature, even 5 at% of H will be soluble in Nb. But below 100 K the solubility limit drops drastically to a few atomic ppm [283].

H is readily absorbed if the temperature of the acid during BCP is allowed to rise above 15 °C, or if hydrogen bubbles during EP are not allowed to escape properly.

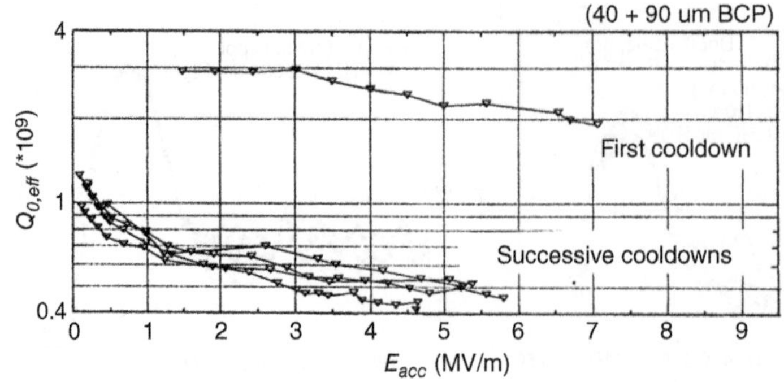

Figure 6.17 *Q* degradation on second and successive cooldowns with 500 MHz cavities operating at 4.2 K. Source: G. Enderlein et al., DESY [280]/JACoW/CC BY 3.0.

In electropolishing, anodic polarization protects the metal from H intake, and the contamination occurs when the niobium is in contact with the HF in the electrolyte without bias.

Niobium-hydride precipitation causing high residual losses is likely for a high-purity (high RRR) niobium cavity. With BCP, the phosphoric acid is known to be the major culprit for H contamination. When it was replaced by lactic acid, the *Q*-disease was reduced [280]. However, this approach to solving the *Q*-disease fell out of favor as the lactic–nitric acid mixture when stored slowly reacts exothermically to cause an explosive mixture with possible harmful consequences to personnel.

According to the phase diagram of the Nb-H system [269, 284], the required concentration of hydrogen to form the hydride phases is very high at room temperature (many hundreds of wt ppm). Therefore, these phases do not form. As the temperature falls, the hydrogen concentration needed to form the hydride phases decreases. Above 150 K the danger of hydride formation is still not very serious, because the concentration required is still relatively high. Hence a cavity can be cooled as slowly as desired to 150 K. Below 150 K, the hydrogen concentration required to form the hydride phases decreases to a dangerous level, so that islands of the hydride phase form even when the concentration is below 1 wt ppm The hydride precipitates at favorable nucleation sites, such as mechanical deformations. The surface hydrides increase the residual loss. These hydrides pose the danger for the *Q*-disease.

Between 150 and 60 K the diffusion rate of hydrogen is significant, so that hydrogen moves swiftly to accumulate to critical concentrations at hydride nucleation sites. Only when the temperature falls below 60 K does diffusion slow down enough that hydrogen can no longer accumulate at hydride centers, and further hydride growth is arrested. When a cavity shows the *Q*-disease, the sharp drop in *Q* (Figure 6.17) indicates that Nb-H islands, initially superconducting at low RF fields (<5 MV/m), become normal conducting at higher fields.

An important aspect of the *Q*-disease is that low RRR (30) Nb cavities do not show the *Q*-disease because interstitial impurities (such as oxygen and nitrogen) in Nb serve as trapping centers for hydrogen, preventing hydrogen mobility, and thereby hydride growth [283]. Herein lies the link between the *Q*-disease, the HFQS and the baking effect. O introduced by the 120 °C baking effect (Figures 6.6 and 6.7) similarly mitigates the HFQS due to hydrides (Section 6.13). Vacancies and dislocations are also effective traps for hydrogen [133, 285].

Formations of Nb-H islands physically dent the Nb surface, as H expands the Nb lattice. The lattice constant in the hydride phase is about 10% larger than in pure niobium. The residual surface relief is observable under a microscope (but not the actual hydrides). On warm-up to room temperature, it is possible to observe relief structures in the SEM. These structures also remain visible after the 800 °C bake (Section 6.12.2). Dents act as nucleation sites for the formation of hydrides on subsequent cooldowns, so that the *Q*-disease degradation repeats. But for the first cooldown, when the hydrides have not yet formed, there are no dents and therefore fewer nucleation sites for hydride formation. Hence the first cooldown does not show a strong *Q*-disease, giving a high *Q*.

Even when the H concentration is in the danger range, the *Q*-disease can be suppressed by fast cooling from above 150 K to below 60 K. The method works because above 150 K the concentration required to form the dangerous hydrides is still very large (see the phase diagram in Figure 6.13) and below 60 K the diffusion rate of H is too slow for hydride island growth. The same features can also be used to aggravate the *Q*-disease by parking the cavity temperature between 150 and 100 K for 24 hours, so as to promote the formation of hydrides. This procedure is often used as a "torture" test for cavities to establish if there is too much H present in the Nb.

The best cure found for the *Q*-disease is to degas most of the H by baking the cavity at 800 °C for a few hours in a vacuum of better than 10^{-6} Torr, or at baking at 600 °C for 10 hours, if the decrease in yield strength from 800 °C bake cannot be tolerated. Most of the bulk H leaves the Nb. Nevertheless, the surface concentration (Figures 6.15 and 6.16) effect shows H is still present at a high level in the first 10 nm.

At Saclay, measurements of the hydrogen concentration [273] near the surface (Figure 6.15) show a large H peak (as high as 5 at.% = 500 wt ppm). The width of the concentration peak is about 40 nm, which is of the same order of magnitude as the penetration depth in niobium. Hence, hydrogen contamination is dangerous to the superconducting properties of niobium. Even the sample annealed at 1000 °C for several hours to degas most of the bulk H continues to show a peak at the surface. Later results [274, 275] showed much higher surface concentrations (up to 25 at%) for chemically treated samples, as well as for 800 °C annealed samples (5–25 at%). The depth of the hydrogen-rich region was about 10 nm.

6.12 Visualizing Niobium Hydrides

6.12.1 Cold-stage Confocal Microscopy

Evolution of niobium hydrides associated dents in samples heavily loaded with H was directly observed with a laser confocal scanning microscope, equipped with

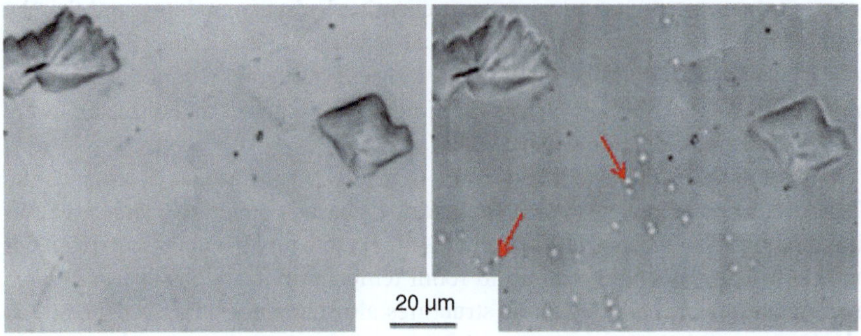

Figure 6.18 Hydrides observed on a mechanically polished sample (a) after holding for three hours at 140 K; (b) after additional cooling to 100 K. Arrows indicate location of dents formed by smaller hydrides. Source: Barkov et al., Fermilab [286]/Fermilab, JACoW/CC BY 3.0.

a cryo-stage [283, 286, 287]. Mechanical polishing was used to heavily load the samples with hydrogen concentration of 20–70 ppm by weight, measured by gas chromatography. For the detection of hydride formation, these studies rely on the optical contrast from morphological changes and provide a lateral resolution of about 1 μm in a temperature range 5–300 K.

In the microscope, samples were given the "torture test" by cooling to 140 K, holding there for 3 hours, then rapidly (~20 K/min) cooling to 5 K. At 140 K several 10 μm hydride islands appeared (Figure 6.18). The islands manifested as bumps on the surface with a characteristic height of about 3 μm. As soon as the temperature reached ~100 K, smaller hydrides appeared in the areas between the larger hydrides grown at 140 K (see Figure 6.18(b)). The smaller hydrides at 100 K are likely due to the precipitation of a second hydride phase, which has a lower solubility limit than the phase formed at 140 K. This second phase is much smaller, likely due to the lower available hydrogen concentration at 100 K, and also to shorter diffusion length of H at a lower temperature. In samples after 120 °C baking the small hydrides did not appear (Figure 6.19). 800 °C baked samples did not show any hydride formation due to the removal of most of the H.

These results allow the tentative hypotheses that large hydrides formed at 140 K are likely responsible for the "*Q*-disease," dropping the cavity *Q* at near 5 MV/m as seen in the manifestations of the *Q*-disease, whereas the smaller hydrides formed at 100 K may be responsible for the HFQS, dropping the cavity *Q* at a higher gradient of 20–25 MV/m, i.e. the onset of the HFQS. The connection of the smaller hydrides to the HFQS is further supported by the result that 120 °C baking suppresses the formation of this phase.

6.12.2 Cold-stage Atomic Force Microscopy (AFM)

The confocal microscope studies [288] above were carried out with samples heavily loaded with H by the artificial means of mechanical polishing. To explore hydride

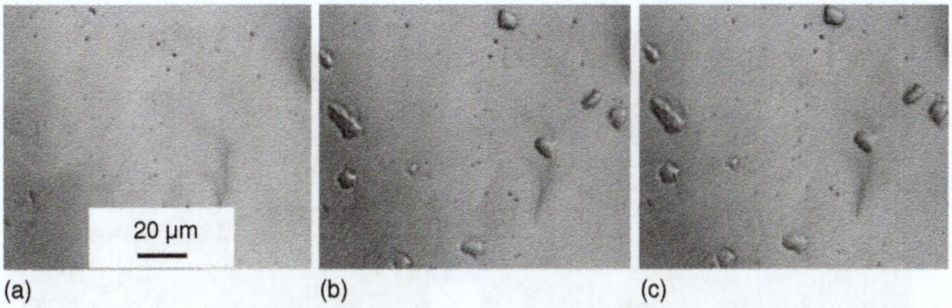

Figure 6.19 Hydrides observed on a mechanically polished sample after additional 120 °C baking for 48 hours: (a) room temperature; (b) after 140 K, two hours; (c) after further cooling to 5 K. Notice the absence of the second phase precipitation unlike the unbaked sample (Figure 6.18). Source: Barkov et al., Fermilab [283]/AIP Publishing LLC/CC BY 3.0.

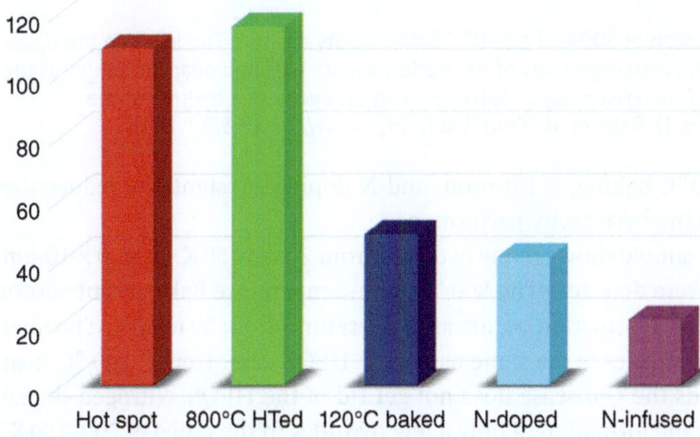

Figure 6.20 Total number of dents (from NbH precipitates) in a 10 × 10 μm^2 unit area during cooling to 10 K. The measurements were carried out on hot spots from cavities treated by EP, 800 °C bake, 120 °C bake, and also from N-doping and N-infusion. Source: Z-H Sung et al., Fermilab [288]/JACoW/CC BY 3.0.

formations for samples cut out from cavities, studies were carried out with a more sensitive imaging method of cold-stage atomic force microscopy (AFM). These samples had much less H content introduced by standard cavity preparation methods. The samples came from cavities exhibiting the HFQS, before and after 120 °C baking, and cavities prepared by N-infusion at 120 °C (Section 7.3), and N-doping. An additional hot-spot cutout came from a cavity prepared by 800 °C bake, followed by BCP. Precipitate formations were counted and compared at 200, 100, and 10 K.

As Figure 6.20 shows, the total number of precipitates/dents found from 300 to 10 K for samples from cavities with HFQS was over 100, whereas there were less than 50 precipitates from cavities that were 120 °C baked, N-doped, or 120 °C N-infused. The average diameter of the precipitates was between 400 and 700 nm for samples with HFQS, decreasing to about 300–400 nm for baked, doped, and infused samples. These results are consistent with the postulate that hydrides are responsible for the

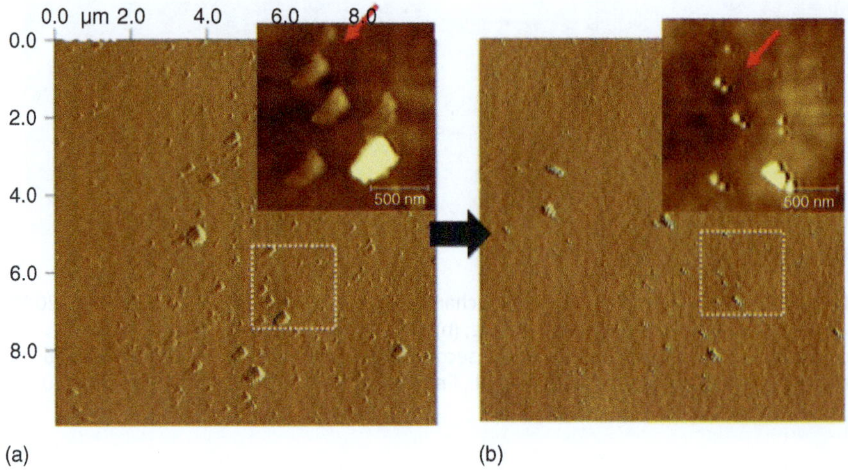

Figure 6.21 (a) AFM scan at 300 K of a 120 °C baked cavity cut-out. The inset on the upper right-hand corner is the expanded copy of the dotted square outlined near the center of the image. (b) AFM scan of the area in (a) at 320 K, showing reduced size hydride in the expanded inset. Source: D. Bafia et al., Fermilab [289]/JACoW/CC BY 3.0.

HFQS, and that 120 °C baking, N-infusion, and N-doping substantially reduce the hydride presence to improve cavity performance.

The HFQS cavity sample shows many hydrides from 200 to 150 K. In a 10 × 10-μm region, 8 hydrides were detected. The 800 °C high-temperature bake cutout shows that most of the hydride formation occurs at temperatures from 50 to 10 K. The total average number of hydrides is the same as for the HFQS case. Hence, 800 °C heat treatment that avoids the *Q*-disease does not get rid of the HFQS. Nitrogen-doped cavity cutouts show the formation of only a few hydrides in the range of 200–150 K. Nitrogen-infused cavity cutouts show even fewer hydrides formed at even lower temperatures. Hence, these two treatments with interstitial N are successful in mitigating the HFQS.

In an unexpected result, a 120 °C baked cavity cutout shows an overall low hydride count, but there are 300 nm hydrides that form close to room temperature. This is surprising as the phase diagram of Nb-H shows phases forming at room temperature only for concentrations above 5 at%. However, surface enrichment studies do show such high concentrations even for 120 °C baked samples (Figure 6.16a).

Figure 6.21 shows two AFM scans performed on the same region of the sample taken at 300 and 320 K. At 320 K, the size of these hydrides decreases considerably. This inspires the idea of warming the cavity to 320 K to possibly dissolve room temperature hydrides followed by rapid cooling to 4 K (Section 7.2.1). We will return to this idea in Section 7.1 on two-step baking.

6.13 Model for HFQS – Proximity Effect Breakdown of Nano-hydrides

Based on the collection of results presented above, Romanenko presented a model for the HFQS and the baking effect [135]. Between 60 K and 150 K, nano-meter sized

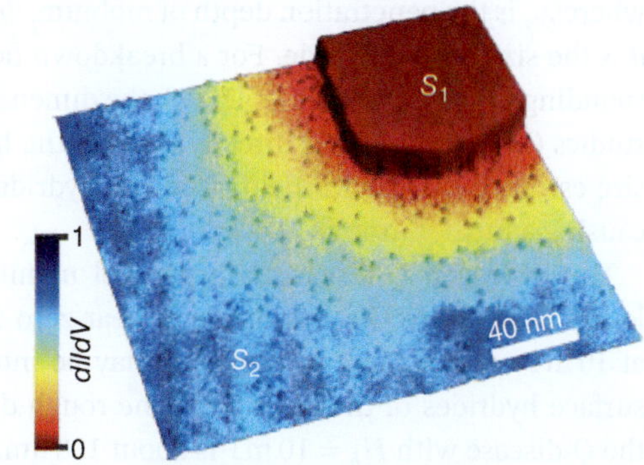

Figure 6.22 Visualizing the proximity effect at 0.3 K between two superconductors. The STM image shows the Pb nano-island S_1 connected to the Pb monolayer S_2. In the vicinity of the Pb island, the Pb monolayer locally develops a finite proximity-induced superconducting order parameter. Source: V. Cherkez et al. [165]/American Physical Society/CC BY 3.0.

Nb-H precipitates form in the H-rich layer at the surface at favorable nucleation sites. The hydrides remain superconducting due to the proximity effect (discussed below) until the onset field of HFQS. 120 °C bake introduces O atoms from the oxide layer to serve as trapping centers that prevent the H from accumulating as NbH at nucleation sites.

The T_c of Nb-H is near 1.5 K [290]. The proximity effect [164] between Nb-H precipitates and the surrounding superconducting Nb renders the nano-hydrides superconducting at 2 K or higher. The basic mechanism of the proximity effect is that the Nb superconducting wavefunction tunnels or "leaks" into the weaker superconducting hydride over a short distance of coherence length. Figure 6.22 [165] illustrates the proximity effect for a superconducting Pb island connected to a weak superconducting Pb monolayer. The monolayer locally develops a finite proximity-induced superconducting order parameter, well above its own critical temperature so that correlations penetrate the monolayer over a much larger distance.

The presence of hydrides in contact with Nb also adversely influences the superconductivity of the good Nb superconductor in proximity. The degradation of superconducting properties is detected as hot spots in cavities, or other degraded properties of the surrounding Nb as seen from Point Contact Tunneling Studies. The degradations may also help account for the MFQS.

Temperature maps on a cavity operating in the HFQS region show significant spatial nonuniformity in the RF losses, which supports the idea that stronger losses come from hydride-rich regions. There is similarly a nonuniformity in the tunneling properties of samples that come from regions with HFQS (Section 6.15).

Proximity superconductivity can be sustained only up to a critical magnetic field value inversely proportional to the smallest dimension of the hydride ($B_c \sim 1/d$). At $B \sim 100$ mT, nano-hydrides start to transition to the normal conducting state, because breakdown of the proximity effect induced superconductivity, starting from the largest hydrides. This is the onset of HFQS [135]. From the physics of the proximity effect, the breakdown field (H_b) is

$$H_b \approx \frac{1}{6}\frac{\Phi_0}{\lambda_N d} \tag{6.2}$$

where λ_N is the penetration depth of niobium, Φ_0 is the magnetic flux quantum, and d is the size of the hydride. For a breakdown field of approximately 100 mT, corresponding to the onset field of HFQS, the dimension d of the hydrides is 10 nm. TEM studies (Figure 6.13) that directly observe the hydride regions support the hydride size estimate. As the RF field rises, the hydrides smaller than 10 nm turn normal, causing a further drop in Q.

The hydrogen Q-disease is a parallel manifestation of the proximity effect at lower fields. Here Q starts out high near zero fields, and then drops precipitously at 10 mT or less. Such a threshold may be interpreted as the breakdown field of surface hydrides of the large size. The rough diameter of hydrides responsible for the Q-disease with $H_b = 10$ mT is about 100 nm.

If the hydrides are formed with a normal distribution of characteristic sizes d, then the distribution of the breakdown fields H_b is also determined by a normal distribution of $1/d$ (i.e. $1/H$) to allow reasonable fits to the Q-drop for the HFQS, as well as for the Q-disease as shown in Figure 6.23. The nano-hydride proximity effect model can be used to fit the Q-drop in the HFQS regime, as well as the Q-drop in the Q-disease phenomenon by choosing appropriate parameters from the model. Thus, the model based on the breakdown of the proximity effect acting on niobium hydrides helps unify the Q-disease with the HFQS. The HFQS is a miniature form of the Q-disease.

In addition to approximate size differences of the hydrides involved in the Q-disease (100 nm) and the HFQS (10 nm), there are other important differences between the two phenomena. Vacuum heat treatments at T = 600–900 °C decrease the bulk H concentration to eliminate the Q-disease, but the near-surface region still exhibits a high hydrogen content (Section 6.10), so that the HFQS is still found after the furnace degassing. As another difference, fast cooldowns from 150 to 4 K can avoid the Q-disease, but it will not prevent HFQS. A simple estimate based on the typical fast cooldown rate and the diffusion constant of hydrogen leads to a diffusion length of 30 nm, which is >10 nm (estimated depth of enriched H layer in Figure 6.16). This means that all the H within the thin segregation layer near the surface can still precipitate into NbH during a fast cooldown, leading to the HFQS. Fast cooldown will avoid the Q-disease but will have no effect on the HFQS.

Since there is a disagreement between the depth of the H-rich surface layer (10 or 50 nm) between the two ERDA results shown in Figures 6.15 and 6.16, it is possible that a fast cooldown may have a subtle effect on the HFQS and the baking effect if the depth is closer to 50 nm. Section 7.2 presents one form of such an effect on cooldown rates.

It should be noted further that if nano-hydrides are present, then they should contribute to RF losses in the medium field Q-slope regime, especially if there is a field dependence of the proximity effect below H_b. The field dependence may also arise from the degradation of the Nb surrounding the hydride precipitates.

6.13.1 Baking Benefit and Proximity Effect Model

As seen from the SIMS studies in Section 6.6, the 120 °C baking benefit is due to diffusion of O from the oxide into the RF layer so that O atoms become interstitials to

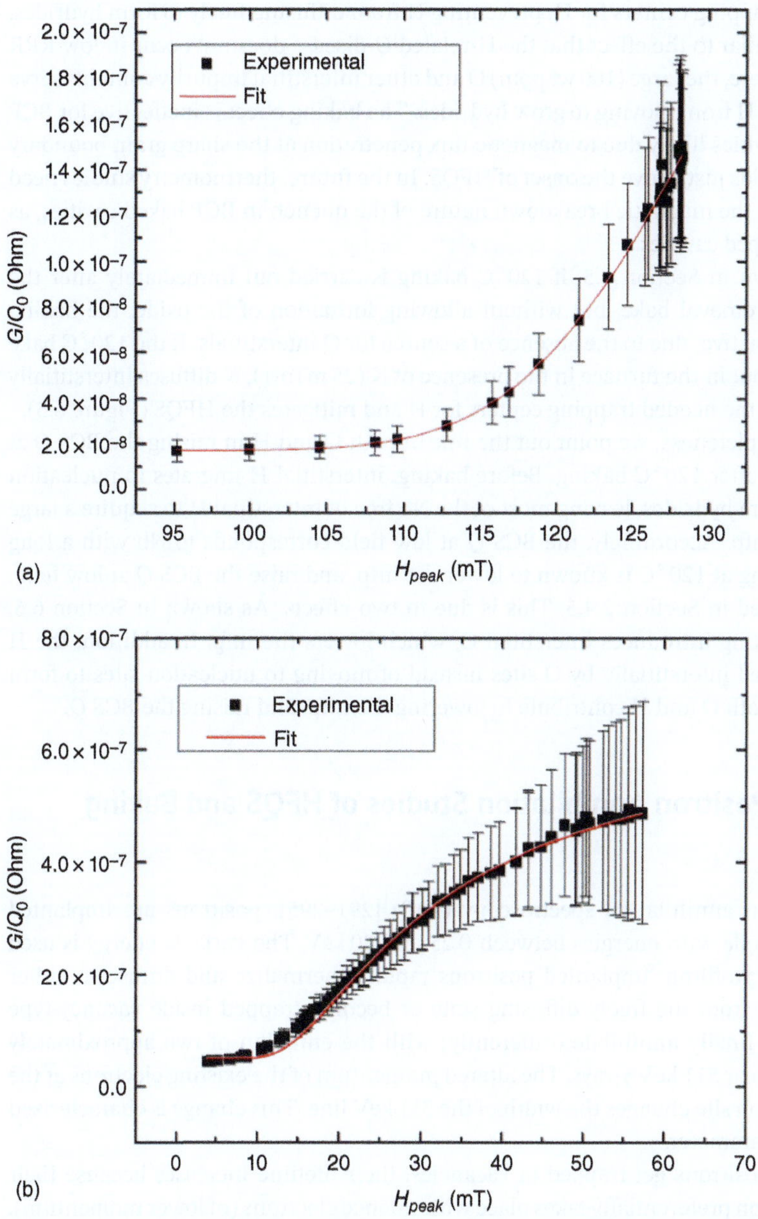

Figure 6.23 (a) Best fit to the hydride proximity effect model for the high-field Q-slope using a normal distribution of hydride sizes. (b) Best fit to the Q-disease with the proximity effect breakdown model. Source: [135]/A. Romanenko et al., Fermilab with permission of IOP Publishing. Both phenomena can be explained by a single model.

serve as trapping centers for H, preventing H from diffusing freely to form hydrides. This is similar to the effect that the H-related *Q*-disease does not occur in low RRR cavities. Here, the large (100 wt ppm) O and other interstitial impurity contents serve to trap the H from moving to grow hydrides. The baking effect is ineffective for BCP treated cavities likely due to magnetic flux penetration at the sharp grain boundary steps at fields just above the onset of HFQS. In the future, thermometry studies need to confirm the magnetic breakdown nature of the quench in BCP baked cavities, as with N-doped cavities.

As shown in Section 6.5, if 120 °C baking is carried out immediately after the 800 °C H-removal bake, but without allowing formation of the oxide, the baking effect is inactive, due to the absence of a source for O interstitials. If the 120 °C bake is carried out in the furnace in the presence of N (25 mTorr), N diffuses interstitially to provide the needed trapping centers for H and mitigates the HFQS (Figure 6.5).

For completeness, we point out the role of both O and H in raising the BCS *Q* at low fields after 120 °C baking. Before baking, interstitial H migrates to nucleation sites to form hydrides, leaving most of the Nb free of interstitial H to acquire a large electron mfp. Accordingly, the BCS *Q* at low field corresponds to Nb with a long mfp. Baking at 120 °C is known to lower the mfp, and raise the BCS *Q* at low field, as discussed in Section 2.4.5. This is due to two effects. As shown in Section 6.6, 120 °C baking introduces interstitial O, which lowers the mfp. In addition, the H gets trapped interstitially by O sites instead of moving to nucleation sites to form NbH. So both O and H contribute to lowering the mfp, and raising the BCS *Q*.

6.14 Positron Annihilation Studies of HFQS and Baking Effect

In positron annihilation spectroscopy (PAS) [291–295], positrons are implanted into a sample with energies between 0.25 and 30 keV. The variable energy is used for depth profiling. Implanted positrons rapidly thermalize and diffuse to either annihilate from the freely diffusing state or become trapped inside vacancy-type defects to finally annihilate (differently) with the emission of two approximately anti-collinear 511 keV γ-rays. The altered momentum of the existing electrons at the annihilation site changes the width of the 511 keV line. This change is characterized by the S parameter.

When positrons get trapped in vacancies, their lifetime increases because their annihilation preferentially takes place with valence electrons (of lower momentum), due to a lack of core electrons (of high momentum), resulting in a smaller Doppler shift and a narrower line as shown in Figure 6.24 for the defect case. Vacancy-type defects result in an elevated S parameter due to the reduced overlap of the positron wavefunction with high momentum core electrons.

Referring to Figure 6.24, the S parameter is the ratio of the number of events in some fixed central region of the peak to the total number, and is very sensitive to small changes in the width. The second parameter, W, is the ratio of surface area under the wing part of 511 keV line to the total surface area. W is related to

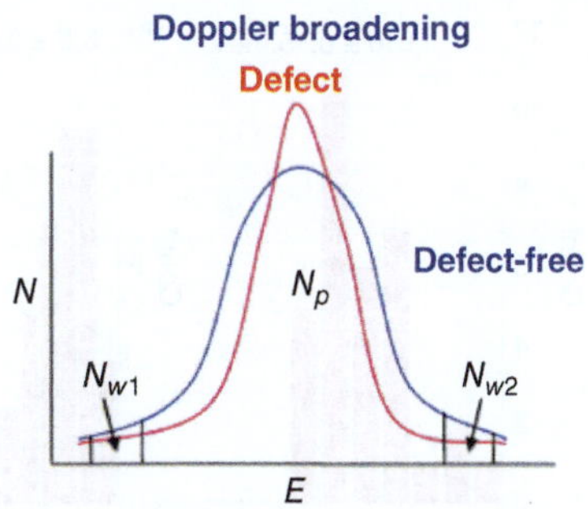

Figure 6.24 Positron annihilation spectroscopy yields two important parameters, *S* and *W*. When vacancy defects are present, the emission lines get sharper (higher *S* value) with flatter tails characterized by *W* [42] Courtesy of A. Grassellino, Fermilab.

annihilation of positrons with high-momentum electrons, and so provides information about chemical environment of the defect.

$$S = N_p/N_{\text{total}}, S_{\text{defect}} > S_{\text{defect-free}} \tag{6.3}$$

$$W = \left(N_{w1} + N_{w2}\right)/N_{total} \tag{6.4}$$

For the PAS studies [291], samples were cut out from an EP cavity from the HFQS hot region, as well as from the 120 °C baked cavity from a cold region. PAS studies show that 120 °C baking dopes a 50 nm surface layer with stable vacancy–hydrogen (V–H) complexes. Hence, besides O interstitials, formation of V-H complexes may also inhibit the formation of Nb hydrides as another mechanism behind the baking effect. Vacancies introduced by 120 °C baking trap near-surface interstitial hydrogen, and suppress niobium hydride formation.

PAS studies also suggest why 120 °C baking is more effective for EP cavities than for BCP cavities. PAS shows that the EP process introduces more vacancies than BCP, which then provide more trapping centers to form V-H complexes during the bake.

6.15 Point Contact Tunneling Studies of HFQS and Baking Effect

As with N-doping, point contact tunneling spectroscopy (PCTS) helps characterize the superconducting behavior of Nb cavity samples [296], extracted from cavities with HFQS prepared by EP or BCP, and 120 °C baked samples from cavities without HFQS [144]. The samples provided to PCTS experts by Fermilab and Jlab were from cavity regions that were found by T-maps to be "hot" due to HFQS, and from regions that were "cold" after 120 °C bake. About 35 tunnel junctions were studied.

The cold sample reveals (Figure 6.25 – upper) both larger gap values between 1.5 and 1.65 meV, and a narrower gap distribution than the hot spot BCP sample (without bake). The large spread (1.0–1.7 meV) in the hot sample (Figure 6.25 – lower) is consistent with the presence of deleterious Nb hydride phases. Baked samples exhibit much sharper gap features in the conductance, characterized by a larger peak height to background and a smaller sub-gap conductance, than the unbaked

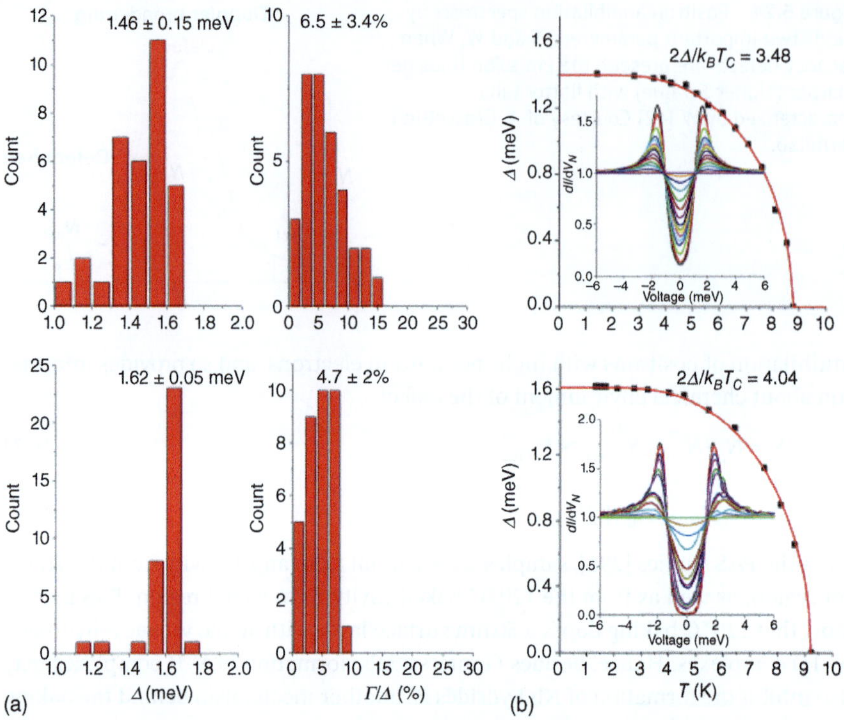

Figure 6.25 PCTS measurements on samples cut out from (upper) a hot spot and (lower) a cold spot. (a) Statistics of Δ in meV and Γ/Δ in % extracted from the fits on the measured conductance curves. The hot sample shows a higher spread of Δ values and a larger value of /Δ. (b) Typical normalized conductance curves, with temperature dependence of Δ. The cold sample shows the higher value of $2\Delta/k_BT_c$ [144]. Source: N. Groll et al. [144]/arXiv.

Nb and with minimal smearing of the density of states (DOS). This result suggests that baking reduces the normal, quasiparticle states inside the superconducting gap that otherwise gives rise to dissipation, and lower Q values.

7

Quest for Higher Gradients: Two-Step Baking and N-Infusion

7.1 Two-Step Baking

An enhancement of the 120 °C bake treatment developed at Fermilab shows encouraging prospects for higher gradients [289, 297]. A 4-hour bake at 75 °C is added to the customary 120 °C baking step. The treatment turns well-performing cavities into exceptional ones. Cavities treated with the 75/120 °C two-step bake exhibit some of the highest gradients with quench fields at or above 46 MV/m (Figure 7.1). More than 50 tests on single-cell cavities fabricated from material from various vendors showed gradients above 40 MV/m with a high yield (Figure 7.2) [12]. 38 tests exceeded 45 MV/m, with one cavity reaching nearly 50 MV/m (213 mT). Three cavities from the batch showed quench fields lower than 28 MV/m, but these have been identified to have defects. Several laboratories around the world have confirmed (Figure 7.3) the near 50 MV/m record-breaking gradient result [289, 298, 299].

One basis for the explanation comes from the companion finding that with the 75 °C baking step there is lower sensitivity to trapped dc magnetic flux (Figure 7.1b). The sensitivity to trapped dc flux drops to 0.2 nΩ/mOe from 0.4 nΩ/mOe for the one-step, 120 °C bake sample. This indicates the presence of a higher density of pinning sites for vortices, or stronger pinning sites, which could be due to additional vacancy–H complexes forming at 75 °C [300]. For example, the additional 75 °C bake could introduce more vacancies to capture more hydrogen. Possibly PAS studies could help verify if this is the reason for the 75 °C benefit.

7.2 Subtle Effects of Two-Step Baking – Bifurcation

The histogram in Figure 7.2 a has a two-peak shape, which suggests one batch of quench fields show an average of 45 MV/m, and another batch an average ofz 48 MV/m. This bifurcation effect is also seen in the *Q versus E* performance of the corresponding cavities (not shown). Trapped magnetic flux was eliminated as the

Superconducting Radiofrequency Technology for Accelerators: State of the Art and Emerging Trends, First Edition. Hasan Padamsee.

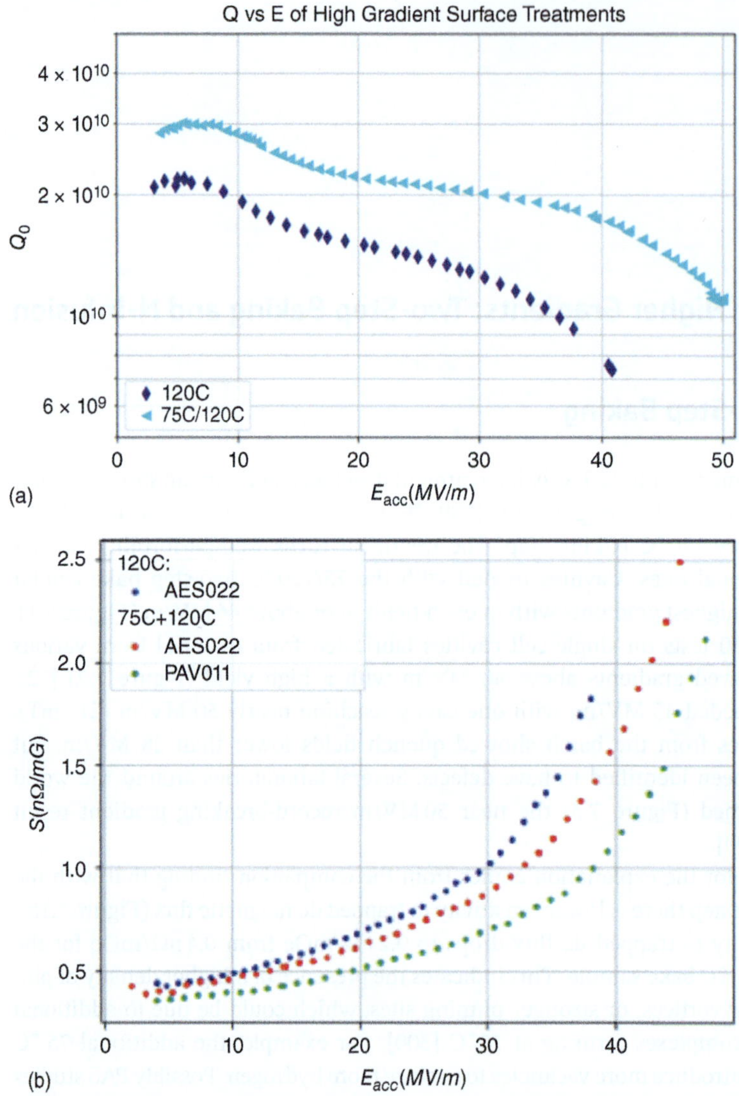

Figure 7.1 (a) Comparing the performance of cavities, after 120 °C bake with 75/120 °C two-step baking. Source: Bafia et al., Fermilab [289]/JACoW/CC BY 3.0. (b) Sensitivity to trapped magnetic flux reduces after two-step baking, indicating stronger pinning, or a higher density of pinning sites [12]. Source: D. Bafia et al., Fermilab [289]/JACoW/CC BY 3.0, Courtesy of D. Bafia.

reason for the bifurcation [12]. The cavities must trap more than 12 mG to cause the observed degree of Q bifurcation, which is not possible since the test dewars are shielded to 5 mG of longitudinal field, and less than 0.5 mG of transverse field. The cause of the branching effect was later ascertained to be related to the protocol followed to cool the cavities.

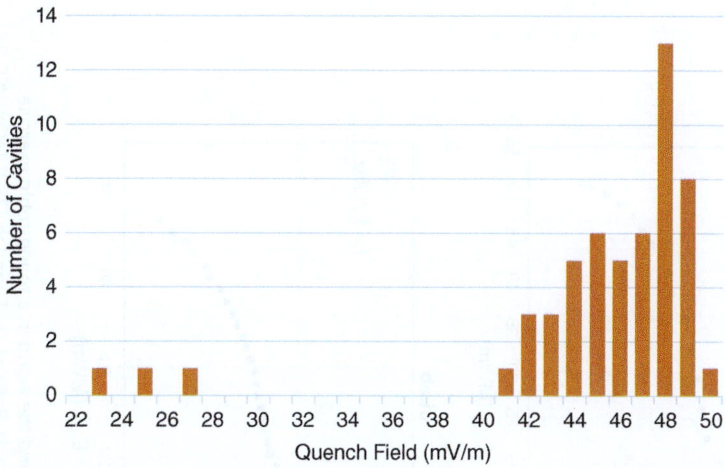

Figure 7.2 Histogram of 2 K quench fields for 50 tests on cavities processed with the 75/120 °C two-step bake. Source: D. Bafia et al., Fermilab [12] Courtesy of D. Bafia.

Figure 7.4 shows the deliberate reproduction of the branching effect by changing the cooling rate from 300 K between fast and slow for a cavity baked at 120 °C. Once again, there is an interesting correlation to the H-Q-disease, which gets aggravated due to slow cooldown. Hydrides grow when cavities linger in a dangerous temperature zone. A similar phenomenon could be occurring for the formation of residual nano-hydrides with slow cooling. If the cavity lingers at higher temperatures during the cooldown, it is likely to end up with larger and more nano-hydrides.

However a puzzle remains. In Section 6.13, we presented the argument that hydride formation in the <10 nm surface enrichment zone of H is too fast to be avoided by the fast cooldown rate, for which the diffusion length is about 30 nm. Nevertheless, the cooldown rate here appears to play a role, (even after the bake) suggesting that some hydrides do form deeper than the 10 nm surface enrichment layer so as to manifest changes with cool-down. Possibly the depth of the H-rich layer is larger than 10 nm (e.g. 50 nm as suggested by Figure 6.15) so that a fast cooldown does reduce some of hydride growth.

Tests at other laboratories (Figure 7.5) confirmed both the record high field cavity performance (50 MV/m) as well as the bifurcation behavior in the performance of cavities with the 75/120 °C two-step bake [298, 299].

7.2.1 Bifurcation Reduction

Guided by AFM results of Figure 6.21 which show that 320 K shrinks the size of room temperature hydrides, an attempt was made to reproduce this effect in cavity performance by cooling cavities starting from 320 °K instead from the usual 295 K. Four 1.3 GHz cavities treated with the 75/120 °C bake were first tested after a "fast" cooldown (~10 K/min) starting from ~295 K. Figure 7.6 a shows the performance of

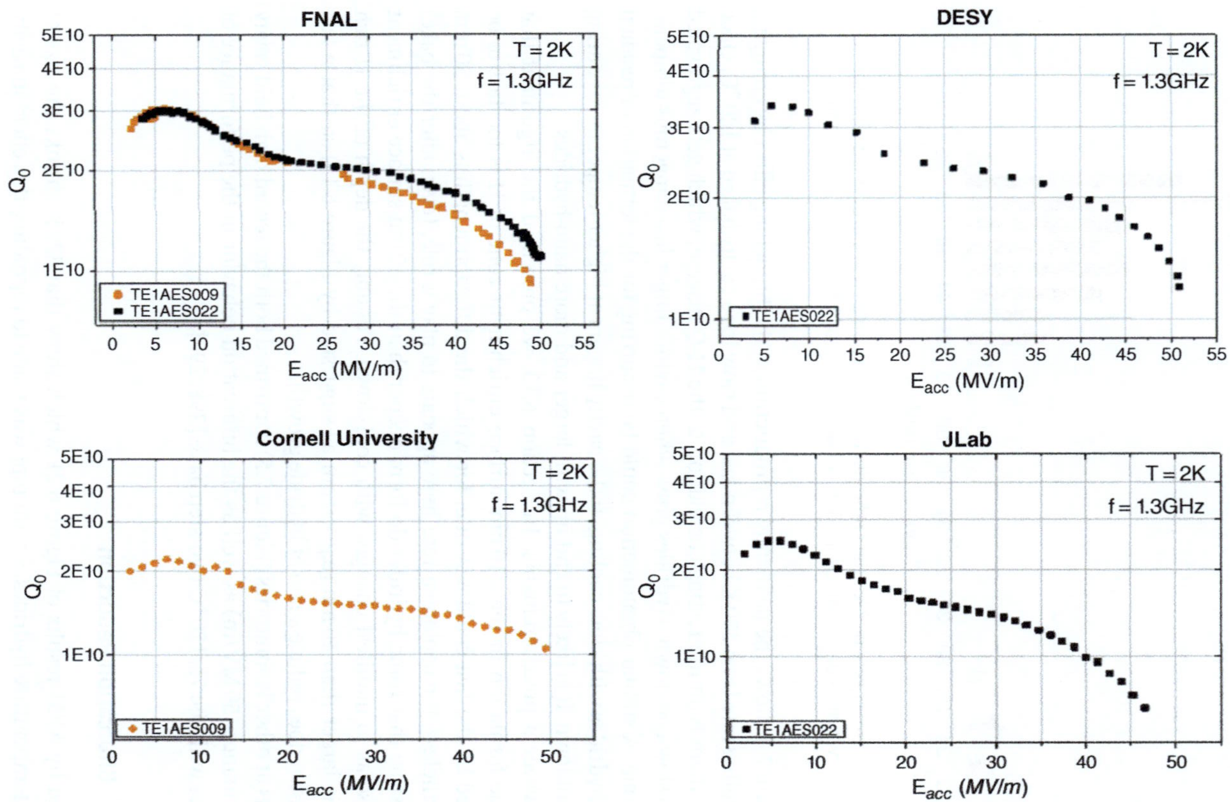

Figure 7.3 Two 1.3 GHz TESLA-shaped cavities, with 75/120 °C bake and tested at FNAL laboratories around the world in a round-robin exercise. Performance from the different laboratories confirms the record fields reached with 75/120 °C bake. Source: D. Bafia et al., Fermilab [289]/JACoW/CC BY 3.0.

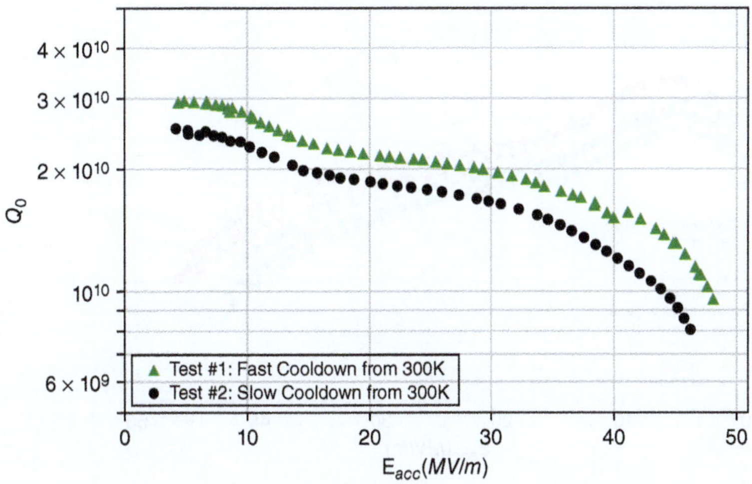

Figure 7.4 An example of the bifurcation effect. RF performance at 2 K of a single cell cavity after 120 C × 48 hours bake, after fast and slow cooldown protocols. The higher performance is for fast cooldown. Source: [12] Courtesy of D. Bafia, Fermilab.

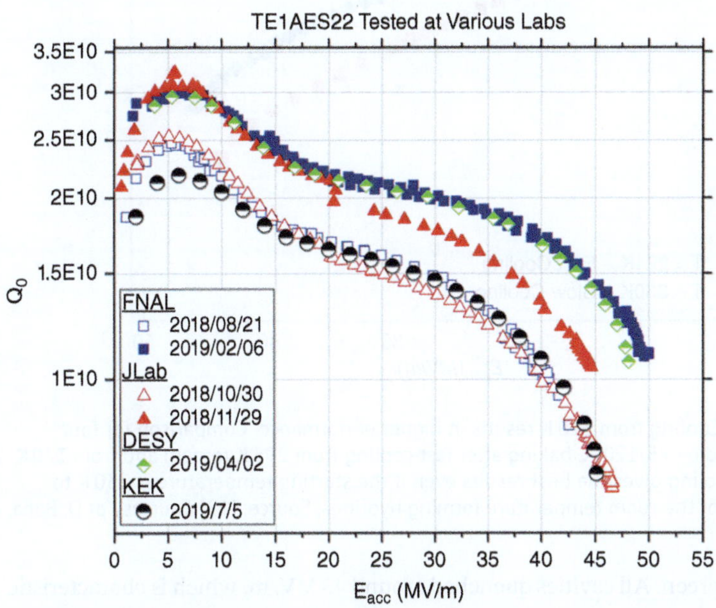

Figure 7.5 Confirmation of near 50 MV/m, as well as for the bifurcation effect. Jlab reproduced the two-branch behavior, but the upper branch curve was compromised by a multipactor-induced quench. DESY confirmed the upper branch reaching 48 MV/m. KEK confirmed the lower branch to reach 45 MV/m [12, 298, 299]. Source: [12] Courtesy of D. Bafia, Fermilab.

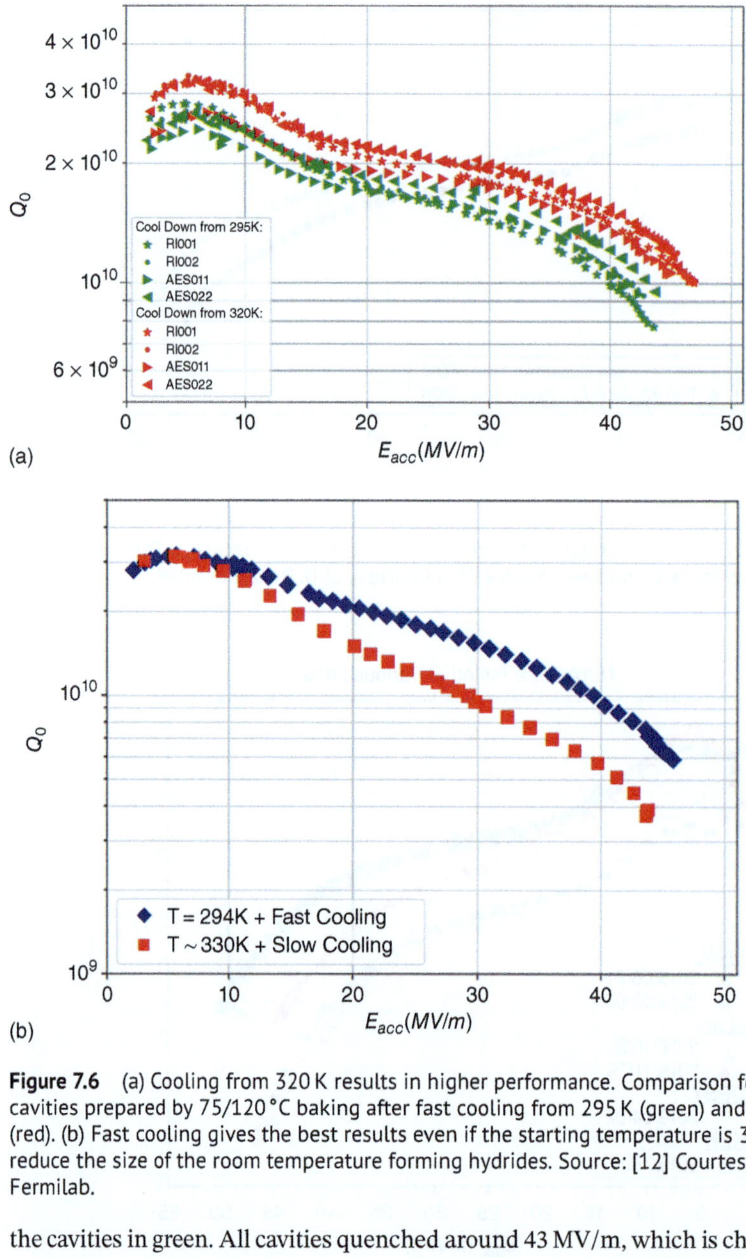

Figure 7.6 (a) Cooling from 320 K results in higher performance. Comparison for four cavities prepared by 75/120 °C baking after fast cooling from 295 K (green) and from 320 K (red). (b) Fast cooling gives the best results even if the starting temperature is 330 K to reduce the size of the room temperature forming hydrides. Source: [12] Courtesy of D. Bafia, Fermilab.

the cavities in green. All cavities quenched around 43 MV/m, which is characteristic of the lower branch, despite the fast cooling. All cavities were then warmed up to 320 K, and fast cooled. The performance of all three cavities improved to the upper branch in red. (Figure 7.6).

However, Figure 7.6b shows that *fast cooling* remains the dominant method to get the best performance, independent of whether cooldown starts at 320 K (to reduce the size of room temperature hydrides) or at 295 K. Once again there is another connection between the H-Q-disease and the HFQS via hydrides.

7.3 N-Infusion at 120 °C

Interstitial N has played an influential role in improving the Q of Nb cavities via N-doping by lowering both the BCS and residual components of RF surface resistance (Chapter 3). When bringing out the role of the oxide as the source for the interstitial O to prevent hydride formation and cure the HFQS (Chapter 6), we touched upon the analogous role of N to provide trapping sites for H (Figure 6.5). This process of baking at 120 °C in N earned the name of N-infusion. Grassellino at Fermilab discovered the N-infusion phenomenon, which this section discusses more thoroughly.

After 800 °C bake, the presence of 25 mtorr N during a subsequent 120 °C bake (without removal from the furnace) cures the HFQS. The HFQS mitigation, in this case, must take place by injecting N into the RF layer, just as O injected from the oxide layer during a 120 °C bake cures the HFQS. Interstitial N presents trapping centers to prevent the formation of hydrides.

The main benefit of N-infusion is higher fields (40–45 MV/m) and higher Q's ($>10^{10}$) (Figure 7.7) than obtained with 120 °C bake which provides O interstitials. In one of the early N-fusion experiments, there is a possibility that some O was accidentally admitted with the N [301], and so O injection also played a role there. However, O alone cannot be responsible for the good results, as the fields and Q's achieved with N-infusion are higher than from a standard 120 °C bake, which demonstrates extra benefits from N trapping over O trapping.

Jlab reproduced the 120 °C N-infusion results to reach the high gradients (Figure 7.8a) [303]. They reached 44 MV/m at a Q of 1.4×10^{10}. Attempts to reproduce the high field/high Q results from N-infusion at 120 °C met with difficulties [304] at other labs (KEK, DESY), most likely due to the lower cleanliness of furnace vacuum [305]. At KEK they successfully overcame furnace contamination issues to reach 38 MV/m at a Q of about 1.45×10^{10} (Figure 7.8b) after N-infusion, and without HFQS [306].

The N-infusion procedure introduces a very shallow (20 nm) N profile as shown by SIMS analysis [68] (Figure 7.9). The N concentration is significantly below the concentration reached during N-doping. These features make N-infusion a delicate procedure due to the very thin N layer formed at 120 °C. It becomes important to keep the thin layer contaminant free, demanding a very clean furnace environment.

An HF rinsing study [308, 309] corroborated the depth of the N infusion benefit, similar to the study conducted for the 120 °C bake (Section 6.4.2). By progressively removing layers of Nb with repeated HF rising and oxidation, the depth of the N diffused layer is estimated to be about 20 nm, as shown in Figure 7.10a. The accompanying Q *versus* E curves of Figure 7.10b show a progressive return of the HFQS with more material removal. Signs of the HFQS onset could be detected after 4 HF rinses, which corresponds to about 8 nm of removed material. HFQS was completely reestablished after 12 HF rinses, after which the Q *versus* E curve remained unchanged with further material removal. These findings agree well with SIMS measurements of N diffusion depth. We conclude that N-infusion at 120 °C allows nitrogen to enter the material and create a diffusion profile of about

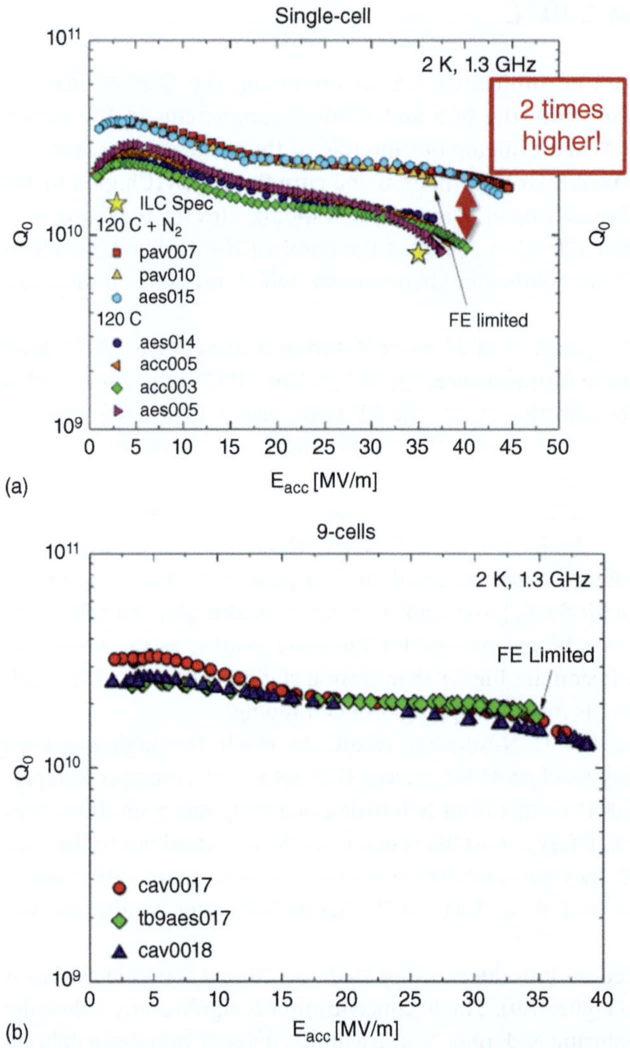

Figure 7.7 (a) N-Infusion with 1-cells show improved gradient and Q over the standard 120 °C bake. Source: Grassellino [261]/IOP Publishing/CC BY 3.0. (b) N-infusion results for 9-cells. Source: [302] Courtesy of M. Martinello Fermilab.

10–20 nm. The short diffusion profile of N (as compared to O) is consistent with the much slower diffusion rate of N. The N interstitial profile (Figure 7.9) for N-Infusion shows a lower concentration of N, and about half the depth as for the mid-T-baked case (Figure 4.2b), which behaves closer to a N-doped sample, but does not reach as high gradients. Reducing the depth of N diffusion for the Mid-T bake by reducing the baking time for example could possibly make it behave more like the N-infused sample and result in higher quench fields.

As the Section 7.4 shows, infusing more N deeper into the Nb by raising the temperature from 120 to 140–175 °C transforms the high gradient behavior of N-infusion to a N-doped like behavior.

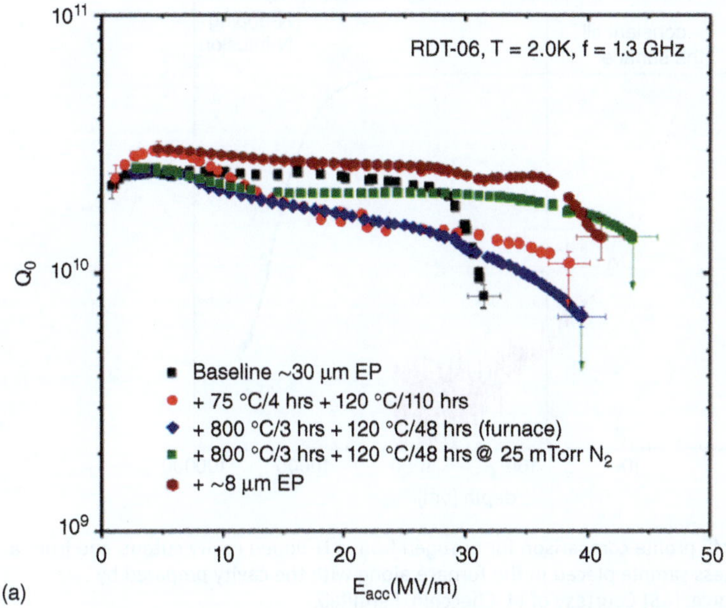

(a)

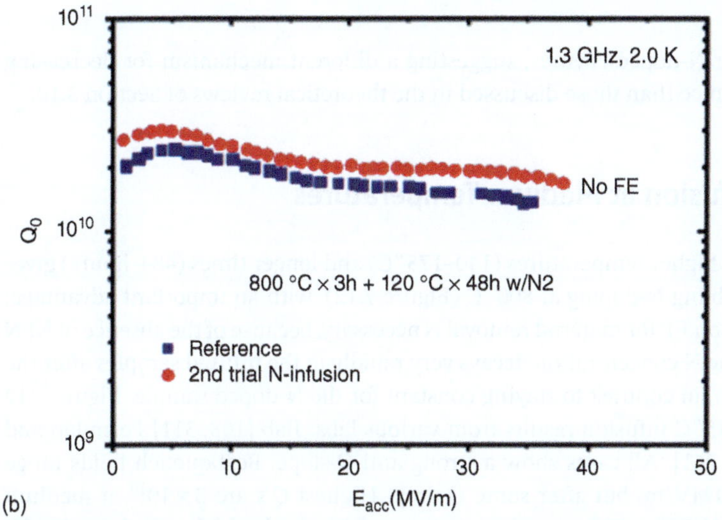

(b)

Figure 7.8 (a) N-Infusion with 44 MV/m gradient results from Jlab (green curve). Source: P. Dhakal, Jefferson Lab [303]/JACoW/CC BY 3.0. (b) 38 MV/m result from KEK (red curve) Source: K. Umemori, KEK [306]/JACoW/CC BY 3.0.

Finally, there are intriguing similarities in the field dependence behavior of R_{BCS} and R_{res} for the high gradient treatments: 120 °C baking, two-step 75/120 °C baking, and N-infusion, and quite different from the R_{BCS} behavior for N-doped cavities. Figure 7.11 shows the decomposition analysis results for the three treatments. In all three cases, the BCS component shows an interesting knee around 20 MV/m, where the resistance starts to *decrease* with field between 30 and 40 MV/m [310]. The decrease is strongest for N-infusion. The R_{BCS} decrease takes place at much higher

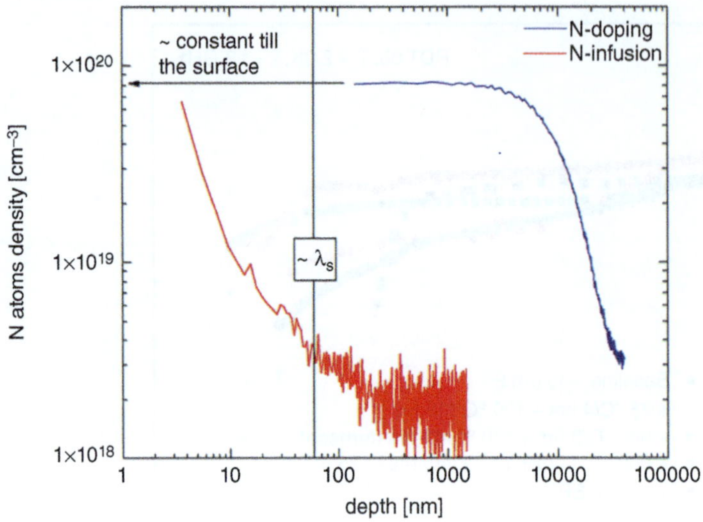

Figure 7.9 SIMS profile comparison for nitrogen from a N-doped cavity cutout, and from a N-infused witness sample placed in the furnace along with the cavity prepared by N-infusion. Source: [68] Courtesy of M. Checchin. Fermilab.

fields than for N-doped cavities, suggesting a different mechanism for decreasing surface resistance than those discussed in the theoretical reviews of Section 3.10.

7.4 N-Infusion at Medium Temperatures

N-Infusion at higher temperatures (140–175 °C) and longer times (48+ hours) gives results resembling N-doping at 800 °C (Figure 7.12), with an important advantage: no post-infusion EP for material removal is necessary, because of the absence of NbN formation. The N concentration decays very rapidly in the infused samples after the first 10–20 nm, in contrast to staying constant for the N-doped sample. Figure 7.12 shows 140–175 °C infusion results from various labs: Jlab [108, 311] Fermilab and Cornell [148, 261]. All cases show a strong anti-Q-slope. Best quench fields range from 35 to 40 MV/m, but after some Q-drop. Highest Q's are 3×10^{10} at medium fields of 20–25 MV/m. The sensitivity to trapped flux is also higher than for 120 °C baked cavities, but not as high as for N-doped cavities, for example, 0.5 nΩ/mOe for the Cornell 160 °C/24 hours. infused cavity. The lower sensitivity suggests stronger pinning from lower mfps.

The Cornell 160 °C infusion performance is comparable directly to N-doping results at about 50 nm mfp (heavier doping). Cornell explored several cases at 160 °C with short mfp (1–10 nm) to reach quench fields of 80–90 mT whereas expected quench fields for such short mfp are <40 mT. Fits of the anti-Q-slope to

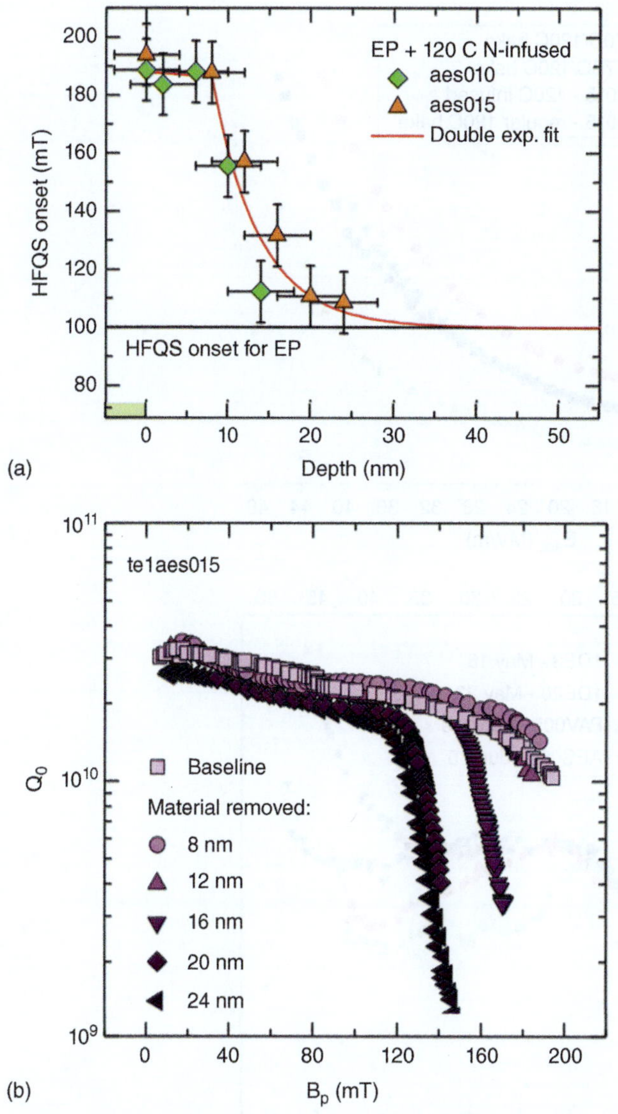

Figure 7.10 (a) Depth of N-infusion benefit determined by repeated HF rinsing and oxidation. (b) Tracking the *Q versus E* performance with successive material removal. HFQS behavior is completely reestablished after removing ~20 nm. Source: M. Checchin et al., Fermilab [308]/JACoW/CC BY 3.0.

the Gurevich theory [153] with low values of the overheating parameter helped to determine the mfp values and lower values of the overheating parameter [148].

A remarkable result is that the low mfp (1–50 nm) with 140 and 175 °C infusion show quench fields higher than B_{c1} at short mfp (Figure 7.7). Figure 3.26 [68]

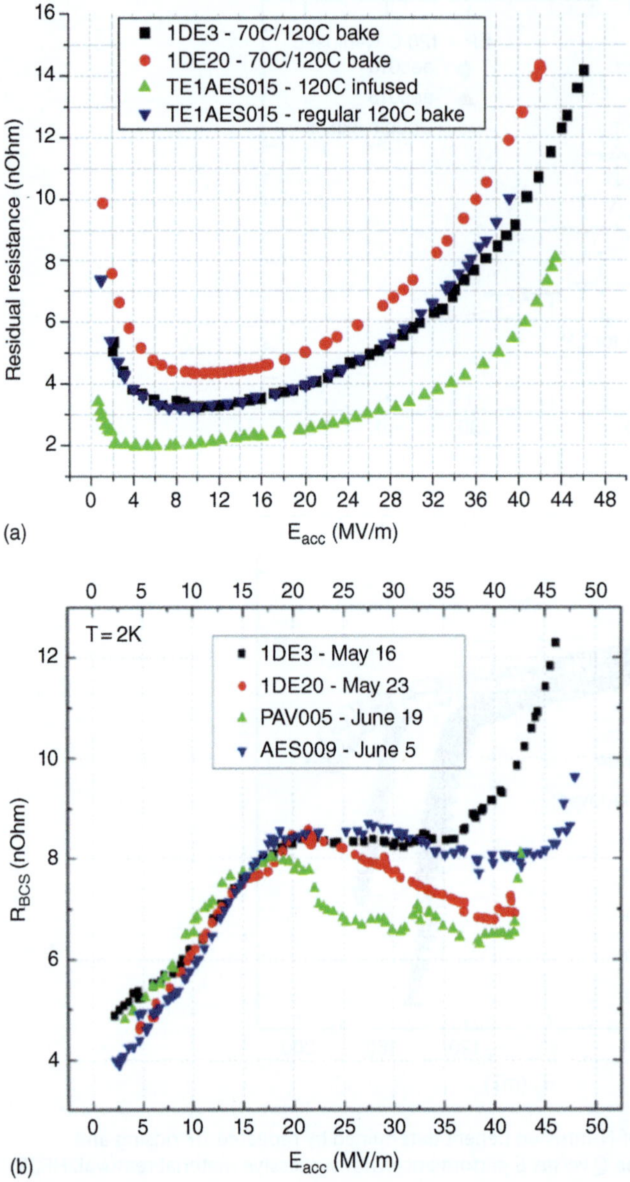

Figure 7.11 Comparison of surface resistance decomposition results for three high-field treatments: 120 °C bake, 75/120 °C bake, and N-Infusion. Source: [261, 310] Courtesy of A. Grassellino et al., Fermilab. (a) R_{res} (b) R_{BCS}.

includes a similar data point for 160 °C infusion, 5 nm/150 mT – low mfp/higher than B_{c1} quench field.

As discussed in Section 3.11, 800 °C heavy N-doping cases with low mfp show quench fields close to B_{c1}. With N-infusion, the doping at lower temperatures is likely to yield fewer defects that are prone to magnetic flux entry, to allow fields in excess of

B_{c1}. As the most likely cause, the lower temperatures for infusion prevent the formation of the nitride phases. We encountered a similar case with N-doping achieved by mid-T baking at 300 °C; here the mfp/quench fields were 50 nm/106–126 mT compared to B_{c1} of 90 mT at 50 nm mfp (Section 4.2), again suggesting fewer magnetic defects with lower surface barrier for flux penetration.

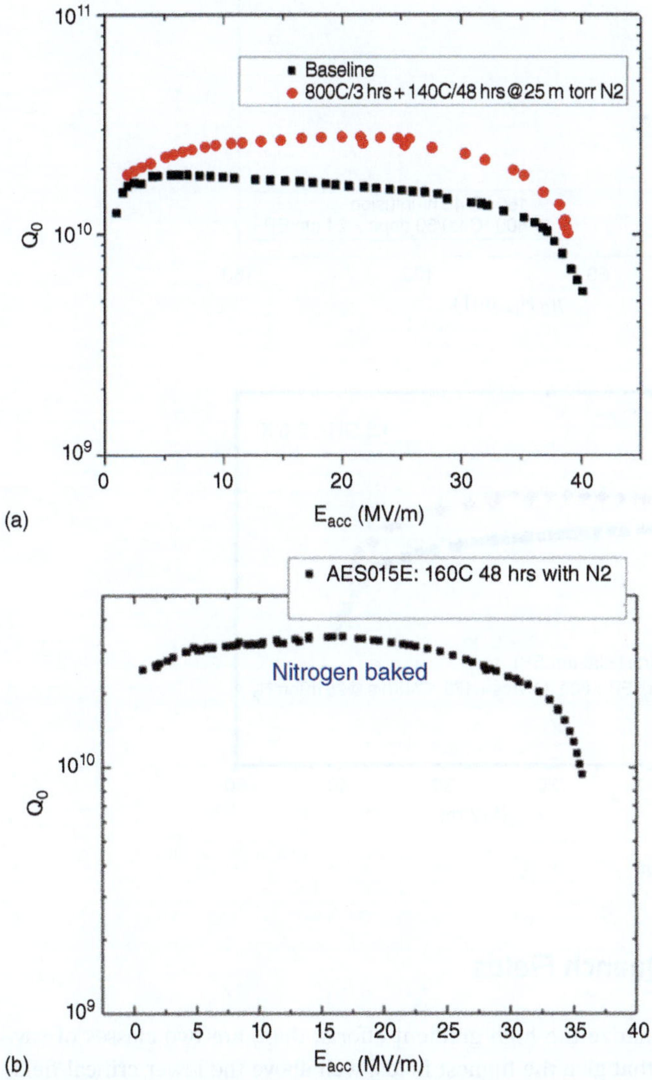

Figure 7.12 *Q versus H* (or E_{acc}) at 2 K for higher temperature infusion: (a) 140 °C N-infusion results from JLAB. Source: P. Dhakal, Jefferson Lab [311]. (b)160 °C N-infusion results. Source: Adapted from [261]. (c) A N-infused cavity at 160 °C, and a nitrogen-doped cavity with a 20/30 protocol at 800 °C followed by 24 μm EP from Cornell. Source: J. Maniscalco, Cornell University, Thesis [148]/Cornell University/CC BY 4.0, Courtesy of J. Maniscalco. (d) 175 °C infusion results from Jlab. Source: [108]/P. Dhakal, Jefferson Lab/with permission of Elsevier/CC BY 4.0.

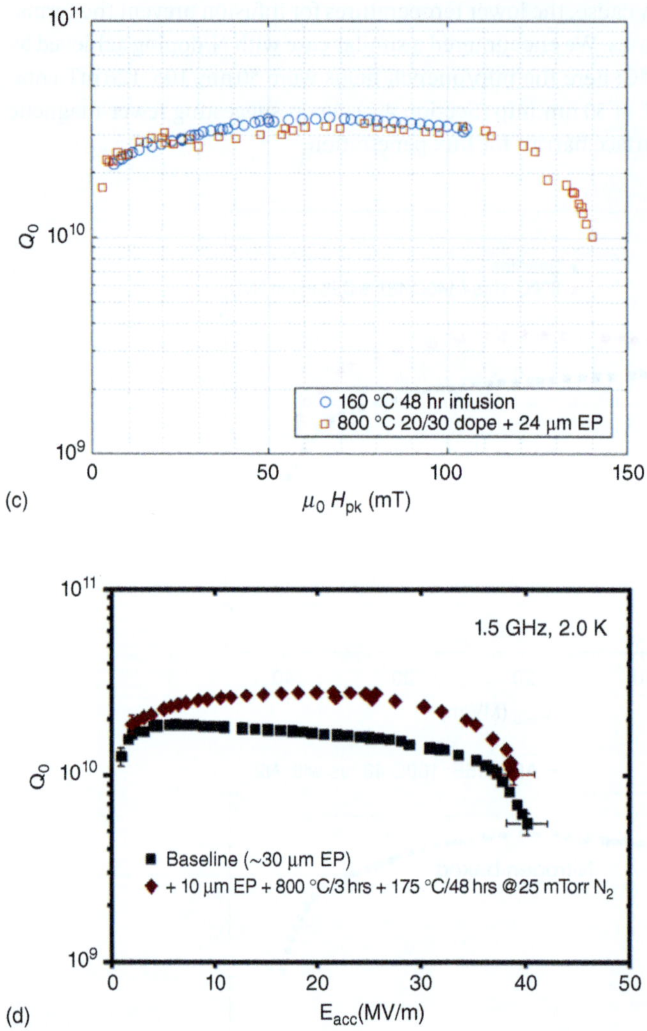

Figure 7.12 *(Continued)*

7.5 Unifying Quench Fields

To review and summarize the high gradient efforts, there are two classes of cavity treatments. One that give the highest fields, well above the lower critical field, and approaching the superheating field (220 mT); these are: EP plus 120 °C baking, two-step 75/120 °C baking, and N-infusion at 120 °C. The maximum field for a 120 °C baked cavity is close to 45 MV/m, corresponding to B_{rf} = 192 mT, and for two-step baking, the maximum gradient is 50 MV/m or 213 mT.

For the other class of treatments such as mid-T baking, 800 °C N-doping, and N-infusion at 140–175 °C, quench fields are lower, and closer to H_{c1}. In several cases, these treatments do exceed H_{c1}, but do not get close to H_{sh}. The second class of treatments generally show an anti-Q-slope starting from low fields.

The fundamental limitation that determines the maximum surface magnetic field of SRF cavities is related to the penetration of vortices into the superconductor at H_{c1}, which brings it out of the Meissner state. At H_{c1}, a single vortex is stable in the bulk of the superconductor. The free energy is lower by having the vortex in the bulk rather than outside. But even though it is energetically favorable for flux to be inside the superconductor at H_{c1}, it does not necessarily penetrate at this field, unless there are defects present for nucleation. The Bean–Livingston energy surface barrier allows the superconductor to remain in a metastable state up to at most H_{sh}, the superheating field [83, 84]. The energy barrier offers a simple way to understand the occurrence of the metastable Meissner state. It represents the energy cost for a vortex to penetrate the surface of a type-II superconductor to reach the bulk. The barrier becomes lower at defects. H_{sh} corresponds to the highest field at which the Meissner state can exist in a metastable condition above H_{c1} [90, 313, 314].

An interplay of two forces at the surface determines the energy barrier: (i) the attractive interaction of the vortex with its image anti-vortex on the opposite side of the surface, and (ii) the repulsive interaction between the vortex and the penetrating field from the surface. The energy barrier decreases for increasing κ so that the efficiency to thermodynamically exclude stable vortices from the material's bulk decreases as κ increases. The lower quench field of N-doped cavities near H_{c1} is explained by the suppression of the vortex nucleation barrier because of a constant κ, and by the presence of defects (such as NbN precipitates). Treating a N-doped cavity surface with cold EP helps to reduce those defects. Similarly field enhancement due to sharp grain boundary steps can also result in a lower quench field for 120 °C baked BCP cavities.

However, the presence of a thin dirty (high κ) layer on the surface can positively affect the vortex nucleation barrier at the surface, increasing its height, as κ of the layer increases. The metastable Meissner state can be stabilized by the presence of the dirty layer. Certain cavity treatments that produce a structure with a layered penetration depth (or layered mfp, or layered κ) promote a metastable Meissner state above H_{c1}, and therefore enhance the maximum accelerating gradient. The persistence of the metastable Meissner state is described by energy balance arguments for vortex penetration at the RF surface [68].

The 120 °C baked cavity after EP provides one example of such a layered structure. Low energy muon spin rotation (LE-μSR) measurements [131] show that 120 °C baked cavities have a low mfp, high κ layer (mfp = 2 nm, κ = 10) near the surface (Figure 6.4). However, deeper in, the mfp is longer, and κ is lower, resulting in the layered structure. From the 2 nm mfp layer, we expect that H_{c1} will be <40 mT, (according to Figure 3.26), far lower than for pure Nb. Instead, the best-observed quench fields are about 190 mT, close to H_{sh}! The layered structure offers a substantial benefit.

In a layered structure, when the effective penetration depth of the bulk is smaller (i.e. the mfp is larger) than the penetration depth in the superficial layer, the vortex repulsive force is enhanced. When there is an abrupt variation in penetration depth, the screening currents at the layer–bulk interface are enhanced, and the force repelling the vortex at the interface becomes larger. A vortex needs to spend more energy to penetrate the surface, so the stability of the metastable Meissner state is increased.

The Bean–Livingston energy barrier [84] is one framework to discuss vortex nucleation at the surface. For a uniform material with constant κ, Section 2.4.9 gave results for the numerical calculation for vortex penetration from the Ginzburg–Landau equations. The calculation shows that as κ increases, the Bean–Livingston energy barrier keeps vortices out of the superconductor. But as H_{c1} falls with decreasing mfp, the barrier gets weaker. If defects are present, the Bean–Livingston barrier is weakened further, or even totally suppressed. Therefore nitrogen-doped cavities (prone to NbN or other defects) tend to quench at or near the lower critical field.

In a similar numerical approach, Checchin at Fermilab modified the Ginzburg–Landau equations [68] to account for a material with a nonconstant κ profile. In particular, he performed the calculation for the scenario in which a dirty layer (high κ) is deposited (or grown) on top of a clean superconductor bulk (low κ). Here κ of the layer is greater than of the bulk, and the thickness of the layer is lower than the penetration depth at the surface. With such a dirty layer present at the RF surface, the field for vortex penetration (and so the quench field) can be enhanced. Cavities with a nonuniform κ profile, such as 120 °C baked and N-Infused cavities, have a stronger barrier to survive in the metastable Meissner state, even in the presence of defects.

The enhancement of the accelerating gradient for SRF cavities by layered structures was first introduced by A. Gurevich [315]. He showed that high κ (and high T_c) superconducting layers separated by insulating layers (SIS structures) deposited at the RF surface can lead to higher gradients. Kubo et al. and others [316–320] continued theoretical investigations of such SIS structures.

T. Kubo also described the "SS structure" [317, 320], i.e. a high κ (dirty) superconducting layer on top of a low κ (clean) bulk superconductor. In the high κ approximation, the London equations show that a dirty layer can enhance the accelerating field even without the insulating layer. In the case for 120 °C baked and 120 °C N-infusion cavities, an SS structure exists, and solutions are found by numerically solving the Ginzburg–Landau equations for arbitrary κ [313, 321]. N-infusion between 140 °C and 175 °C also yields low mfp values (1–10 nm), with quench fields well above H_{c1}. Similarly, mid-T (300 °C) baked cavities produce a low mfp (50 nm) layer over Nb resulting in the combined benefits of higher Q's and fields above H_{c1}.

Section 9.3.4 further discusses LE-uSR results for SS structures where higher T_c, higher κ materials, such as Nb_3Sn and MgB_2, are deposited onto Nb surfaces to successfully enhance the Nb dc critical field from H_{c1} to H_{sh}. The section also explores SIS structures with higher T_c materials.

7.6 Quench Detection by Second Sound in Superfluid Helium

To continue progress on the high gradient frontier it is important to continue to understand the causes of quench by determining the quench location with T-maps, or with other techniques, and to analyze the nature of the quench by surface analysis techniques on cutouts. We end this chapter on quench fields by discussing an important development in quench detection and location.

The High-Energy Physics Laboratory (HEPL) at Stanford University originated a novel technique based on second-sound propagation in superfluid helium in the 1970's [322]. Below the lambda point (2.18 K at 38 Torr bath pressure), liquid helium transitions to the superfluid state. The two-fluid model describes the microscopic properties. When a heat pulse such as that released at the quench site disturbs the equilibrium of the normal/superfluid components, the pulse propagates in a wave-like motion, instead of the usual diffusion. The name "second sound" denotes a wave-like propagation of entropy and temperature. At 1.8 K, the propagation velocity is about 20 m/s.

HEPL used an array of 14-resistance thermometers distributed along the length of the 7-cell structure to measure the time of arrival of the heat pulse initiated by breakdown. The point of origin could be established within ±1 cell of the structure. Argonne National Laboratory (ANL) [323] used second sound to locate quench spots in low-β resonators. They used several germanium resistance thermometers inside the niobium tube of split-ring resonators to narrow down the quench location.

Cornell further developed the method by measuring the time of arrival of the second-sound wave at three or more detectors (Figure 7.13), to locate the quench with a spatial resolution of about (~1 cm) [312]. The detectors were oscillating

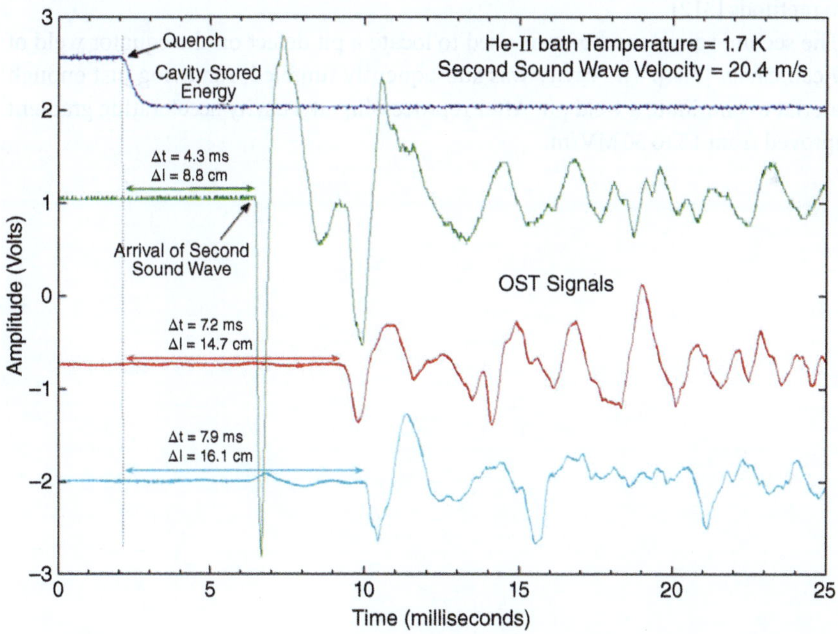

Figure 7.13 A typical quench event observed with three different transducers. The top trace is the amplitude of the RF field. The lower three traces are the second-sound signal measured with three distinct OSTs. The arrival time is well correlated with the propagation distance to each transducer. Source: Z. Conway et al., Cornell University [312]/JACoW/CC BY 3.0.

super-leak transducers (OST) which measure the fluctuating wave-like superfluid helium counterflow velocity to detect the time of arrival of 2nd sound waves [312, 324–326]. The OST elements are parallel-plate capacitors with one rigid plate and one flexible porous plate or membrane. With a dc voltage applied between the membrane and the rigid plate, the OST works similar to a condenser microphone. The pore diameter is chosen to clamp the flow of the normal fluid while allowing the superfluid, with zero viscosity, to pass through freely. The arrival of a second-sound wave at the OST causes the membrane to move with the normal fluid as the second-sound wave passes. The changing capacitance of the detector is continuously monitored to measure the arrival of the second-sound wave. A typical OST arrangement uses 8 transducers evenly distributed around a cavity. This became a cost-effective and simple method to determine quench locations.

From a given propagation velocity, the arrival times of the wave signals are translated into the respective distances between the quench location and the OSTs. Thus the position of the quench is triangulated. The advantage of the second sound method over conventional temperature mapping (Section 2.1) is the small number of OSTs compared to the many hundreds of carbon thermometers normally in use. In addition, the OSTs are permanently attached to the cavity holding insert, while temperature mapping sensors are affixed to the cavity, and have to be assembled for every cavity test. In principle, just a single quench is needed for determining the quench location. There is generally good agreement for the quench location by the two methods [312].

The second sound method was used to locate a pit defect on the equator weld of a 9-cell cavity [324]. The cavity was subsequently tumbled, removing just enough material to eliminate a weld pit. After reprocessing, the cavity accelerating gradient improved from 15 to 30 MV/m.

8

Improvements in Cavity Preparation

8.1 Comparisons of Cold and Warm Electropolishing Methods

Section 3.12 on N-doping for LCLS-HE cavities discussed how electropolishing (EP) at cold temperatures [327] (T < 16 °C) for the final EP stage improves cavity performance. After removing the first few μm impregnated with nitrides, the performance improves significantly, for example from 20 to 24 MV/m. Figure 8.1 gives statistics [168] to show how the number of cavities that exceed 23 MV/m drops from 100% to below 10% when the EP temperature rises from 12 to 18 °C.

Figure 8.2 a shows performance improvement for a 650 MHz N-doped cavity with cold EP. The maximum field achieved increases from about 20 MV/m to about 25 MV/m [157]. One possible explanation is that in removing the nitrides, cold EP prevents preferential etching around the nitrides to yield a smoother surface. Cold EP reduces the etching phase.

Figure 8.3 a shows the standard EP setup for a single cell, with the cooling water arrangement (Figure 8.3b) to enable cold EP.

It is useful to trace some of the long history of EP to provide context for the cold EP changes. Sulfuric acid acts as an oxidizing agent, forming an insulating niobium pentoxide (Nb_2O_5) layer. HF dissociates the pentoxide. Under the influence of an electrical potential, both processes provide a high rate of material removal (>10 μm/h). As EP proceeds, a dielectric layer of niobium salts [329] develops at the niobium (anode) surface. It takes time for HF to diffuse through this layer to break down additional Nb_2O_5 buildup. With local depletion of HF concentration within a cycle, the current drops until the HF concentration at the pentoxide rebuilds over the diffusion time. The electrochemical cycle then repeats, provided the bulk electrolyte solution can supply the required amount of HF, which decreases as the Nb content from removed material rises. The repetitions cause oscillations in the current flow.

Siemens researchers first described the EP procedure in the early 1970's [330]. They obtained best results by starting with large current oscillations that diminished after polishing for several minutes. Several minutes of stirring restored oscillations. Therefore, the etching process was interrupted frequently. Researchers at KEK and Nomura plating [331] advanced to continuous EP with large amplitude oscillations by rotating niobium cavities, using rotary seals to confine the acids inside the cavity.

Superconducting Radiofrequency Technology for Accelerators: State of the Art and Emerging Trends, First Edition. Hasan Padamsee.

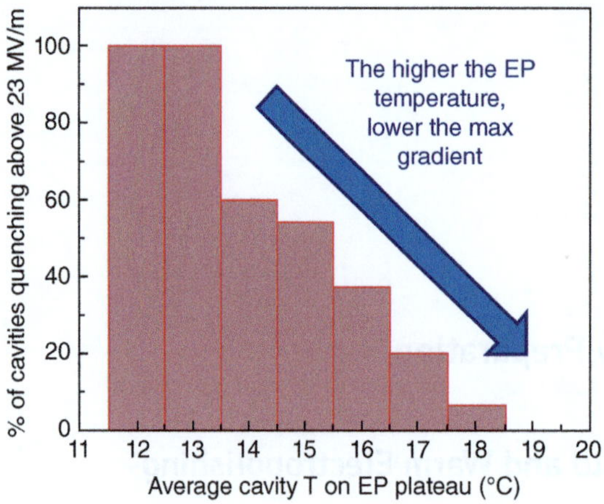

Figure 8.1 Statistics of gradient performance with acid temperature for electropolishing. Source: M. Checchin, Fermilab [328].

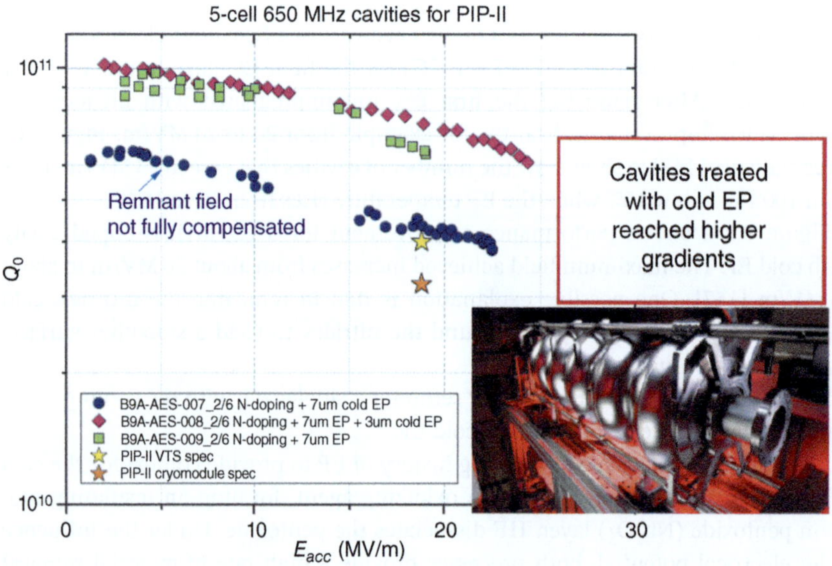

Figure 8.2 Gradient increase for a cavity with cold EP. Source: M. Martinello, Fermilab [157].

The material removals at the equator and iris are relatively uniform without the need for shaping the cathode to conform to the cell shape. Allowing the electrolyte temperature to rise to 32 °C at the equator results in a removal rate of about 12 μm per hour (Table 8.1).

For the cold EP method, invented by Crawford at Fermilab, the most critical factors are low concentration of water in the electrolyte, low temperature at the cell surface (12–22 °C), and low electrolyte flow velocity. Water interferes with the insulating quality of the dielectric layer. As water concentration increases, the depth of the current oscillations decreases. Hence the electrolyte is modified to reduce the quantity of water. The temperature of the exterior cavity wall is controlled by cold water flow.

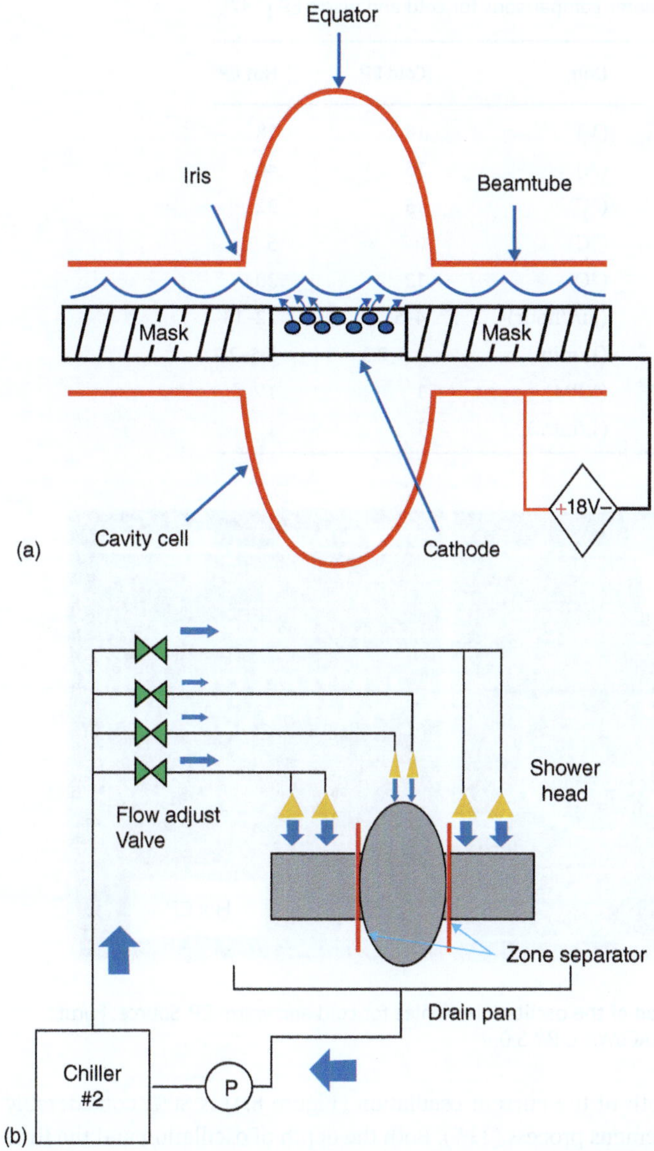

Figure 8.3 (a) Standard cathode arrangement at Fermilab for 1-cell EP. (b) Outside cooling schematic for cold EP. Source: C. Crawford et al., Fermilab [327] Elsevier/CC BY 4.0.

The electrolyte is chilled to 8 °C and continuously introduced through the cathode holes. By adjusting the flow rate of the 5 °C external cavity cooling water independently for each zone, the temperature of the equator is maintained at 16 °C while the beam tubes are at 6 °C. The temperature of the electrolyte is 10 °C as it exits the cavity. Table 8.1 compares EP parameters for warm and cold EP.

The current oscillations are useful to keep track of the process and EP quality. At 18 V, current oscillation is regular and repeats until the HF in the electrolyte

Table 8.1 Main parameter comparisons for cold and warm EP [332].

Parameters	Unit	Cold EP	Hot EP
EP voltage	(V)	18	18
EP current	(A)	15	40
Equator temp	(°C)	15	32
Beam tube temp	(°C)	0	5
Acid temp	(°C)	12	20
Removal rate	(μm/hour)	4–5	12–13
Acid circulation	(L/min.)	1.5–2.3	1.5–2.3
Cavity rotation	(RPM)	1	1
Nitrogen gas flow	(L/min.)	1	1

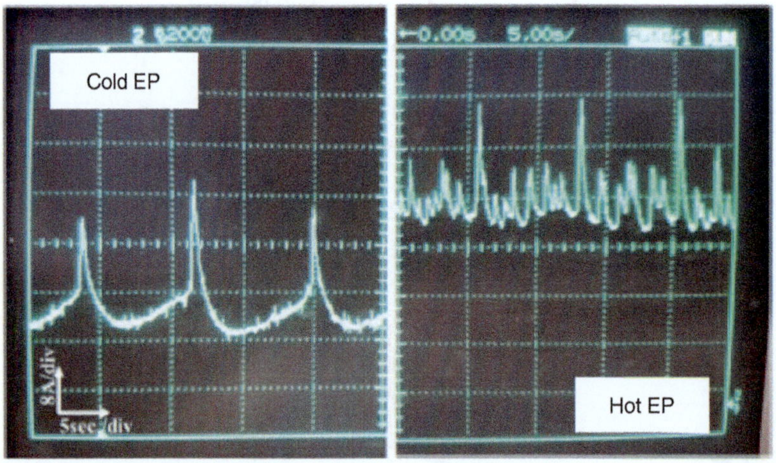

Figure 8.4 Comparison of the oscillation profiles for cold and warm EP. Source: Furuta, et al., Fermilab [332]/JACoW/CC BY 3.0.

is depleted. The depth of the current oscillation (Figure 8.4) is 80%, considerably larger than in the Siemens process (33%). Both the depth of oscillation and the time period for the waveform increase with decreasing cavity wall temperature. Figure 8.4 compares the current oscillations for cold with warm EP. There is a minimum temperature at which oscillation stops. The temperature is a function of the HF concentration in the electrolyte.

The rate of material removal in cold EP for the parameters of Table 8.1 is approximately 5 μm per hour. This rate is suitable for precision material removal of 10 μm or less, as required for the final EP stages, but the rate is too slow for the larger initial "bulk" removal (approximately 150 μm) required for a new cavity. Therefore, the first EP is carried out at 30 °C followed by a cold EP stage at reduced temperature for the final 10 μm. In most cases, only 5 μm is required after N-doping.

It is useful at this stage to summarize some of the customary steps for successful EP with reduced danger of field emission. After EP, niobium some oxide crystals adhere to the surface [333] to become a source of field emission. To dissolve these crystals, the electrolyte is allowed to circulate through the rotating cavity for 30 minutes after conclusion of polishing. The flow rate is 3.5 liters/minute and the rotation speed is 4 revolutions/minute. The electrolyte flow functions as a slow, dilute "HF rinse" for the cavity. Another important aspect is to avoid in H take-up. During EP, the electric potential provides a barrier to prevent H+ ions from entering the Nb. During the final electrolyte rinse phase, a small negative potential (−0.125 V)_ is maintained at the cathode to continue protection.

8.2 Chemical Soaking

In an attempt to reduce the performance spread of N-doped cavities, Palczweski at Jlab explored the results of soaking a cavity in 50%–70% nitric acid after the nitride removal with EP [334]. He followed the nitric soak with an HF rinse, or by light EP

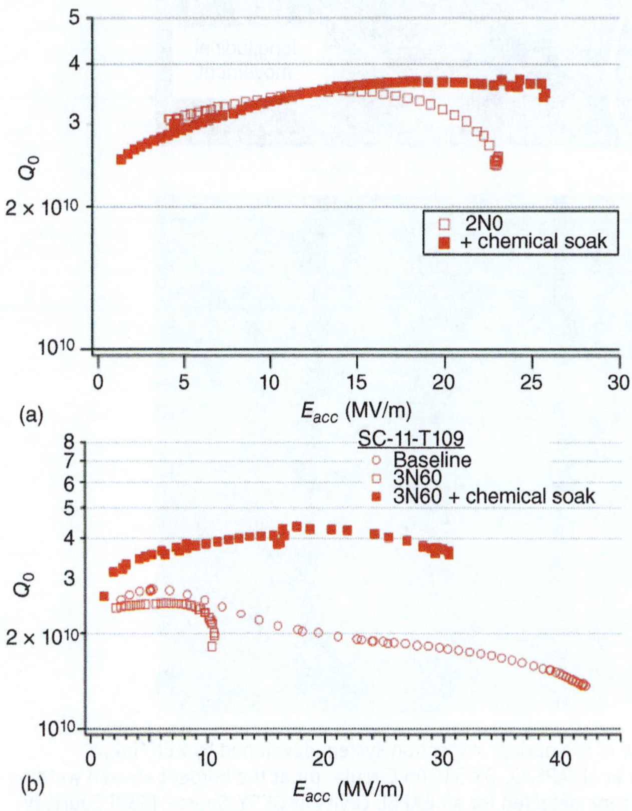

Figure 8.5 Cavity performance improvements by soaking in nitric acid. Source: [334]/Palczewski et al., Jefferson Lab with permission of Ari Palczewski.

(1 μm) to mimic the HF rinse. As shown in Figure 8.5 (a) the quench field increased by 3 MV/m, and the Q at max field improved to nearly 4×10^{10}. In a more spectacular second case (Figure 8.5b) they carried out 3/60 N-doping on a cavity that started with an excellent baseline (>40 MV/m) performance achieved by the standard treatment of EP and 120 °C bake. The doping treatment failed, with Q drops at 9 MV/m. After a 70% nitric acid soak followed by 10 × HF rinse, the cavity reached excellent performance, with maximum $Q > 4 \times 10^{10}$, and maximum field of about 30 MV/m.

The reason for the improvements with soaking needs more investigation. One possible cause suspected for the improvement is the removal of carbides that grow on the surface at 800 °C [148] or possibly other furnace contaminants.

(a)
(b)

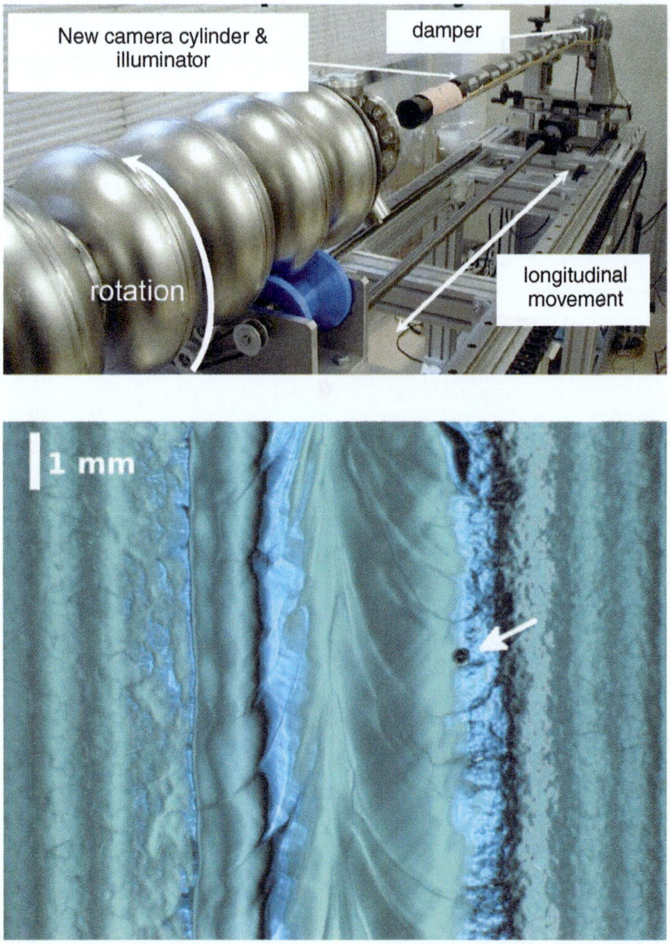

Figure 8.6 (a) Overview of the optical inspection system developed by KEK/Tokyo University [28]. Iwashita et al./APS/CC BY 3.0. (b) Circular pit at the border between welding seam and heat affected zone detected for an EXFEL cavity at DESY. Source: [338] Courtesy of S. Aderhold, DESY.

8.3 Optical Inspection System and Defects Found

One of the effective strategies to raise the quench field in the past has been to optically examine the defective regions of a cavity surface after T-maps reveal the quench locations. But a simple borescope system does not provide the resolution nor the illumination needed for close studies of the cavity surface, especially at welding seams, or weld-related defects. Tokyo University [28] developed a special optical inspection tool (Figure 8.6a) which was widely used for 9-cell ILC and EXFEL cavities. It uses a high-resolution camera and specially designed lighting equipment. Once the defect is located, imaging systems take pictures with a resolution better than 10 μm. Using such a tool, several 9-cell cavities were found to show pit-like defects (Figure 8.6b), some from cavities supplied by new vendors, still gaining experience. The heat-affected zone near the weld shows a tendency to form large (100–200 μm) voids, likely originating from high-stress regions [335]. These voids may start smaller, but grow to 100–200 μm during EP, and retain their sharp edges [336]. Such defects can be removed by global centrifugal barrel polishing (CBP) or by a local grinding tool [337].

DESY adapted and improved the Tokyo inspection system [338], as seen in Figure 8.7. Housed in a pipe inserted into the cavity, the camera looks in the forward direction with a mirror directing the field of view towards the cavity surface. The illumination system is an array of stripes located on the outside of the tube around the opening for the camera. There is a central light inside the pipe to illuminate the surface in direction of viewing. The light of different stripes reaches the surface at different angles. Since the cavity surface is not flat, especially at the welding seams and at defects, the striped illumination adapts to the surface topology. In particular, the illumination of a defect from different directions allows a measurement of the wall angle, to determine whether there is a pit or a bump, together with an estimate of the depth or height. Figure 8.8 shows a severe aluminum inclusion defect found in an EXFEL cavity.

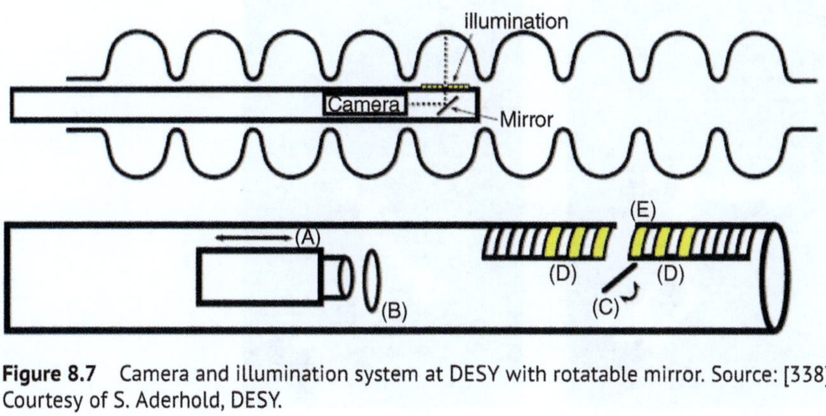

Figure 8.7 Camera and illumination system at DESY with rotatable mirror. Source: [338] Courtesy of S. Aderhold, DESY.

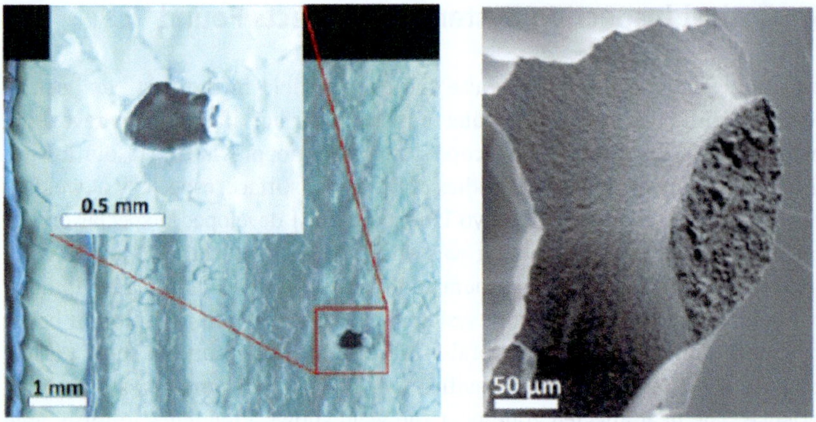

(a) (b)

Figure 8.8 (a) Optical image of the quench area in a cavity with the marked defect and (b) SEM image from a sample cut out from the quench area. The defect consists of aluminum as identified by EDX and AES analysis. The inclusion caused quench at 13.7 MV/m. Source: S. Aderhold, DESY and U. of Hamburg, Thesis [338] Courtesy of S. Aderhold, DESY.

8.4 Robotics in Cavity Preparation

Automation has been applied extensively in cleanrooms for the photovoltaic, semiconductor, and pharmaceutical industries. Their aim is to increase the production yield by increasing the reliability of operations. The FRIB project at MSU (Section 11.2.7) was first to introduce a robot in the SRF cleanroom for operations such as high-pressure rinsing [339, 340] to reduce processing times and labor. Many hundred rinse cycles have been successfully completed with high reliability and more than 95% uptime. The automated system has versatility for multiple

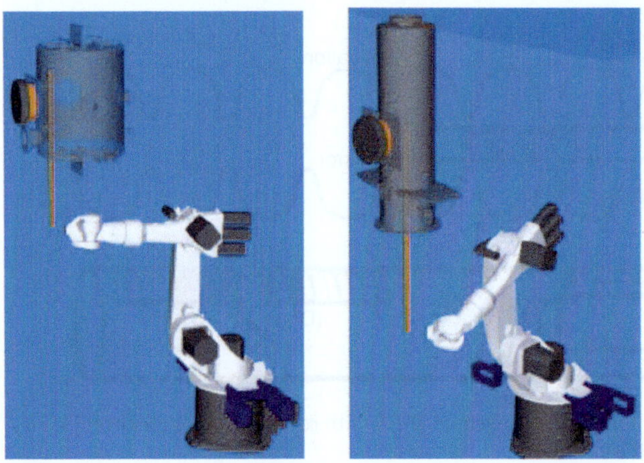

Figure 8.9 Schematic for the robotic system developed at MSU for automated high-pressure water rinsing on QWR and HWR resonators for FRIB. The same robot is reprogrammed for rinsing two different types of cavities. Source: L. Popielarski [339]/Facility for Rare Isotope Beams, Michigan State University, U.S. Department of Energy Office of Science, Courtesy of L. Popielarski.

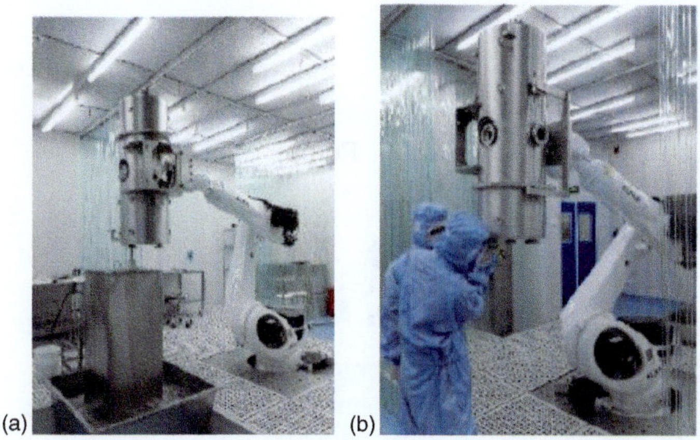

Figure 8.10 (a) Robot-assisted HPR for an HWR Cavity at IHEP. (b) Robot for assembly at IHEP. Source: Guo et al. [341]/JACoW/CC BY 3.0.

geometries (Figure 8.9) as needed for the FRIB project. Robot reprogramming allows fast optimization.

Robots could also be used effectively in cavity string assembly and coupler assembly. Now several laboratories are contemplating or already using robotics for cavity high-pressure rinsing, cavity assembly in a clean room, or monitoring of cavity operations for X-rays due to field emission. These techniques are likely to develop further with large-scale projects and will lead to less contamination due to fewer personnel, and more reliable preparation.

Operators and their movements or improper practices could increase particle contamination in the cleanroom. Robot operation reduces these risks. IMP in China has introduced robots for assisting with HPR and assembly operations (Figure 8.10). For assembly, the flexibility and maneuverability of the robot arm improve the ease of alignment between flanges. With a reduction in the number of operators needed, HPR and assembly cleanliness can reach higher levels. For CIADS at IMP (Section 11.3.5), the robot-assisted string showed much higher onset levels for field emission and a higher working gradient.

8.5 Plasma Processing to Reduce Field Emission

Plasma processing has been shown to effective against field emission, when it is caused by surface hydrocarbons, or against multipacting and, in one case, against quench due to artificial surface contamination.

A new procedure called in situ plasma processing [342] was developed at Oak Ridge National Lab (ORNL) and applied to reduce field emission to raise gradients for cavities installed in spallation neutron source (SNS). Field emission has multiple causes: metallic particles, surface defects, hydrocarbon contaminations, and adsorbates [4, 5]. Hydrocarbons (C_xH_y) lower the work function (Φ) of the cavity surface, increasing field emission. Plasma processing removes C_xH_y type surface contaminants to restore higher Φ and to increase the threshold for field emission, enabling the cavity to operate at higher accelerating fields [343].

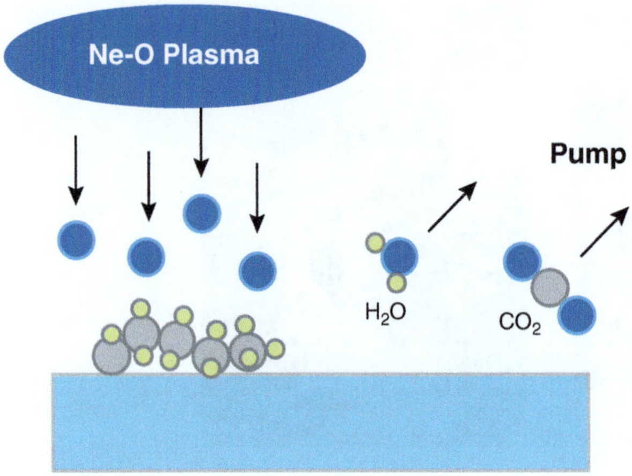

Figure 8.11 Conceptual process for plasma cleaning to break up surface hydrocarbons that lower the work function, and enhance field emission [344]. Source: S. Massahi et al./with permission from SPIE and Courtesy of S. Massahi, National Space Institute, Technical University of Denmark.

Figure 8.11 [344] outlines the schematic of the process. Free electrons with energy >10 eV crack the hydrocarbons into free molecules. Plasma processing uses free oxygen produced in the plasma to break down the hydrocarbons on the surface. The residuals of water, carbon monoxide, and carbon dioxide are removed from the cryomodule as part of the process gas flow.

The process is carried out at room temperature with a mix of inert gas (argon or neon) and a few percent oxygen. The gas pressure in the cavity is usually between 50 and 200 mTorr. A pair of modes from the fundamental passband are superposed to maximize the electric field in one cell so as to ignite the plasma in that cell.

Unlike helium processing [345–347] which is based on erosion of adsorbed gases and sputtering of particulates by ion bombardment, plasma processing likely does not help with particulate field emission. Sample studies showed that plasma processing increases the work function of the niobium surface systematically by about 0.5–1 eV. The process can also reduce the secondary emission coefficient, which helps reduce multipacting.

The procedure was first proved with cavities in off-line cryomodules at the SNS (Section 11.3.1), and then applied to cavities in several cryomodules in the tunnel. The RF power is <1 kW, and the processing time up to 8 hours. The average improvement in E_{acc} obtained for 38 cavities is 2.4 MV/m (21% increase). The individual cavities improve about 0.2–6 MV/m. Figure 8.12 shows the range of improvements per cavity for 10 cryomodules. There was no performance degradation observed from plasma processing. For one high-β cavity, a multipacting limitation was removed after plasma processing confirming the reduction of the secondary emission coefficient [342].

Fermilab is exploring the technique to reduce field emission that may compromise the behavior of LCLS-II cavities [349]. Such a remedy will likely be needed when commissioning starts in 2023 if field emission limits gradients. They use higher order

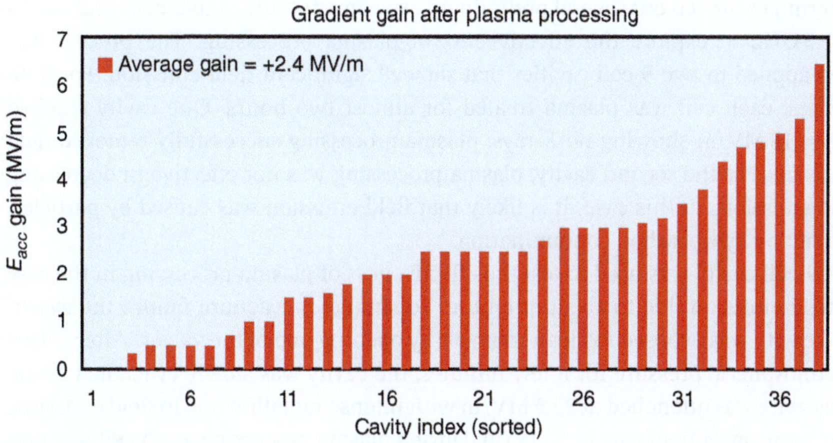

Figure 8.12 38 cavities plasma processed at SNS with an average E_{acc} increase of 2.4 MV/m. Many cavities were processed online. Source: [342] Courtesy of M. Doleans, Spallation Neutron Source, Oak Ridge National Laboratory.

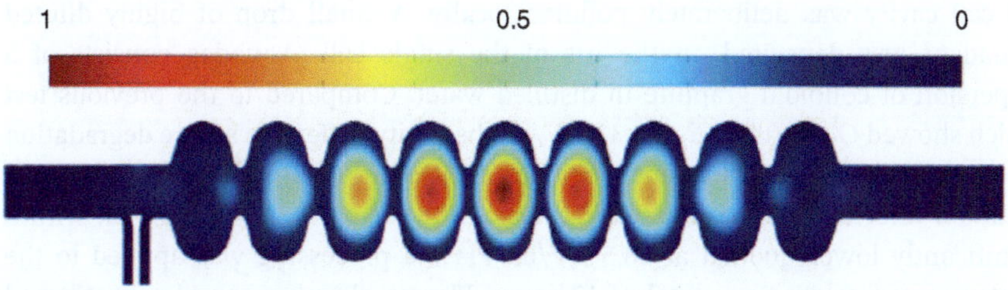

Figure 8.13 Normalized electric field of the first mode of the second dipole passband. This mode is used for igniting and growing plasma. Source: B. Giaccone, Fermilab [349]/JACoW/CC BY 3.0.

modes (HOM) instead of the fundamental passband modes, used by SNS. At room temperature (with $Q_0 \approx 10^4$) the coupling between the cavity and the fundamental power coupler is extremely low (with $Q_{ext} \approx 7 \times 10^5$, [350], demanding a high level of forward power to ignite the glow discharge. Also, the peak field enhancement at the coaxial coupler raises the risk of plasma ignition at the antenna, which is better to avoid.

The first and second dipole HOM passbands are selected for plasma ignition and plasma transfer between cells. The HOM couplers for these modes present good coupling at room temperature. Plasma ignition needs only few watts! Glow discharge is ignited one cell at a time to maximize the electric field in that cell. Using the electric field of the first mode of the second dipole passband (Figure 8.13), glow discharge processing starts in the central cell. On completion, it is possible to transfer the plasma to the rest of the cavity through adjacent cells via superposition of selected higher order modes. A convenient diagnostic comes from the drop in the dielectric constant where the plasma is present. The plasma location in a multicell cavity can be determined from the frequency shift of the resonance peaks.

Fermilab carried out several preliminary experiments with single-cells and 9-cells at 1.3 GHz to explore the effectiveness of plasma processing. The process was then applied to two 9-cell cavities that showed significant field emission. For both cavities, each cell was plasma treated for almost two hours. One cavity reached E_{acc} = 18 MV/m showing no X-rays; plasma processing successfully removed field emission. For the second cavity, plasma processing was not effective in decreasing field emission. In this case, it is likely that field emission was caused by particles, and not by hydrocarbon contamination.

A 9-cell cavity was used to test the effectiveness of plasma processing in the case of field emission due to vacuum failure. To simulate a vacuum failure the evacuated cavity was exposed to clean room air by opening one of the valves. After sitting at atmospheric pressure for a few minutes, the cavity was slowly evacuated again. The cavity was quenched at 22.5 MV/m with intense radiation due to field emission. The X-ray onset field was 18.5 MV/m. During plasma processing, each cell was processed for almost two hours. The quality factor of the cavity did not change, and the radiation was completely removed. The X-ray intensity drop showed that the processing removed field emission as shown in Figure 8.14a.

In another experiment to explore the effects of the plasma on local contaminants, a 1-cell cavity was deliberately polluted locally. A small drop of highly diluted Aquadag® was deposited on the iris of the single cell. Aquadag consists of a dispersion of colloidal graphite in distilled water. Compared to the previous test which showed $Q_0 > 10^{10}$ at E_{acc} = 33 MV/m, the cavity suffered a severe degradation in the quality factor and quench field by the pollution. But there was no field emission after contamination. The Q_0 was 1.7×10^{10} at E_{acc} = 16 MV/m, with a significantly lower quench at 18.5 MV/m. Plasma processing was applied to the contaminated cavity for a total of 17 hours. Plasma cleaning completely restored the initial accelerating field, reaching quench at E_{acc} = 33.5 MV/m, resulting in a 15 MV/m increase. And plasma removed the aquadag stain. Although this test does not prove the effectiveness against field emission, it shows benefits of removing surface contamination.

At Jlab [351], both HOM and fundamental passband modes were used for plasma processing. A camera was used to verify the plasma location. The RF power level to ignite plasma in one cell was 1–600 W depending on the available system, gas species, pressure, and cell. They maintained the plasma for 30–60 minutes in each cell. The plasma was moved from cell to cell by changing the RF frequency. Exhaust gas monitored with an RGA showed cracked hydrocarbon residuals of CO_2, CO, and H_2O. The overall success was marginal, perhaps because the processing times were low compared to Fermilab. Field emission onset gradient increased by 1 MV/m (Figure 8.14b). However, the operating gradient with 100 mRem/h of field emission radiation was increased from 16.8 to 19.5 MV/m, so there was some benefit.

In situ plasma processing a cryomodule (CM) was not very successful. One CM (C100–10) had severely reduced operating gradients probably due to a catastrophic gate valve O-ring failure on the cavity at the end of the cryomodule. The average increase in field emission onset due to plasma treatment was 0.4 MV/m. The average increase in 100 mRem/hr. There was probably too much particle contamination

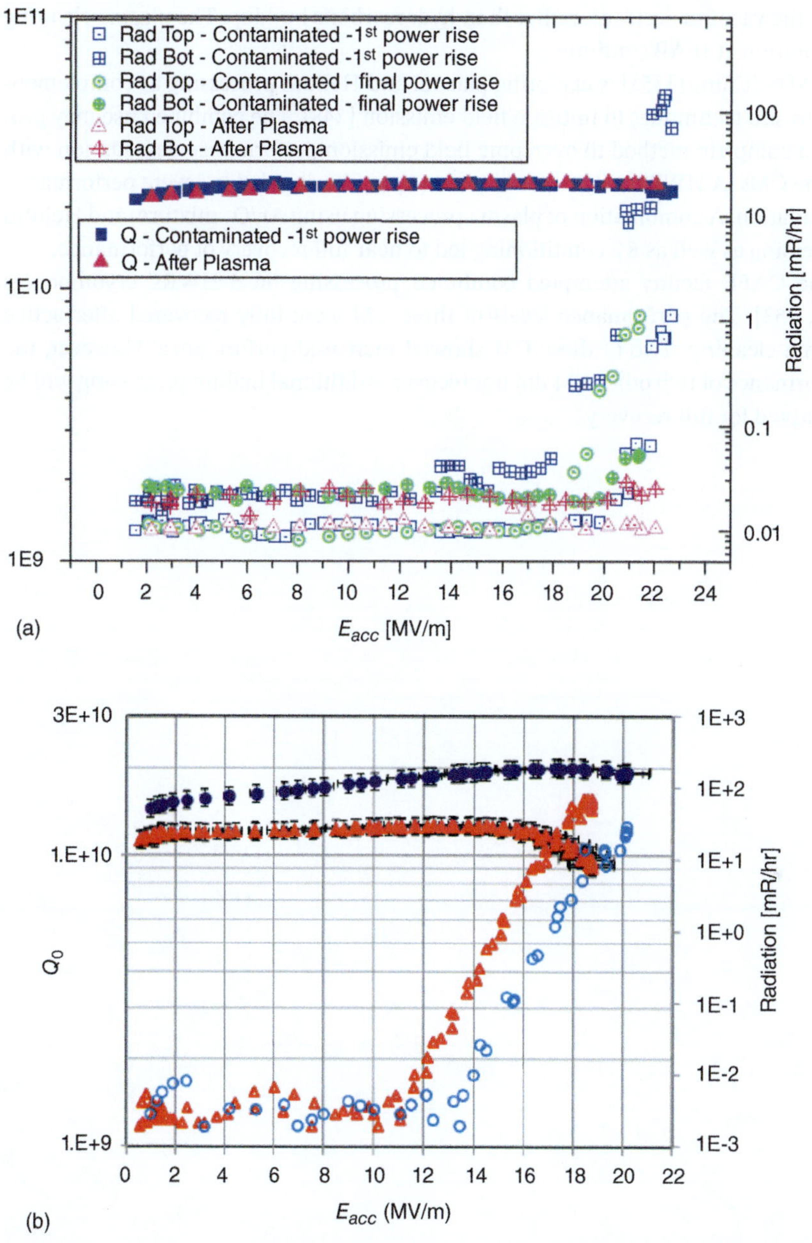

Figure 8.14 (a) Plasma processing successfully removed field emission introduced by vacuum contamination. Source: B. Giaccone, Fermilab [349]/JACoW/CC BY 3.0. (b) Impact of plasma processing on a multicell cavity at Jlab. The higher Q values on the second test were due to the rapid cooldown and magnetic flux expulsion, not plasma processing. Source: T. Powers et al., Jefferson Lab [351]/arXiv.

from the vacuum accident, rather than hydrocarbons buildup. The plasma cleaning exploration at JLAB continues.

CiADS (China) [352] is exploring plasma and Helium processing as complementary in situ techniques to mitigate field emission [348]. The combined use may provide a complete method to overcome field emission performance degradation with online CMs. A HWR cavity exposed to air (3 μm filtering) led to severe performance degradation. A combination of plasma processing using Ar/O_2 mixture, and Helium processing as well as RF conditioning led to near full recovery of performance.

The CAFE facility attempted combined processing on 5 HWRs' cryomodules [348, 353]. The performance levels of three CM were fully recovered after active plasma cleaning. Two of these CM showed increased performance! However, the performance of two other CM did not recover. Additional helium processing will be attempted for full recovery.

9

Pursuit of Higher Performance with Alternate Materials

Since the best Nb cavities with gradients of 45–50 MV/m (H_{pk} = 190–210 mT) approach the fundamental superheating field H_{sh} = 220 mT limit of the surface magnetic field, new materials with higher superheating field will eventually need to enter the picture in the quest for ever higher gradients and *Q*'s. The development time for full-scale cavities with new materials is likely to be very long (decades). Hence it is important to explore new material avenues as early as possible. Many institutions are engaged in a variety of efforts, some of which we will cover in this chapter. At this stage, a single successful cavity that exceeds Nb performance in new materials is still far from realization.

It is likely that the new material ventures will use various coating methods, some of which have been developed for thin (μm) Nb films on Cu cavities. The effort to produce good quality Nb films on Cu cavity substrates has been in progress for many decades. Hence we start with an update on the status of Nb–Cu cavities before we move on to new materials.

9.1 Nb Films on Cu Substrates

Nb-coated Cu cavities are cost-effective, especially for low (<800 MHz) frequency (large scale) cavities due to the low cost of Cu compared to sheet Nb. The cost of Cu material is only about 10% of the cost of bulk Nb. The high thermal conductivity copper substrate enhances thermal stability to avoid quenches. Cu has thermal conductivity of 300–2000 Wm^{-1} K compared to Nb of about 75 Wm^{-1} K at 4.2 K. But the fabrication and surface preparation of the Cu substrate turns out to be as difficult as for bulk Nb, in order to achieve high quality Nb film depositions.

An unexpected advantage of thin film Nb is that the residual surface resistance is almost insensitive to trapped flux from the dc ambient magnetic field (Figure 9.1a), reducing, or even eliminating, the need for complex cavity magnetic shielding in the cryostat. The sensitivity to magnetic field is 0.3 nΩ/mOe at 1GHz for a bulk Nb cavity and about 0.005 nΩ/mOe (about 100 times) smaller for sputtered Nb films [215], which typically have a very small mean free path, comparable to the BCS coherence length for magnetron sputtered cavities [354]. This occurs at a film RRR of about 10.

Superconducting Radiofrequency Technology for Accelerators: State of the Art and Emerging Trends, First Edition. Hasan Padamsee.

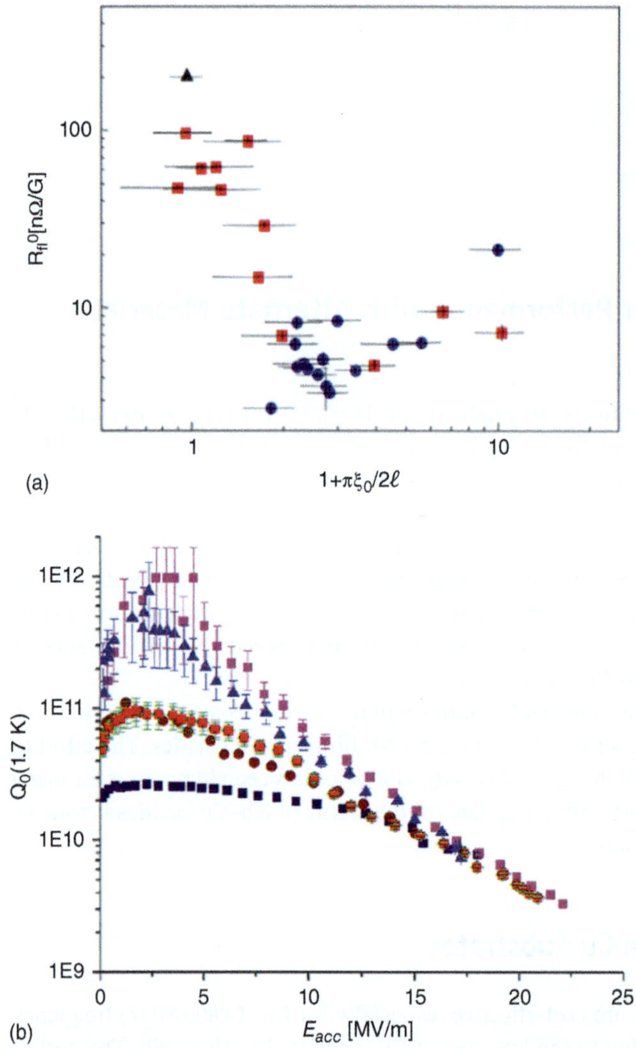

Figure 9.1 (a) Sensitivity of sputtered Nb films to trapped flux vs electron mean free path of the films. Source: [215, 355] Courtesy of S., Calatroni, C. Benvenuti et al. (b) Best performance at 1.7 K for several 1.5 GHz Nb/Cu cavities sputtered coated with Kr gas at CERN. The symbols correspond to Nb film deposition attempts. Source: [356]/C. Benvenuti et al., with permission of Elsevier.

Note how the sensitivity follows a bell-shaped curve (plotted here versus 1/mfp) similar to the behavior for N-doped cavities (Chapter 3).

Despite the important advantages, Nb–Cu is not yet ready for accelerators requiring highest accelerating gradient, or the lowest surface resistances. Even though at low E_{acc} the surface resistance starts out very low (compared to bulk Nb cavities) due to the short mean free path of the films (Figure 9.1b), the surface resistance of

Nb films increases strongly with E_{acc}. The increase in field has been a long-standing problem with Nb–Cu cavities, with much effort to understand and suppress it.

The origin of the field-dependent resistance has been the subject of many studies [354, 356–361] over a vast range of R&D, but is still far from being fully understood. The quality of the substrate copper cavity – in particular at the seams – is recognized as one of the contributing causes. Sputtered films of Nb generally do not adhere to the copper substrate well enough to withstand the standard full pressure (100 atm) during HPR to remove field emission. This suggests that the interface layer may have poor contact with the substrate, ensuing hot spots [362]. More recent studies [358] show that the thermal boundary resistance between the Nb film and the copper substrate is not the main contribution to the field-dependent losses.

There are several new coating techniques under exploration based on energetic condensation to address the Q-slope problem, with initial signs of some success. The development will be very important for future low frequency (<500 MHz) applications, such as FCCee (Section 12.2) or CepC (Section 12.3).

In the case of FCCee for example, successful R&D on the energetic condensation techniques will make Nb/Cu at 400 MHz and 4.5 K competitive to bulk Nb at 800 MHz and 2.0 K in terms of cryogenic consumption. Estimates show that the cost for a CM with 400 MHz Nb/Cu cavities is about 20% less than for a CM with 800 MHz bulk Nb cavities [363]. In order for these technologies to mature and fulfill their promise for future superconducting accelerators, there need to be considerable research and development investments.

9.1.1 Direct Current Magnetron Sputtering

Bias sputtering or magnetron sputtering are necessary for niobium because the sputter rate is low and films can get easily contaminated by the residual impurities in the coating chamber. Bias sputtering helps to remove contaminants as the film grows, while magnetron sputtering increases the rate of deposition to lower the relative contamination content. Direct current magnetron sputtering (dcMS) was successfully used by CERN on a large scale (250 cavities) for the 4-cell 352 MHz elliptical cavities in the large electron–positron (LEP) collider [364], and the single-cell 400 MHz elliptical cavities installed in the LHC [365]. Typical dcMS sputtering conditions [215] require a dc power of 1 kW in Kr plasma, resulting in 1.5 μm thick films in 15 minutes of coating time. Coatings are performed at 150 °C.

9.1.2 DC-bias Diode Sputtering at High Temperature (400–600 °C)

DC bias diode sputtering is more suitable for the complex QWR geometry. CERN has prepared thirty-two 101 MHz QWR Nb-coated Cu cavities for HI-ISOLDE with two β values [366–372]. The welded copper cavity substrates were coated with sputtered Nb at about 500 °C. Figure 9.2 gives the performance of the fifteen cavities for the first three cryomodules [373]. Best performance was obtained after slow cooldown to avoid trapped flux due to thermocurrents arising from temperature gradients in welded regions of the composite cavity where there is poorer contact between film and substrate.

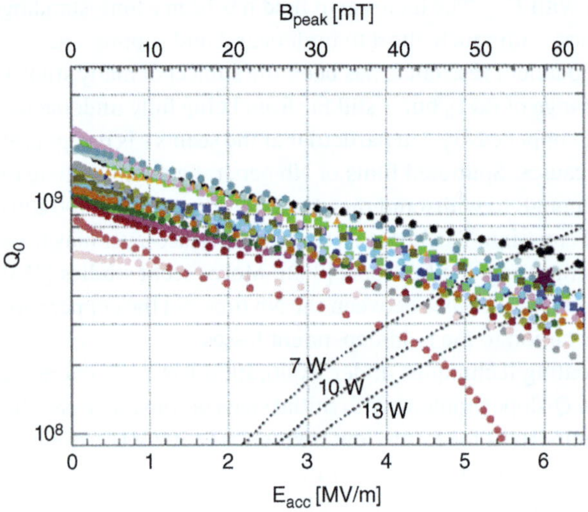

Figure 9.2 Performance of 15 Nb–Cu QWR cavities at 4.5 K for HI-ISOLDE, Nb films on welded Cu substrates. Source: S. Lopez et al., CERN [373]/JACoW/CC BY 3.0.

9.1.3 Seamless Cavity Coating

Recently a 101 MHz QWR fully machined (seamless) copper cavity without welds was coated with dc bias diode sputtering [374] to show a substantially lower Q-slope (Figure 9.3), especially after avoiding trapped dc magnetic flux by reducing the external field to 50 mOe [375]. The cavity reached $Q = 10^9$ at 13 MV/m and 2.3 K instead of similarly sputtered cavities with welded seams that were limited to about 5–6 MV/m for 10 W power consumption (Figure 9.3). The study shows that the residual resistance of films sputtered by this method have a higher sensitivity (0.011 nΩ/mOe) to trapped dc magnetic fields than the magnetron sputtered films (Figure 9.1a), so that in a high external dc field of 1000 mOe (comparable to earth's field and no shielding) the performance shows a strong Q-slope due higher trapped flux losses (Section 5.5). Flux gate sensors show that the external magnetic field is fully trapped, as expected for sputtered films with strong pinning. Historically, Nb–Cu cavities operate without magnetic shielding due to their low sensitivity as seen in Figure 10.1a (0.003–0.005 nΩ/mOe for dc magnetron films with RRR = 11, mfp = 30 nm). When the external field for the seamless coated cavity was lowered to 50 mOe, the Q-slope at 2.4 K is greatly reduced to give a Q close to 2×10^9 at 7 MV/m (Figure 9.3b). However, at 4.5 K there is still the familiar strong Q-slope with field, indicating that the underlying Q-slope behavior of sputtered thin films is still present [375]. With the elimination of trapped flux losses, the observed reduction in Q-slope with temperature from 4.5 to 2.4 K suggests that the explanation for the Q-slope for the dc bias sputtering method depends partly on the BCS component of the surface resistance.

Due to absence of thermal contact problems at welds, the seamless cavity shows more uniform temperatures during cooldown and therefore eliminates thermocurrent-induced flux during cooldown. The higher magnetic field sensitivity

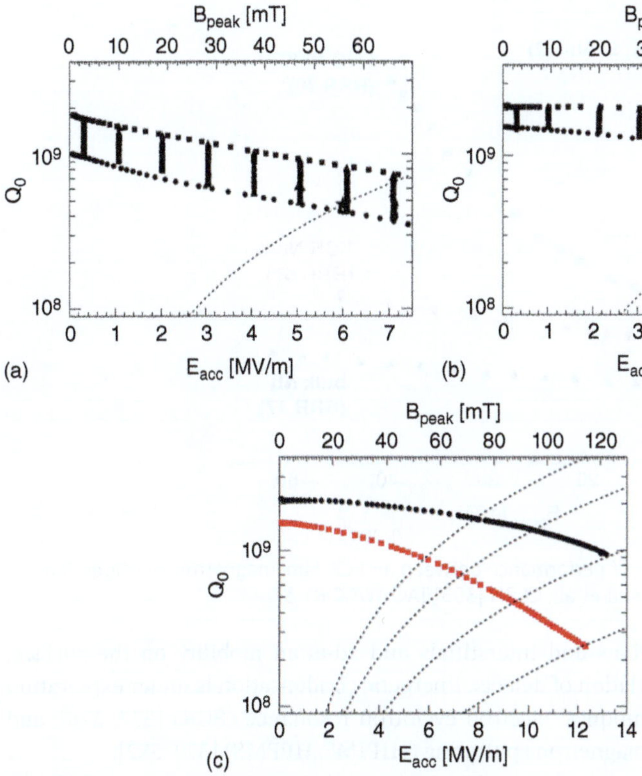

Figure 9.3 Performance of seamless QWR cavities coated with bias dc sputtering at about 500 °C (a) in an external magnetic field (1000 mOe) comparable to no shielding. The circle points show 4.5 K and square ones show higher Q2.4 K data. The dashed line shows 10 W power consumption and the star is the nominal operating point. (b) Improved performance of seamless cavity in a lower magnetic field of 50 mOe. (c) Performance of a seamless cavity at 4.5 and 2.4 K at higher fields in an external magnetic field of 50 mOe. Source: A. Miyazaki and W. Delsolaro, CERN [375]/American Physical Society/CC BY 4.0.

of the dc bias sputtered films over the dcMS films is likely due to a lower density of pinning sites, as well as possibly a higher mean free path than for magnetron sputtered films. From the sensitivity vs. mfp curve (Figure 9.1a), the mfp is about 60 nm, compared to the mfp (30 nm) for dc magnetron sputtered cavities. The improved *Q versus E* performance (lower slope) may also be a manifestation of these same properties. LE-uSR measurements [376] confirm that the dc bias sputtered cavities have a shorter penetration depth (and therefore a longer mfp) than the magnetron sputtered films.

9.1.4 Nb–Cu Films by ECR

Energetic condensation is a deposition process where a significant fraction of the condensing species are ions with higher than thermal deposition energies (10 eV). Ions are implanted into the surface giving high mobility to atoms in the affected volume. Large amplitude thermal vibrations facilitate diffusion, especially the

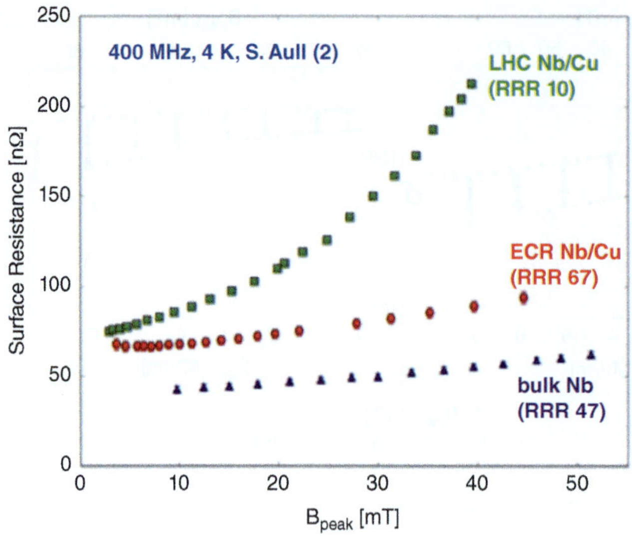

Figure 9.4 Comparison of performance between an ECR film, magnetron sputtered film, and bulk Nb. Source: S. Aull et al., CERN [383]/JACoW/CC BY 3.0.

migration of dislocations and interstitials and ad-atom mobility on the surface. This favors the annihilation of defects. Energetic condensation is under exploration in two groups of techniques: electron cyclotron resonance (ECR) [377, 378], and high-power impulse magnetron sputtering (HIPIMS/HPPMS) [379–382].

In the ECR technique, an electron beam gun produces neutral vapor of Nb that is introduced into the ECR chamber. ECR does not involve a working gas. With a perpendicular magnetic (about 900 G) and RF (e.g 2.45 GHz) electrical fields applied to the chamber, the residual electrons are accelerated to their cyclotron resonance conditions and ionize the neutral Nb vapor. The ions generated are singly charged and have an inherent energy of about 60–70 eV. An applied bias voltage accelerates the ions towards the substrate. The bias voltage controls the incident ion energy. The ECR Nb films typically exhibit enhanced adhesion to the substrate, and the *Q*-slope with field is reduced. ECR samples are now showing glimpses of potentially achievable good RF performance. Figure 9.4 shows best results with ECR compared to dc magnetron sputtering, and bulk Nb [383]. The coating was deposited on a Cu sample which was measured in a host Nb cavity. The big step of ECR coating a full cavity has yet to be accomplished.

9.1.5 Nb–Cu Films via High-Power Impulse Magnetron Sputtering (HIPIMS)

HIPIMS also called high-power pulsed magnetron sputtering, is thin film deposition from standard magnetrons using pulsed plasma discharges, where a large fraction of the material arrives to the substrate as ions. The plasma is generated by a glow discharge with high current density in Kr. The Kr plasma ionizes the Nb from the target, allowing denser films to be formed.

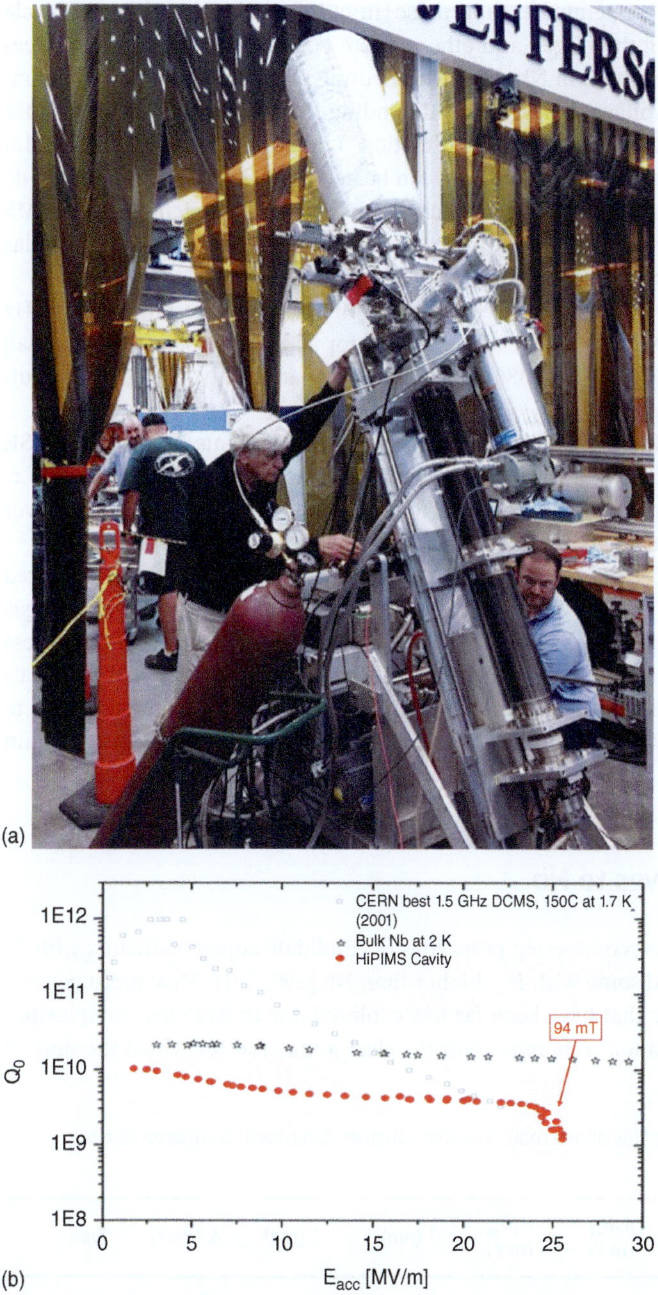

Figure 9.5 (a) HIPIMS apparatus at Jlab. Source: [387]/Cryogenic Society of America, Inc, Photo courtesy of Jefferson Lab. (b) Comparing the performance of Jlab HIPIMS-coated cavity with best CERN magnetron sputtered cavity at 1.5 GHz and bulk Nb cavity. Source: [386] Courtesy of A-M Valente-Feliciano, Jefferson Lab.

HIPIMS is typically operated in short pulse (impulse) mode with a low duty cycle to avoid overheating the target, and other system components. HIPIMS produces multiply charged ions of Kr or Nb along with neutrals. The deposition energy is controlled with a bias voltage to provide good bonding. The technique creates a dense sputtering plasma with high ionization fraction. Literature shows that better film morphology, denser and smoother films, can be achieved compared to standard dc magnetron sputtering (dcMS) coating technology [384]. Early attempts at HIPIMS gave cavity results with the familiar increase of surface resistance with E_{acc}, similar to what is found for cavities coated by dcMS [215].

Both CERN [385] and Jlab are developing HIPIMS for elliptical cavities on 1.5 GHz single cells. Figure 9.5 shows an excellent result for a HIPIMS-coated cavity at Jlab with high Q to nearly 25 MV/m [386]. Results are compared with dc magnetron sputtering, as well as bulk Nb.

RF frequency shift analysis near T_c from cavity measurements, as well as LE-μSR results show that HIPIMS films have a significantly shorter penetration depth, (i.e. longer electron mean free path) compared to standard dcMS [376] likely due to lower impurity concentrations as expected from the energetic ion condensation.

The possibility of "gasless" HIPIMS is also under exploration. Here the sputtered atoms would fill in the role of the process gas. Thus the need for Kr or other gas is eliminated, and the issues related to noble gas inclusion can be avoided. Gasless self-sputtering has been demonstrated for HIPIMS using high sputter yield materials such as copper [388]. In the case of Nb [389], the sputter yield is not sufficient to produce enough "Nb gas". But the process can be started with noble gas to obtain HIPIMS pulses and then tapered into gasless mode.

9.2 Alternatives to Nb

Table 9.1 shows superconducting properties of candidate superconductors with T_c higher than Nb, and some with H_{sh} higher than Nb [390, 391]. Most are superconducting compounds that have been far less explored due to technical complexities of compound formation. It is important to select a material for which the desired

Table 9.1 Key properties of alternate superconductors discussed, compared to Nb [91, 390, 391].

Material	T_c(K)	$\mu_0 H_{c1}$ (mT)	$\mu_0 H_{sh}$ (mT)	λ (nm)	ξ (nm)	Δ (meV)	Type
Nb	9.22	174	255	39	38	1.5	II
NbN	17.1	23	214	200–350	<5	2.6	II
NbTiN	17.3	30		150–200	<5	2.8	II
Nb3Sn	18.3	35–50	425	80–100	<5	<5	II
MgB2	39	43	170	140	5	2.3/7.2	II = 2gaps**

compound phase is stable over a broad composition range. The attractive candidates are the A15 superconductors (e.g Nb_3Sn), the cubic NbN intermetallic, the high T_c (39 K) superconductor MgB_2, and overlayers of the above materials on a substrate of niobium. The very high T_c superconductors (cuprates) are not included here because these are D-wave superconductors with nodes in the gap. Such nodes with zero gap value would cause very high RF losses. All conventional BCS superconductors including Nb have S-wave gap symmetry.

The candidate materials all have a very low H_{c1}, but this may not be a problem since the energy barrier prevents vortices from penetrating (Section 2.4.9). This has been proven for Nb_3Sn with κ about 40, H_{c1} 30–40 mT, as well as for 120 C baked Nb with low mean free path and κ about 5 (Figure 3.27, [392]). For 120 °C baked Nb RF flux penetration has been avoided to fields near H_{sh}. However, defects can be weak points for early vortex penetration, as discussed in Section 2.4.9. The alternate materials have low (<10 nm) coherence length, which means they are sensitive to <10 nm size defects. It may not be possible to reach H_{sh} in Nb_3Sn cavities because of the small coherence length (<5 nm), so that even small defects can act as nucleation sites for flux penetration.

9.2.1 Nb_3Sn

A15 compounds are intermetallic and brittle in the bulk form, so SRF structures are produced as a thin layer on the inner surface of an already formed structure. Nb_3Sn is the most explored compound with best results but does not give as good performance as Nb. The A15 phase is in the composition range of 18–25 at% Sn. The superconducting properties T_c, Δ, and H_c, depend strongly on the Sn content [393]. Perfect ordering in the stoichiometric phase is achieved close to stoichiometry (at 24.5 at%) where H_{sh} is 425 mT as compared to Nb's H_{sh} of 220 mT at 0 K. Accordingly, we can expect the upper limit of the gradient to be 400 mT or near 95 MV/m for a TESLA geometry cavity. But the key question is whether films can be sufficiently defect-free to approach H_{sh}?

Nb_3Sn films of a few microns thickness have been deposited on the inner surface of Nb cavities by exposure to Sn vapor (10^{-3} mbar) in a UHV furnace at temperatures between 1050 °C and 1250 °C. Table 9.2 gives the properties of a Nb3Sn film extracted from RF measurements on a coated cavity [395]. In general, such films exhibit good material quality with Sn content of about 25%, T_c from 16 to 18 K, and Δ from 2.7 to 3.2 meV [394]. Coating results are typically reproducible for the same Nb cavity substrate but have been seen to vary between different cavities. PCT studies [394] show a large spread in Δ and Γ/Δ values (Figure 9.6) confirming the difficulty of producing high-quality films over large areas. Figure 9.7 shows a large spread of H_{c1} values for various film attempts.

Some of the limitations of Nb_3Sn arise from the sensitivity of the thermodynamic critical field H_c (and therefore for H_{sh}) to the exact Sn concentration. For example, a Sn depletion of 3% reduces H_c by 75%. Other difficulties are the high surface roughness at Nb_3Sn grain boundaries possibly causing local field enhancement. Thinner (1 μm) layers give smoother surfaces and best results (Figure 9.8a).

Table 9.2 Measured and calculated properties of the Nb_3Sn film from one cavity prepared at Cornell [21, 90, 93, 94, 395].

Property	Value	References
$\lambda_L(0)$(nm)	89 ± 9	[94]
$\xi_0(0)$(nm)	7.0 ± 0.7	[94]
T_c(K)	18.0 ± 0.1	[395]
$\Delta/k_B T_C$	2.25 ± 0.12	[395]
l(nm)	4.8 ± 2.0	[395]
R_{res}(nΩ)	8.5 ± 1.2	[395]
$\lambda(0)$(nm)	139 ± 23	[21]
$\xi(0)$(nm)	3.4 ± 0.5	[93]
$\kappa(0)$	$41 \pm 1:9$	[21]
$\mu_0 H_c(0)$(T)	0.49 ± 0.10	[21]
$\mu_0 H_{c1}(0)$(mT)	36 ± 3	[94]
$\mu_0 H_{c2}(0)$(T)	28 ± 9	[21]
$\mu_0 H_{sh}(0)$(T)	0.41 ± 0.09	[90]

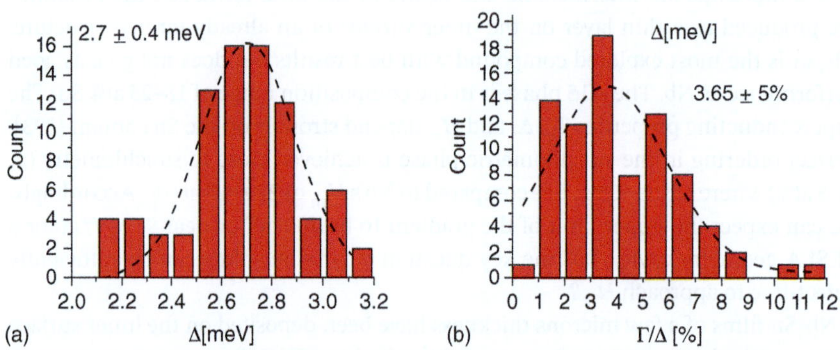

Figure 9.6 PCT results show a spread in values for (a) Δ and (b) Γ/Δ suggesting significant variations in surface superconducting properties. Source: [394]/C. Becker et al., with permission of AIP Publishing.

Most practitioners of Nb_3Sn have encountered a Q-slope problem with gradient limits. Initially there was suspicion about vortex entry due to exceeding H_{c1}. Estimates of H_{c1} from extracted values of mfp and coherence length, as well as mfp from μSR measurements, show high Q values for RF fields far in excess of H_{c1} (Figure 9.7), indicating that vortices do not start entering at H_{c1}.

Progress continues. At Fermilab [397], Posen reduced the thickness of the film down to one μm (Figure 9.8) to achieve a flat Q *versus* E curve out to 23–24 MV/m (Figure 9.9a). The performance at 4.2 K is also very attractive showing $Q_0 > 10^{10}$ at gradient of 18 MV/m. 1 μm films (Figure 9.8) show a factor of two smaller surface

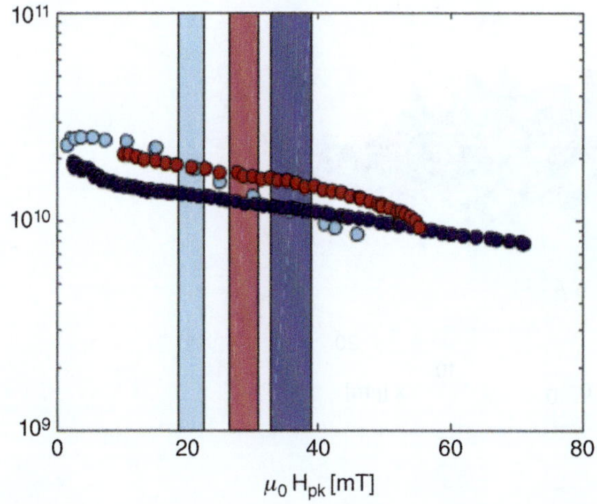

Figure 9.7 *Q versus B* for selected cavities. Bands show H_{c1} estimates for each cavity. Each test that significantly exceeds H_{c1} is without a strong *Q*-slope showing that H_{c1} does not limit performance. Source: [395, 396] Courtesy of S. Posen, Cornell University.

roughness *versus* 2–3 μm films, and smaller grain size (0.7 versus 1.2 μm). The thinner film could be more thermally stable on top of the higher thermal conductivity Nb substrate. Careful material science is yet required to understand and confidently control Nb_3Sn crystal growth dynamics so as to reproducibly produce low-loss and defect-free surfaces.

High-power pulsed RF measurements (Figure 9.9b) at Cornell on a Nb_3Sn cavity show encouraging trends [396]. At high temperature (T > 15 K), the results track the superheating field, extrapolating to 300 mT (E_{acc} > 80 MV/m) at zero temperature. But at lower temperatures, thermal limitations take over to limit the highest field to about 100 mT (24 MV/m), which is close to the cw result of 22 MV/m.

The data confirm that Nb_3Sn cavities can operate in a flux-free Meissner state above H_{c1}. The quench is triggered by localized vortex penetration so that there is negligible preheating [399]. This could be due to local suppression of the superheating field at coating flaws as reported in [400], which prevents cavities from reaching higher fields.

Application to multicell cavities has started. Best 9-cell Nb_3Sn cavities coated at FNAL [401] have reached 15 MV/m with *Q* values of 10^{10}. 5-cell cavities coated at Jlab [402, 403] have reached 13 MV/m (Figure 9.10b).

9.2.2 MgB_2

MgB_2 has a high T_c (39 K) but lower H_{sh} (170 mT) than Nb. The superconductivity comes from phonon-mediated Cooper pairs, similar to the low-T_c superconductors, which distinguishes MgB_2 from the higher T_c (>70 K) cuprates, like YBaCuO, for which the origin of superconductivity is still a mystery.

MgB_2 is at a disadvantage because it is anisotropic [405, 406] and has two gaps (2.3 and 7.1 meV). The RF surface resistance is dominated by the lower gap, which gives

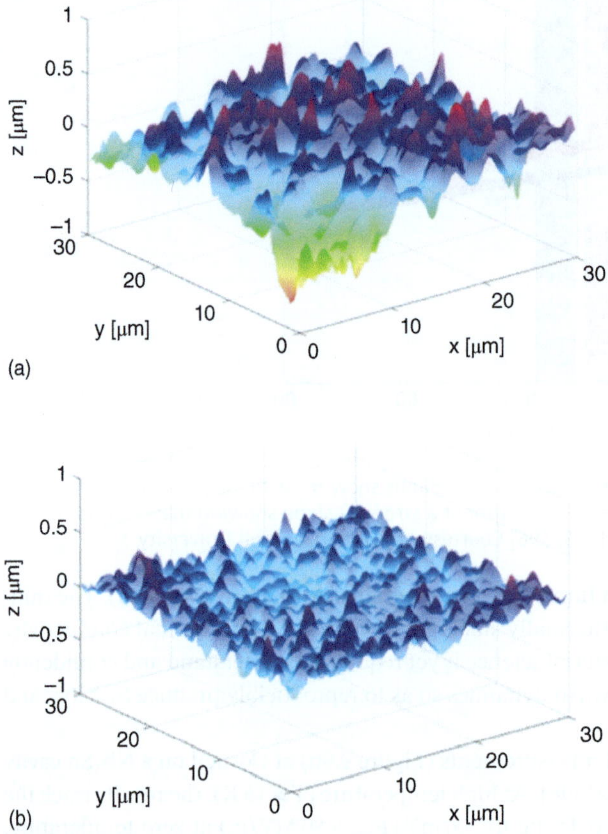

Figure 9.8 Comparison of the smoothness and grain size of (a) thick and (b) thin coatings of Nb_3Sn. The grain size of the thin films was found to be smaller than 0.7 *versus* 1.2 μm for the thick film, and with a lower spread 2015 . The thinner film shows a factor of 2 smaller roughness. Source: [397, 398] Courtesy of S. Posen et al., Fermilab.

higher surface resistance. The lower gap is still higher than the gap of Nb (1.5 meV), so the promised Q performance is better than for Nb. However, the lower gap is less than the Nb_3Sn gap (3.1 meV). Also, the superheating field for MgB_2 (170 mT) is less than that of Nb_3Sn (420 mT). Other than the higher T_c, the anisotropy, lower gap, and lower superheating field make MgB_2 less attractive than Nb_3Sn.

Several techniques have been used to deposit MgB_2 as reviewed in [390]. One that is amenable to cavity coatings is hybrid physical chemical vapor deposition (HPCVD) reacting boron hydride (B_2H_6) with Mg at 550–760 °C [407]. A high-pressure Mg vapor is needed for MgB_2 phase stability. The MgB_2 film is susceptible to contamination with O and C. Thus, a reducing environment during coating is needed to avoid oxidation, which is provided by H_2. Another challenge for MgB_2 is the degradation of the film properties with exposure to moisture (the resistance increases, and T_c degrades).

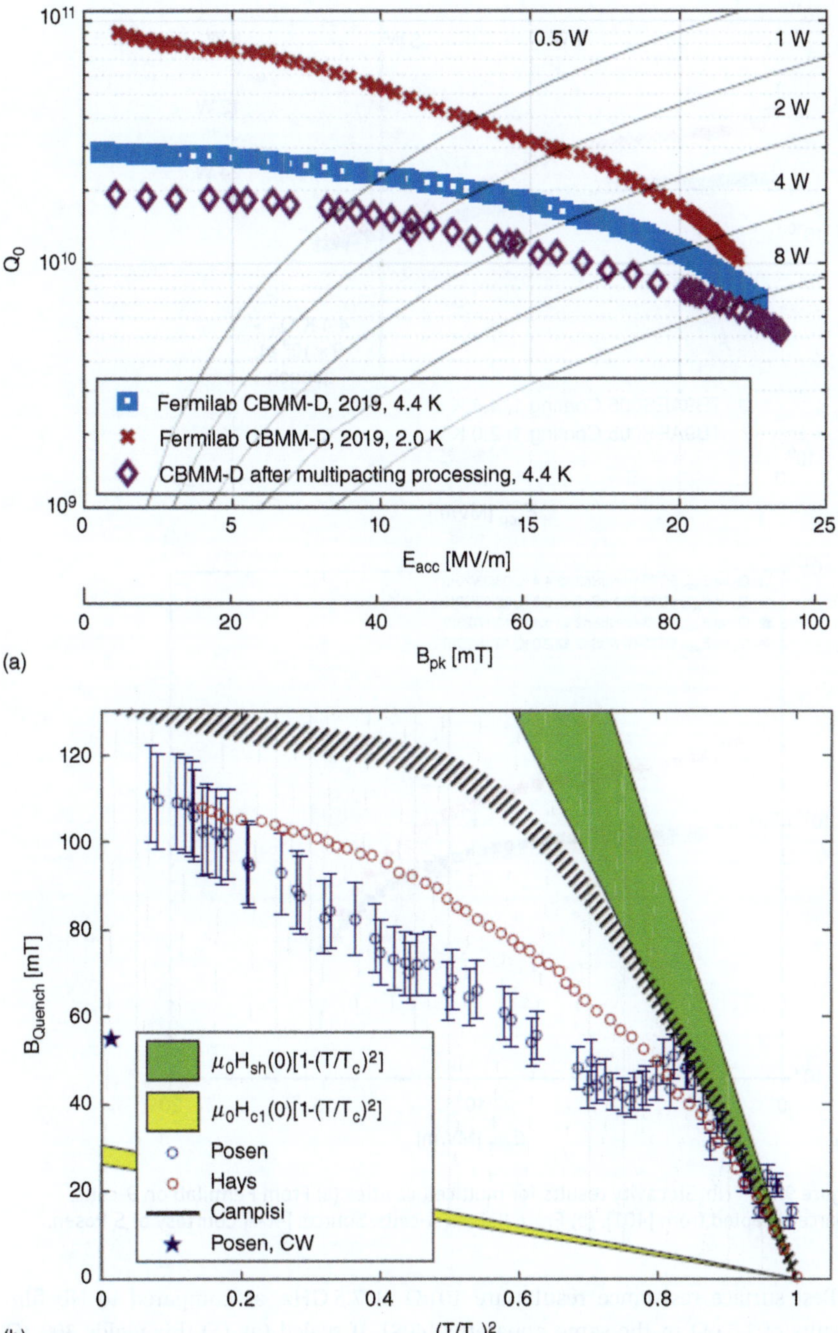

Figure 9.9 (a) Record cw behavior for Nb_3Sn coated at Fermilab reaching E_{acc} = 23 MV/m. Source: [397, 398] Courtesy of Posen et al., Fermilab. (b) Measurements of the critical RF field using high power pulsed RF. The high temperature results extrapolate to a maximum surface magnetic field of 300 mT, which would translate to E_{acc} = 70 MV/m for a TESLA shape cavity. Source: S. Posen et al., Cornell University [392, 396]/IOP BY 4.0.

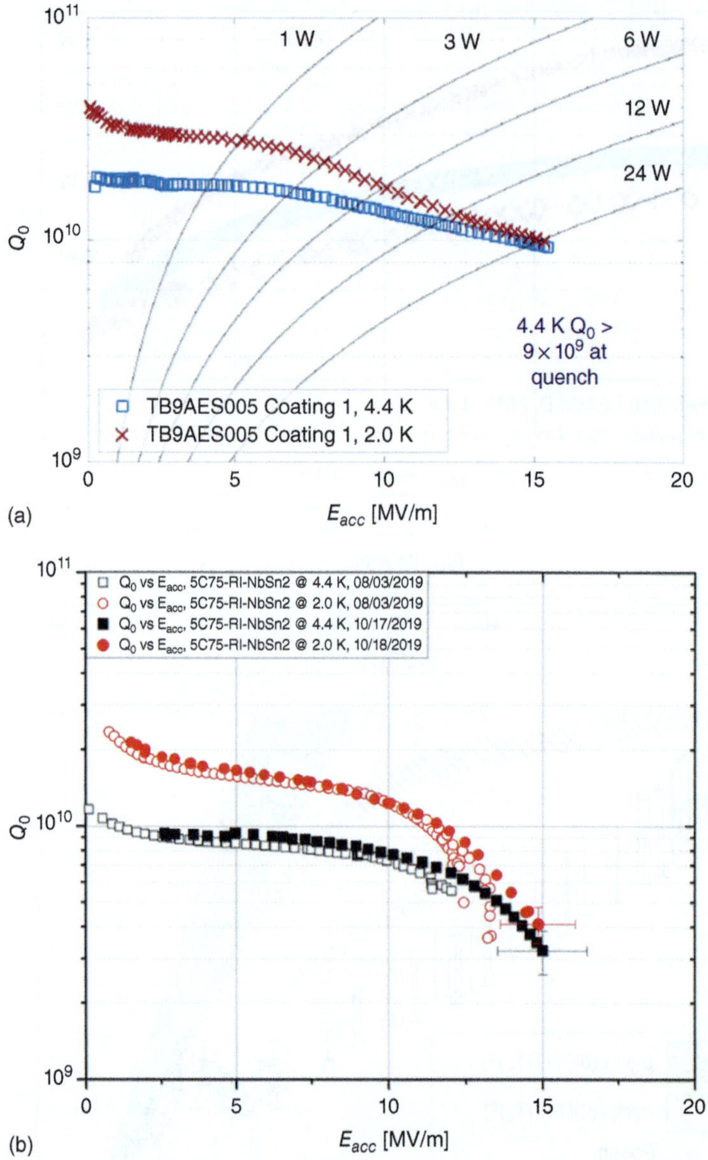

Figure 9.10 Nb_3Sn cavity results for multicell cavities (a) From Fermilab on 9-cells. Source: Adapted from [401]. (b) From Jlab on 5-cells. Source: [404] Courtesy of S Posen.

Best surface resistance results are 10 μΩ at 7.5 GHz, as compared to Nb film results of 1.5 μΩ in the same apparatus [408]. If scaled (as f^2) this yields 300 nΩ at 1.3 GHz for MgB_2 as compared to the Nb film result, which scales to 45 nΩ at 1.3 GHz. For comparison, the best Nb cavities at 1.3 GHz yield far less than 10 nΩ, which puts into question the scaling with f^2. Much improvement is still necessary for MgB_2.

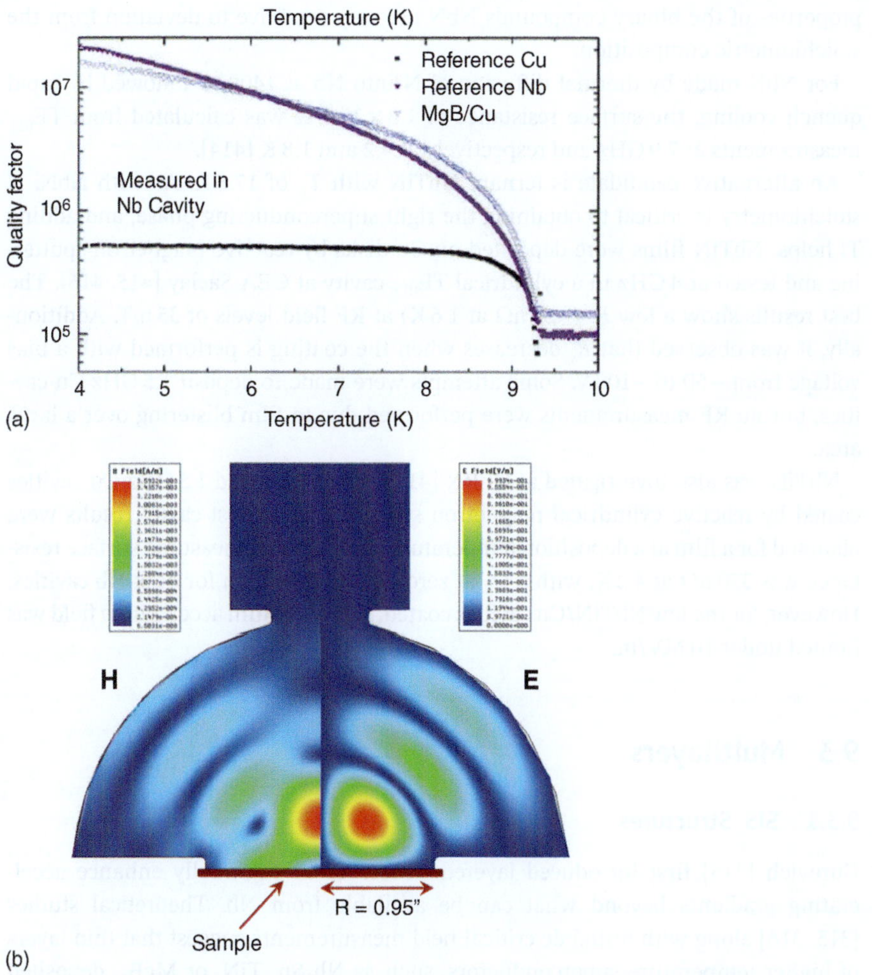

Figure 9.11 (a) Best results for MgB_2 in an RF cavity compared to Cu and Nb Source: [409, 411]/P. Welander et al., SLAC with permission of IOP Publishing [410]. (b) Magnetic (H) and electric field distributions for the TE_{013} host Nb cavity showing high magnetic field on the sample. Source: P. Welander, SLAC [409]/JACoW/CC BY 3.0 and W. Withanage et al. [411] /IOP Publishing BY 4.0.

Figure 9.11 shows some of the best results on a MgB_2 sample with $T_c = 37$ K measured in a TE_{013} Nb cavity [409, 410] at 11.4 GHz as compared to Cu and Nb samples in the same cavity. With the MgB_2 sample, the Q of the host cavity is 30 times higher than with a Cu sample. With a Nb sample, the Q was 40 times higher.

9.2.3 NbN and NbTiN

The phase diagram of the binary system Nb–N system includes many different phases [116, 412], each characterized by different T_c [413]. The NbN superconducting phase of interest is the cubic δ-phase, with $T_c = 17.3$ K. The superconducting properties of the binary compounds NbN are very sensitive to deviation from the stoichiometric composition.

For NbN made by thermal diffusion of N into Nb at 1400 °C followed by rapid quench cooling, the surface resistance of $1.3 \times 10^{-6}\,\Omega$ was calculated from TE_{011} measurements at 7.9 GHz and respectively at 4.2 and 1.8 K [414].

An alternative candidate is ternary NbTiN with T_c of 17.8 K. As with NbN, N stoichiometry is critical to obtaining the right superconducting phase, and adding Ti helps. NbTiN films were deposited on Cu disks by reactive magnetron sputtering and tested at 4 GHz in a cylindrical TE_{011} cavity at CEA Saclay [415, 416]. The best results show a low R_s (<50 nΩ at 1.6 K) at RF field levels of 35 mT. Additionally, it was observed that R_s decreases when the coating is performed with a bias voltage from −50 to −100 V. Some attempts were made to deposit 1.5 GHz Cu cavities, but no RF measurements were performed due to film blistering over a large area.

NbTiN was also investigated at CERN [417] on samples and 1.5 GHz Cu cavities coated by reactive cylindrical magnetron sputtering. The best cavity results were obtained for a film at a deposition temperature (265 °C). The measured surface resistance was 330 nΩ at 4.2 K, with a Q_0 at zero field higher than for bulk Nb cavities. However, for the few NbTiN/Cu cavities coated, the maximum accelerating field was limited under 10 MV/m.

9.3 Multilayers

9.3.1 SIS' Structures

Gurevich [315] first introduced layered structures to potentially enhance accelerating gradients beyond what can be available from Nb. Theoretical studies [315, 316] along with initial dc critical field measurements suggest that thin layers of higher temperature superconductors, such as Nb_3Sn, TiN, or MgB_2, deposited on the surface of Nb cavities with an intermediate low-loss dielectric layer could lead to substantially higher gradients than possible with Nb. The thickness of the superconducting layer has to be less than the penetration depth of the chosen high T_c superconductor, which is greater than the penetration depth of the substrate superconductor (Nb). The intermediate low-loss dielectric material layer has to be thick enough to prevent Josephson coupling and lossy Josephson currents between the two superconducting layers.

Theoretical models [315, 317] predict that appropriately fabricated nanometric SIS' multilayer films can delay vortex penetration in Nb surfaces allowing them to sustain higher surface fields than any pure material. The higher T_c layer screens the magnetic field from the bulk superconducting substrate.

The phase space for such new development is quite extensive. The BCS resistance should also be reduced due to the larger gap of the higher T_c materials. Theoretically, the effect can be furthered by adding one or more S-I layer sequences. The optimum number of SI layers is determined by a compromise between vortex dissipation reduction and superconductivity suppression at the S-I interfaces.

9.3.2 Theoretical Estimates

T. Kubo [316, 317] and S. Posen et al. [318] explore the behavior of the SIS structure. Kubo solves the Maxwell and London equations in the I and S layers, and in the bulk superconductor with continuity conditions of electric and magnetic fields at boundaries to solve the magnetic field in the three regions I, II, and III (Figure 9.12). The solutions reduce to the expression for the semi-infinite superconductor $B = B_0 e^{-x/\lambda_1}$ when the S layer and the bulk superconductor are the same material ($\lambda_1 = \lambda_2$) and the I layer vanishes ($dI \to 0$). Kubo provides predictive contour plots for maximum achievable peak surface field without vortex dissipation, for various layer thicknesses, for NbN, and Nb_3Sn structures on an ideal Nb surface.

If the thickness d of the overlayer is small, most of the external field reaches the base superconductor [391], so there is not much gain in maximum achievable field. If d is thicker, H_{sh} of the composite is lower, but screening is more effective. Therefore, there is an optimum thickness with a maximum screening field [391]:

$$H_m = \left[H_s^2 + \left(1 - \frac{\lambda_0^2}{\lambda^2} \right) H_{s0}^2 \right]^{1/2} \tag{9.1}$$

where the maximum screening H_m is achieved at the thickness an optimum thickness d_m

$$d_m = \lambda \ln(\mu + \sqrt{\mu^2 + k}), \quad \mu = \frac{H_s \lambda}{(\lambda + \lambda_0) H_{s0}} \tag{9.2}$$

Detail numerical calculations [316, 391] give typical examples under perfect film conditions, and with a 10 nm thick insulating layer.

(a) For dirty Nb, optimum thickness of 90 nm, the maximum field possible with Nb base superconductor is 259 mT.
(b) For an optimum thickness of 90 nm of Nb3Sn, the optimum field for Nb base rises to 400 mT, a substantial gain.
(c) For an optimum thickness of 100 nm of good NbN, the maximum field for the underlying Nb of 240 mT is possible.

SIMS results (Figure 6.7) show that 120 C baking provides a "dirty" layer of less than 70 nm via oxygen diffusion, which results in maximum fields of about 210 mT, which is roughly close to the prediction (a). Realistically, the film layers will have defects, nonuniform thickness, and surface roughness to reduce the above optimistic estimates by an unknown suppression factor which could be large.

9.3.3 Results

First results on single-layer NbN magnetron sputtered films of increasing thickness show dc values of H_{c1} at 4.2 K rising with thickness from 20 to 75 mT till 100 nm, after which there is saturation [391].

Achieving low surface resistance with dirty overlayers is a further challenge. The surface resistance of 4 layers of NbN measured in a 3.88 GHz cavity was 6 μΩ falling to 0.8 μΩ at 2 K [391]. ECR deposited single layer (150 nm) of NbTiN as part of SIS multilayer showed surface resistance of 20 μΩ at 2 K with a 7.5 GHz TE011 cavity [418].

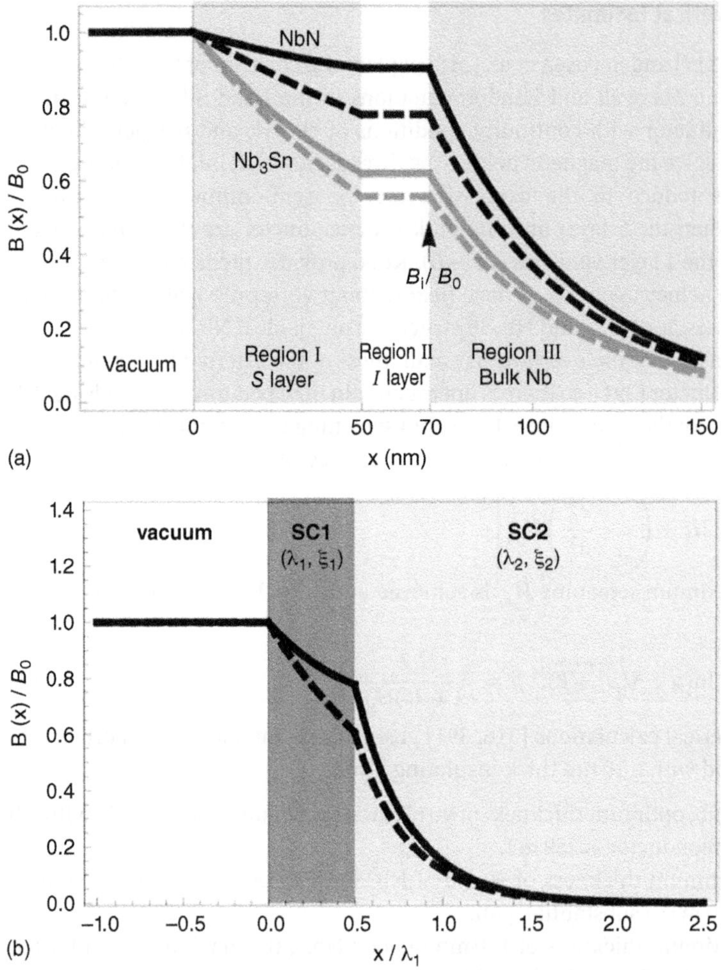

Figure 9.12 (a) Examples of magnetic field attenuations in a typical SIS' structure for NbN and Nb_3Sn, from T. Kubo et al. [316] with permission of AIP Publishing. Solid curves are from calculations. Dashed curves show the naive estimates with $B_0 e^{-x/\lambda_1}$. Black curves and gray curves correspond to the material of the S layer: NbN ($\lambda_1 = 200$ nm) and Nb_3Sn ($\lambda_1 = 85$ nm), respectively. The bulk superconductor is Nb ($\lambda_2 = 40$ nm). The thickness of the S layer and the I layer are fixed at 50 and 20 nm. (b) The magnetic field distribution in the SS' nanolayer coating without insulator layer. The solid curve is for the calculated magnetic field distribution Source: T. Kubo [319, 320] ArXiv & JACoW/CC BY 3.0. The dotted curve represents a naive exponential decay. Source: T. Kubo, [316, 319, 320]/AIP Publishing BY 4.0 /arXiv/JACoW/CC BY 3.0.

At Saclay [419–421], 25 nm layers of NbN (superconductor) and MgO (insulator) were deposited on 250 nm thick Nb film substrates via magnetron sputtering to form an SIS' system. The dc magnetization response was studied using SQUID and local magnetometry. A third harmonic local probe determined the first penetration fields for a reference Nb film sample, the single NbN/MgO nanolayer sample (SL), and for the 4-period multilayers (ML4) sample.

H_{c1} was measured using the third-harmonic response of an ac magnetic field applied from a solenoid coil positioned above a superconducting sample to one side of the sample. The third-harmonic response changes drastically when a vortex penetrates the superconductor [422].

Around 6 K, the first penetration of vortices in the SL structure appears at a field value about twice the reference one, while the 4-layer sample (ML4) is about 10 times higher. RF surface resistance measurements of a NbN/MgO/Nb in a TE011 cavity structure show a multilayer resistance <1 μΩ.

Jefferson Lab [390] deposited NbTiN and AlN films on Nb by reactive magnetron sputtering at 450 °C. Flux penetration field measurements by SQUID magnetometry were performed on 150 nm thick NbTiN films deposited simultaneously on MgO and AlN ceramic. The sample deposited on MgO with a high T_c demonstrated a first flux penetration field enhancement up to 200 mT from 30 mT as measured for a 2 μm bulk-like NbTiN/MgO. The surface resistance of SIS structures is still high at about 20 μΩ at 7.5 GHz.

There is by now an impressive worldwide effort to demonstrate the potential of multilayers. DESY is attempting ALD-deposited NbTiN for studying multilayers [423]. U. of Siegen [424] and collaborators are trying SIS structures with NbN using DCMS, SIMS, and ALD (Atomic Layer Deposition). U. of Wisconsin at Madison [425] is trying SIS multilayers with Nb_3Sn and A_2O_3 over Nb. STFC in UK [426] is attempting multilayer structures with Nb_3Sn, NbTi, and NbTiN. Cornell [427] has a large effort to explore NbTiN/AlN/NbSIS structures along with Nb_3Sn, V_3Si, MgB_2, and NbTiN, via a host of methods such as vapor diffusion, electroplating, chemical vapor deposition, sputtering, plasma-enhanced ALD (atomic layer deposition) and reactive vapor deposition.

9.3.4 SS' Structures

T. Kubo [316, 317] and S. Posen et al. [318] explored the behavior of the SS' structures [319], i.e. a high κ (dirty) superconducting layer on top of a low κ (clean) bulk superconductor. (e.g a Nb_3Sn or MgB_2 layer over Nb). Kubo [316, 317] showed that a dirty overlayer can enhance the accelerating field even if no insulating layer is present. Figure 9.12b shows the model calculations for a SC layer with a penetration depth λ_1 and a coherence length ξ_1 over a SC substrate with a penetration depth λ_2 and a coherence length ξ_2.

Even without the insulating layer, the SS' structure can provide an additional barrier to prevent penetration of vortices. It would not be as robust as the SIS structure, but it also contributes to pushing up the onset of vortex penetration. The superconducting layer thickness is larger than both coherence lengths. The magnetic field distribution is identical to that of the multilayer superconductor with zero insulator layer thickness. The boundary between the two superconductors introduces a force that pushes a vortex in the direction of the material with larger penetration depth. For example, if a thin (<10 nm) superconducting layer with penetration depth 200 nm is deposited on top of s substrate superconductor with 100 nm penetration depth, the field of first vortex penetration ideally rises from 180 mT to over 300 mT.

Checchin [68] discusses how the benefit of the higher κ overlayer comes from increasing the barrier to vortex penetration from layered κ structures (Section 9.3.2). The improved performance of 120 °C N-infused cavities could similarly come from this effect.

At TRIUMF and PSI, μSR studies in the presence of local magnetic field [259, 376] provide encouraging data for the SS' multilayer approach. The technique, introduced in Section 3.5.1 and Section 6.4.3, is a sensitive probe to detect a local magnetic field, and can thus be used to detect the transition from the Meissner to the vortex state. The field for first entry increased from H_{c1} for a "clean" Nb sample to 10% above H_{c1} for a 120 °C baked Nb sample, consistent with the benefits of baking from higher κ overlayer. MgB_2 and Nb_3Sn layers on top of Nb push the field of first penetration to above H_{c1} and to a value consistent with H_{sh} of Nb, independent of the layer thickness, as shown in Figure 9.13 [428, 429]. The field of first flux penetration is enhanced by about 30% for the Nb_3Sn overlayer, and by 40% for the MgB_2 overlayer. Coatings thicknesses between 50 and 2000 nm have been tested. The Nb_3Sn coatings for the μSR studies were produced by vapor diffusion at Cornell University. The MgB_2 coatings were carried out at Temple University HPCVD technique.

Another possible reason for the improvement seen is the impact of the proximity effect from a high T_c film to smooth out the depressed order parameter near defects, and so to raise the field of first penetration [428]. With a simple 1-D model of a thin normal layer (to simulate the presence of defects) a superconducting overlayer can increase the field of first penetration by 10% for normal layer thickness of 20 nm. Thus the proximity effect can recover the stability of the superheating field, for small defects. Point contact tunneling experiments [165] show that the proximity effect between two superconductors can extend over tens of nm (Section 6.13).

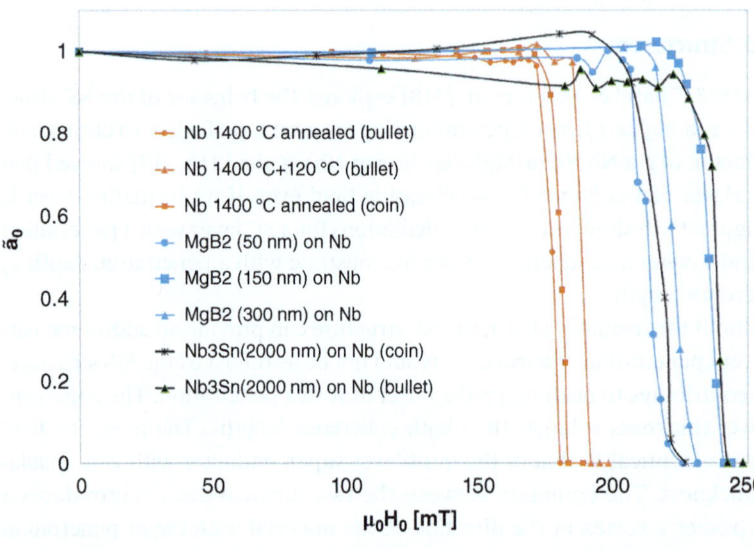

Figure 9.13 The volume fraction probed by the muons, which is in the field-free Meissner state as a function of applied field for various samples. Source: [428]/Junginger et al., TRIUMF with permission of IOP Publishing.

9.4 Summary

The initial results from multilayer studies on samples are very encouraging, but transferring the techniques to a cavity surface is still far away. The phase space for new development paths in the categories of new materials and overlayers is quite extensive. We can expect that the enthusiasm of the proponents in each area will continue to drive efforts in their favorite areas. The danger is that the overall effort becomes too dilute, and results with cavities remain always far away.

To summarize the efforts to push beyond the capabilities of Nb, many institutions around the world that are devoting significant efforts to explore thin films and superconductors with higher T_c are encountering vast challenges. Harder technical challenges will need to be overcome for practical application to cavities. The road to an accelerating cavity with gradients higher than possible with Nb is likely to be very long. Success could dramatically reduce capital and operating costs for future accelerators.

Part IV

Applications

10

New Cavity Developments

A fundamental cavity mode system is the mainstream application of SRF to accelerators. SRF technology has expanded to deflecting/crab cavities, harmonic RF systems, and SRF photoinjectors, which all require specialized cavity designs with unconventional shapes and other uncommon elements.

10.1 Crab Cavities for LHC High Luminosity

To obtain the highest luminosities, high energy particle colliders populate beams with large numbers of bunches, then cross the beams at angles for acceptable bunch separation. But when bunches cross at an angle there is a geometrical loss of luminosity as there is not a good overlap between colliding bunches. A crab-crossing scheme for head-on bunch collision becomes necessary to correct for the geometric loss (Figure 10.1). A crab cavity that provides a transverse kick to the beam rotates the bunches. After collision, a second crab cavity brings the bunches back to their original orientation for continued passage through the beam line. Palmer [430] in 1988 first proposed the crabbing concept to recover luminosity. KEKB [431–434] implemented the first SRF crabbing system. Crabbing will be used in Large Hadron Collider (LHC), in electron-ion colliders such as Relativistic Heavy Ion Collider (RHIC), and in the future for the ILC, as well as for short pulse X-rays at Argonne National Lab (ANL).

Deflecting and crabbing cavities fall into two categories: TM_{110}-type or TEM-type/TE_{11}-like cavities based on their electromagnetic field profiles. A transverse E field, a gradient of the axial E field, or a transverse B field, via $v \times B$, provide the bunch tilt. Some applications use both E and B fields.

In designs using a TM_{11}-like mode (Figure 10.2a), the main kick comes from the transverse B field. In a cylindrical-shaped cavity, the TM_{110} mode is degenerate, with two polarizations. A squashed geometry separates the differently polarized modes. The squashed-elliptical cavity has a lower order mode (LOM), which is the fundamental TM_{010} monopole, accelerating mode. In pure, cylindrical-shaped TE-type cavities (Figure 10.2b) the net deflection from the perpendicular electric field is completely canceled by the net deflection from the magnetic field as a consequence

Superconducting Radiofrequency Technology for Accelerators: State of the Art and Emerging Trends, First Edition. Hasan Padamsee.

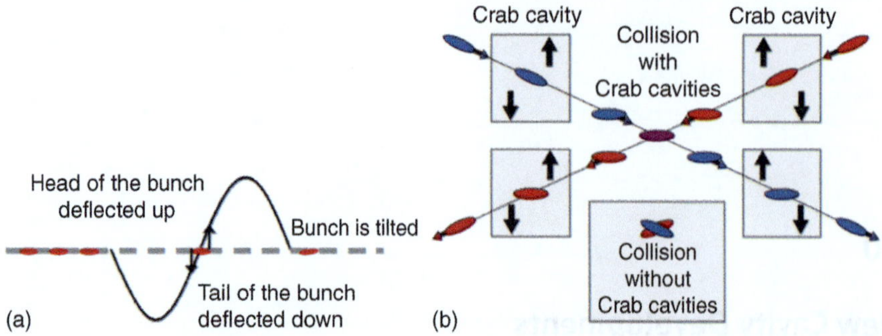

Figure 10.1 (a) At the zero crossing of the B field in the cavity, the RF kicks the tail of the bunch down and the head of the bunch up so as to tilt the bunch for the crab crossing. (b) After collision, the second crab cavity tilts the bunch back with opposite kicks. Source: S. DeSilva, Ohio Dominion University. [435]/JACoW/CC BY 3.0.

of Panofsky–Wenzel theorem [436]. It is necessary to introduce a deformation to create an imbalance, such as a ridge in the TE_{11} waveguide (Figure 10.2c)

Coaxial geometries support transverse electro-magnetic (TEM) modes. When the plates at the two ends are grounded, a standing wave occurs with vanishing E on the end plates. In this TEM-type, or TE_{11}-like crab cavity, the kick comes from both E and B, but the primary contribution is from the transverse E field. TEM-type/TE_{11}-like cavities are compact designs favorable for low-frequency operation. These designs also have high shunt impedance and low surface peak surface field ratios. The QWR is a coaxial device that operates in the TEM mode and provides a compact solution for acceleration or deflection of particle beams. In a coaxial line, the TEM mode has a radial electric field, and an azimuthal magnetic field (Figure 10.3a). The QWR provides a deflecting voltage when the particle beam traverses the cavity, across the orthogonal E field lines. The single QWR also provides an unwanted, residual accelerating voltage. In the double quarter wave resonator (Figure 10.3b), two quarter wave cavities face each other to eliminate the accelerating voltage by symmetry, and also provide a transverse electric field for crabbing. Figure 10.3c shows the E fields for the TEM mode in a four-rod crab cavity design, which is equivalent to four quarter wave cavities, facing each other two by two. The transverse momentum obtained depends on the gradient of the longitudinal electric field along the beam axis, as in the case of the TM_{110} mode.

KEKB is the only facility that has implemented full crab crossing to date. KEKB adopted the squashed cell shape for the pioneering KEKB crab cavity operating in the TM_{110} mode at a frequency of 509 MHz, as shown in Figure 10.4. The cavity is elongated in one direction to separate the crab mode from its degenerate partner, by pushing the degenerate mode frequency up to 700 MHz. The squashed cell shape cavity was studied extensively at Cornell in 1991 and 1992 for CESR-B under a KEK-Cornell collaboration [441, 442]. The cavity delivered 0.8–1.6 MV crabbing voltage. Both HOMs and LOMs were successfully damped.

The crab cavities at KEK [443] successfully delivered horizontal RF kicks to both beams to restore head-on collisions. The tilt of the bunches was recorded by streak

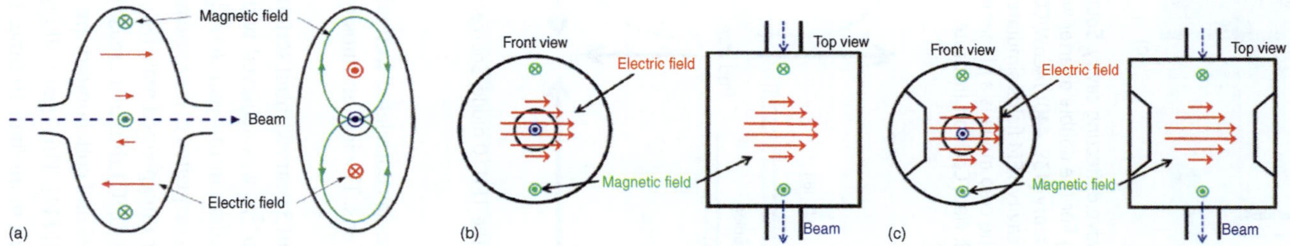

Figure 10.2 Electric and magnetic fields for (a) the TM_{110} mode cavity. Here the kick comes from the gradient of E field and the B fields on the beam axis. The crab mode is a higher order mode with frequency higher than the fundamental TM_{010} accelerating mode. Source: S. DeSilva, Ohio Dominion University. [435]/JACoW/CC BY 3.0. (b) For a pure TE_{11} mode cavity, the kick from the E and B fields cancel. Source: S. DeSilva, Ohio Dominion University. [437] Courtesy of S. DeSilva, Ohio Dominion University. (c) With a ridge-like distortion of the TEM-type waveguide, the fields are TE_{11}-like, but there a net kick, and the deflecting mode is fundamental. This is the basis of the RF-dipole crab cavity. Source: S. DeSilva, Ohio Dominion University. [435]/JACoW/CC BY 3.0.

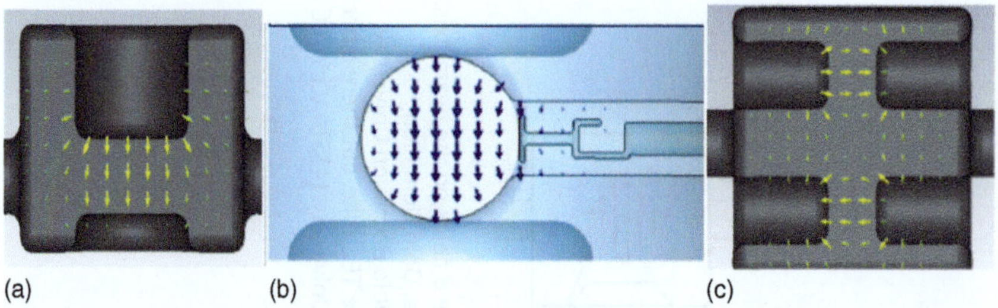

Figure 10.3 Electric fields for (a) Quarter wave deflecting cavity. Source: B. Hall, Cockroft Institute, Thesis [438] Courtesy of B. Hall. (b) For the double quarter wave cavity. Source: S. Verdú-Andrés, Brookhaven National Laboratory [439, 440]/JACoW/CC BY 3.0. (c) TEM 4-Rod or 4 QWR resonators. The 4-Rod crab cavity TEM fields simulate the TM_{110} fields. There is a lower order fundamental mode. The crab mode is a higher mode. Source: B. Hall, Cockroft Institute, Thesis [438] Courtesy of B. Hall, Cockroft Institute.

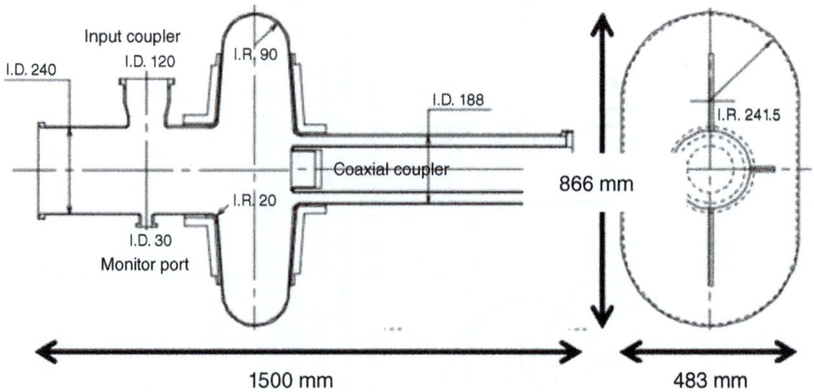

Figure 10.4 KEK-B crab cavity operates in the TM110 mode. Source: K. Hosoyama et al./KEK [432]/JACoW/CC BY 3.0.

cameras located in both rings; the tilt angles correlated well with the RF phase of the cavities, in accordance with simulations. The lost luminosity at low current was regained.

The crab cavities at KEK operated about 3 years without serious problems. A maximum peak luminosity of 21.1×10^{33} cm^{-2}/s was achieved with the crab cavity in operation, which set the record for colliders in the year 2009 [443]. However, the measured beam–beam tune shift of 0.088 (expected to increase with luminosity) was lower than the expected value of 0.15 for unexplained reasons.

The LHC high luminosity upgrade (HL-LHC) aims to increase the peak luminosity by a factor of 2.5 to 5×10^{34} cm^{-2}/s. CERN and collaborators are developing the crab cavities to be installed in LHC by 2025 [444]. This goal will enable the integrated luminosity over the twelve years to reach more than 10 times that during the first ten years of the LHC life. Studies show that 65% of the increase will come from the contribution of crab crossing at 400 MHz, with compensation of a crossing angle of 0.59 mrad [445].

Each interaction point (IP) for the two detectors at the LHC will need both types: crab cavities to induce the rotations, and "anti-crab" cavities to remove the tilt after collision. There will be a total of 16 cavities in 8 cryomodules. Beams cross horizontally at one IP, and vertically at the other IP. Due to the nature of the IP's, the beam pipes are very close together - situated less than 20 cm apart. Each beam pipe has an inner radius of 42 mm. This requires the crab cavities to have a maximum outer radius of about 152 mm. For a 400 MHz, TM110 KEKB-like elliptical cavity, the expected outer radius of 375 mm would be unacceptable. The total crabbing voltage required is 10 MV per beam per side to deflect on one side and cancel the tilt on the other side. Precise control of voltage and phase is necessary. High crabbing voltage and limited space along the beam pipe require SRF technology.

The collaboration developed three crab cavity designs based on [438, 446, 447] the double quarter-wave resonator (DQW) [448], the RF dipole cavity (RFD) [449], and the four-rod cavity (4RCC) [450]. All cavities are at least three times smaller than an elliptical crab cavity at 400 MHz. Prototyping the three designs successfully

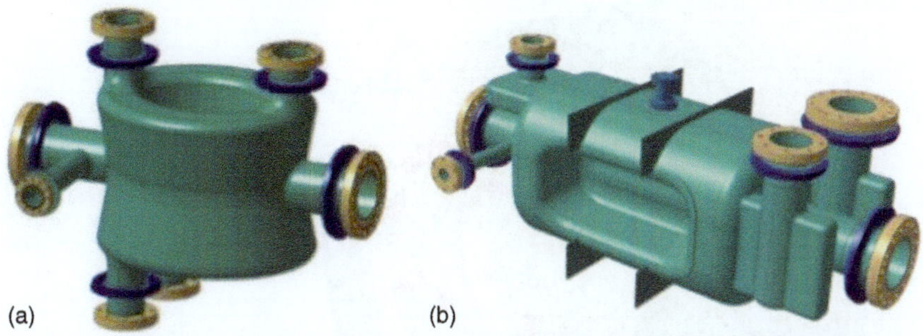

(a) (b)

Figure 10.5 (a) DQW cavity with coupling ports for vertical crabbing and (b) the RF dipole cavity with coupling ports for horizontal crabbing. Source: F. Carra et al. [451]/JACoW/CC by 3.0.

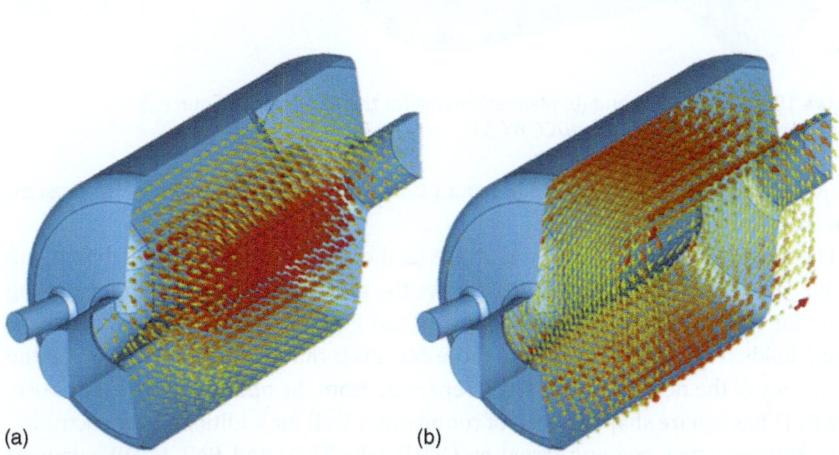

(a) (b)

Figure 10.6 (a) Electric and (b) Magnetic fields for the RFD cavity. Source: S. DeSilva. ODU [452]/JACoW/CC BY 3.0.

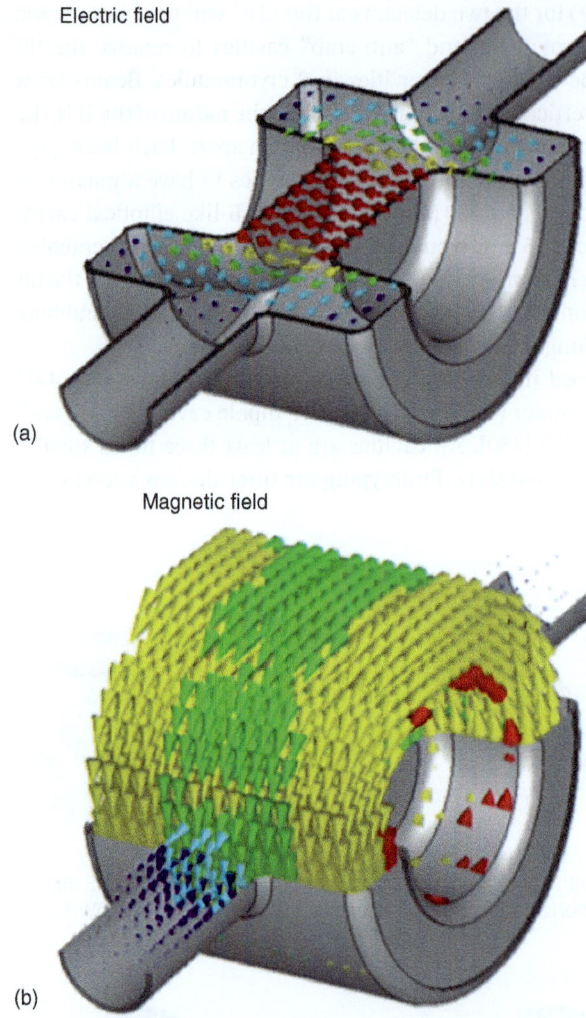

Figure 10.7 (a) Electric and (b) Magnetic fields for the DQW cavity. Source: S. Verdú-Andrés, BNL [453]/JACoW/CC BY 3.0.

demonstrated a kick voltage of 3 MV per cavity. Multipacting barriers were present, but easily processed.

The final technology choice was made as the RFD for horizontal crabbing and as the DQW for vertical. Figure 10.5 shows the two final geometries with coupling ports for fundamental and HOM couplers, and Figures 10.6 and 10.7 shows the E and B fields. The main advantage of these designs is that there are no LOMs and the frequency of the nearest HOM is well separated from the operating frequency. Also, the RFD has square shape instead of round outer wall for additional compactness.

Prototypes developed and tested by ODU/Jlab (RFD) and BNL (DQW) demonstrated the design performance. 4 DQW and 2 RFD cavities are prepared by BCP and

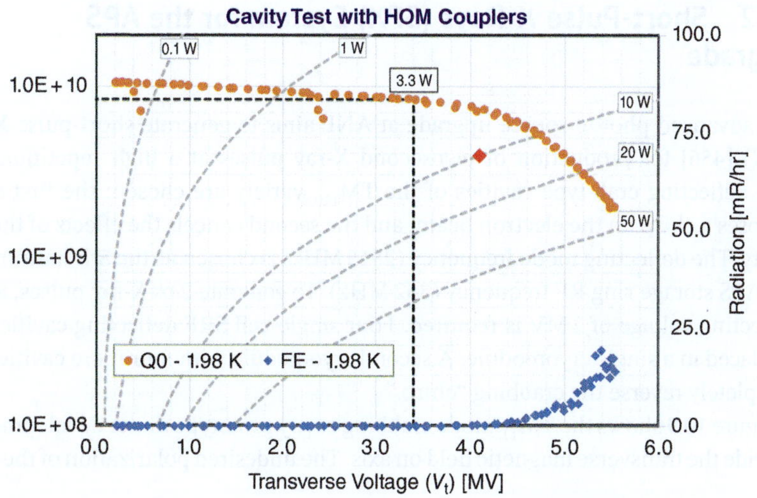

(a)

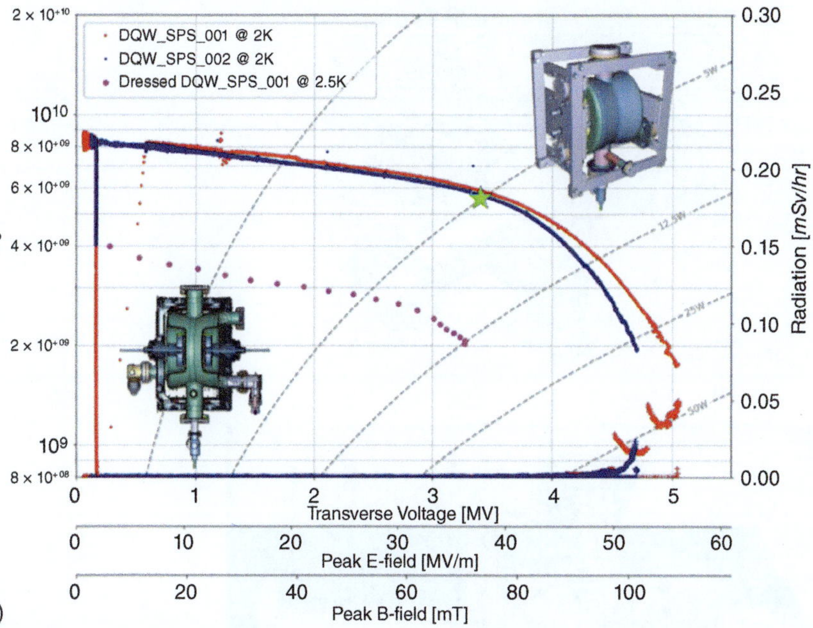

(b)

Figure 10.8 (a) RFD cavity test at JLAB. Source: S. DeSilva [455]/JACoW/CC BY 3.0. (b) DQW cavity test at CERN. Multipactor barriers were observed at low fields and successfully processed. Source: S. De Silva [455]/Old Dominion University/JACoW/CC BY 3.0.

tested to reach kick values of 4.4–5.8 MV as compared to design for 3.4 MV (corresponding $E_{pk} < 40$ MV/m and $H_{pk} < 70$ mT). The dynamic heat load expected is five watt per cavity. Figure 10.8 shows test results for both types. The cavity production effort for the RFD is pursued by Fermilab under the *LHC* accelerator research program (LARP collaboration), and by CERN for the DQW [454].

10.2 Short-Pulse X-Rays (SPX) System for the APS Upgrade

The advanced photon source upgrade at ANL aims to generate short-pulse X-rays (SPX) [456] for production of picosecond X-ray pulses at a high repetition rate. Two deflecting crab-type cavities of the TM_{110} variety are chosen: the first cavity imposes a chirp on the electron beam, and the second cancels the effects of the first cavity. The deflecting mode frequency (2815 MHz) is chosen as the 8th harmonic of the APS storage ring RF frequency (352 MHz). To generate 2 ps X-ray pulses, a total deflecting voltage of 2 MV is required. Four single-cell SRF deflecting cavities will be placed in a single cryomodule. A second cryomodule with four more cavities will completely reverse the crabbing "chirp."

Figure 10.9 shows the TM_{11} mode crabbing cavity prototype under development to provide the transverse magnetic field on axis. The undesired polarization of the TM_{11}

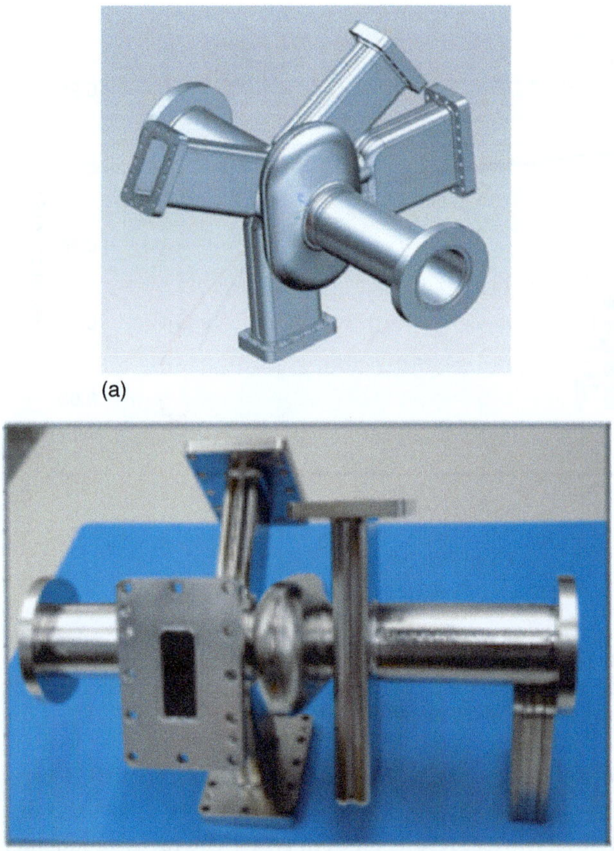

(a)

(b)

Figure 10.9 SPX deflecting cavity design (a) 3D Schematic (b) Photograph of prototype. Source: A. Nassiri et al. [459]/JACoW/CC by 3.0.

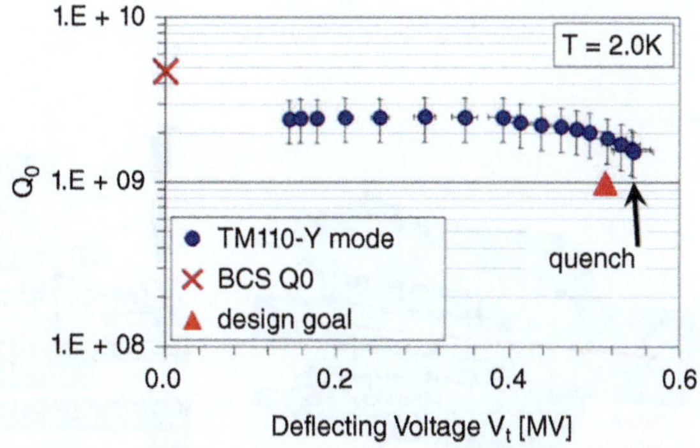

Figure 10.10 Performance of prototype deflecting cavity for SPX. Source: J. Fuerst and A. Nassiri, ANL [458, 459]/ JACoW/CC BY 3.0.

mode is split by polarizing the cavity. A LOM coupler removes the beam-induced power from the fundamental mode. Strong damping of parasitic HOM is necessary to avoid multibunch instabilities. Waveguides will be used for coupling HOMs out from the cells. A Y-shaped waveguide end group has been adopted, similar to that developed at Jefferson lab for high-current cavities [457]. The LOM is damped by a waveguide coupler on the beam pipe (Mark I). ANL and Jefferson Laboratory (JLAB) are collaborating on the development. Vertical cavity tests on the prototype cavity reached 120 mT surface field at $Q = 10^9$, exceeding the design requirements (Figure 10.10) [458].

10.3 QWR Cavity for Acceleration

Institute of High Energy Physics (IHEP) needs a very low frequency (166 MHz) cavity for acceleration to be used in the 6 GeV storage ring HEPS in China (Section 11.5.1). The low-frequency choice is driven by the long bunch length (30 mm) and the limited space of 6 m for two cavities. They have chosen the unusual QWR design for acceleration, driven by the desire to develop a compact cavity. The QWR is usually adopted for low-β accelerator applications, but here the beam goes through the QWR via the center conductor. The electric field from the center conductor to the outer wall is the accelerating field. Figure 10.11 shows the concept of the cavity with the beam direction, and Figure 10.12 shows the electric and magnetic fields for the accelerating mode [460, 461]. The peak surface fields are $E_{pk}/E_{acc} = 3.2$ and $H_{pk}/E_{acc} = 5.1$ mT/MV/m.

The first cavity did very well in the vertical test reaching surface fields $E_{pk} = 82$ MV/m and $H_{pk} = 130$ mT, to provide 3 MV acceleration compared to the design specification of 1.5 MV. Multipacting barriers proved easy to process. After dressing the cavity with He vessel, couplers, and tuners, the horizontal test showed some degradation due to overheating at the fundamental power coupler port.

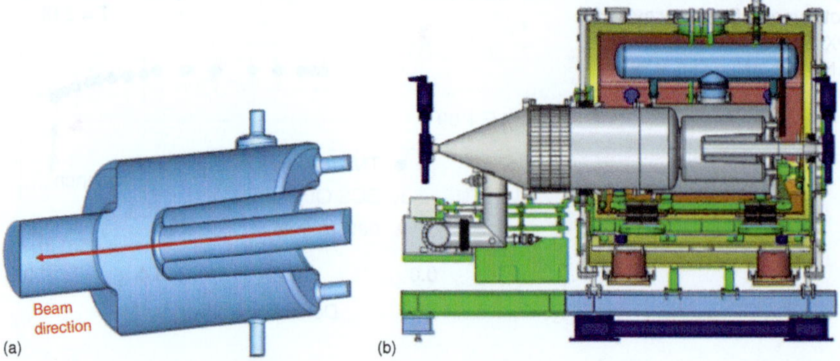

(a) (b)

Figure 10.11 (a) The QWR as an accelerating cavity for high beta electrons in a storage ring [461]. (b) Schematic for the cavity in its cryomodule and the beam pipe HOM damper. Source: P. Zhang [461]/JACoW/CC BY 3.0.

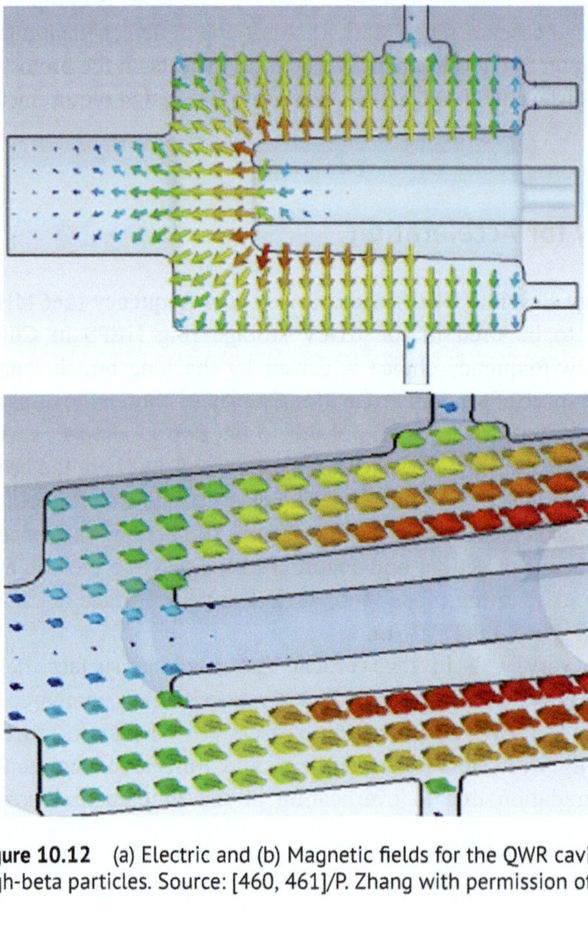

(a) (b)

Figure 10.12 (a) Electric and (b) Magnetic fields for the QWR cavity used for accelerating high-beta particles. Source: [460, 461]/P. Zhang with permission of Feisi He.

10.4 Traveling Wave Structure Development

With standing wave (SW) SRF structures approaching the theoretical critical field limit of H_{sh}, and with alternative materials to niobium remote from maturity, another approach to higher accelerating gradients is to shift from SW to traveling wave (TW) structures. TW structures offer two main advantages: substantially lower peak magnetic and peak electric field ratios, and substantially higher (x2) R/Q for lowering cryogenic losses. Superconducting TW structures with a small phase advance per cell (Figure 10.13) would significantly increase (up to 48%) accelerating fields [463–465]. Considering the best accelerating field for a TESLA single-cell cavity is about 50 MV/m, we can optimistically anticipate near 70 MV/m accelerating field after necessary developments to reach best performance. With the Geometry Factor ($G = QR_s$) of 1.86, the cryogenic losses will be lower by 1.45, when compared to the TESLA structure. Figure 10.13 shows geometry parameters for a half-cell, and Table 10.1 shows the calculated RF properties [466].

Figure 10.14 compares a one-meter TW structure to a TESLA 9-cell cavity. The TW structure requires a Nb feedback waveguide for redirecting power from the end of the structure back to the front to avoid high peak surface fields in the accelerating cells that would form with a standing wave. This requires careful tuning to compensate for reflections along the TW ring to obtain a pure traveling wave regime at the desired frequency.

Many significant challenges remain for development of practical TW niobium structures. There is increased complexity due to doubling the number of cells (e.g. more welds). The cavity fabrication and surface processing fixtures and procedures must be modified and qualified. The fabrication and processing will also be more

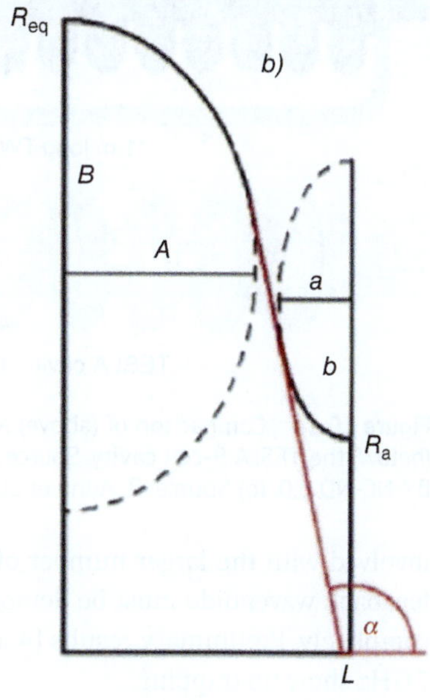

Figure 10.13 Parameters defining the cell shape. Source: V. Shemelin et al., Cornell University [466]/APS CC BY 4.0.

Table 10.1 Parameters of optimized cells with limiting surface fields: $E^*_{pk} = 120$ MV/m and $B^*_{pk} = 200$ mT; $L - A = 5$ mm, aperture radius $R_a = 25$ mm.

Optimization	**120/200**
Phase advance θ (deg)	90
A (mm)	23.826
B (mm)	36.4
a (mm)	4.512
b (mm)	7.52
E_{pk}/E_{acc}	1.727
B_{pk}/E_{acc} (mT/[MV/m])	2.878
R_{sh}/Q (Ohm/m)	2127
α (degrees)	90.91
R_{eq} (mm)	98.950
v_{gr}/c	0.01831
E^*_{acc} (MV/m)	69.5
$E^*_{acc} \times 2L$ (MV)	4.00

E^*_{acc} is the accelerating rate when the limiting surface fields are achieved. The cell shape parameters are defined in figure.
Source: V. Shemelin et al., Cornell University [466]/Americal Physical Society CC BY 4.0.

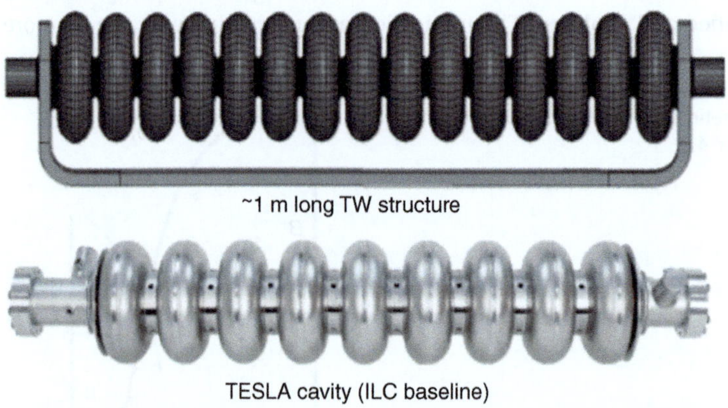

Figure 10.14 Comparison of (above) a one-meter long TW structure [467, 468] with (below) the TESLA 9-cell cavity. Source: (a) R. Kostin et al. [468]/IOP Publishing/CC BY-NC-ND 3.0. (b) Source: B. Aune et al. [25]/American Physical Society/CC BY 3.0.

involved with the larger number of cells per meter. High circulating power in the feedback waveguide must be demonstrated. HOM damping must be studied more completely. Preliminary results [466] show that the first 10 monopole modes up to 7 GHz show no trapping.

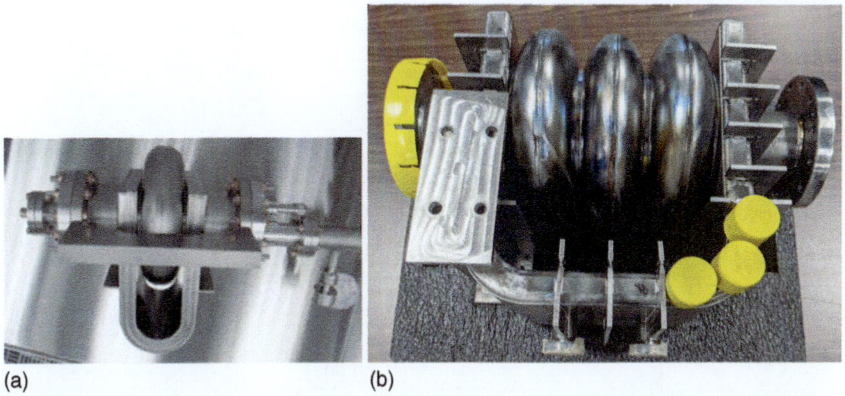

Figure 10.15 (a) 1-cell [468] and 3-cell TW structures out of Nb [468]. The 1-cell has received the first test to reach 31 MV/m. Source: R. Kostin et al. [468]/IOP Publishing/CC BY-NC-ND 3.0.

The TW approach is in early development stages. A single-cell prototype has been built and tested [465] (Figure 10.15a). It reached a gradient equivalent to 31 MV/m in a TESLA-shaped cavity. A three-cell structure has been fabricated and is in preparation for testing (Figure 10.15b).

11

Ongoing Applications

11.1 Overview

Steady advances in SRF performance have enabled, and continue to enable, a large variety of SRF-based accelerators for applications in materials science, nuclear physics, nuclear astrophysics, and high-energy physics. Accordingly, this chapter will discuss the status of major ongoing applications for both high-beta and low-beta accelerators. The updates naturally capture the state of the art for SRF performance for the corresponding accelerator structures as of January 2022 TESLA Technology Collaboration Meeting. No doubt progress will continue to be reported in subsequent conferences. The following Chapter 12 will summarize plans for far-future accelerator projects.

As evidenced by the wide range, SRF applications have indeed become ubiquitous. The total installed voltage has risen from 7 GeV in the year 2000 to 25 GeV installed and operating by 2020. In another decade the total is expected to rise to 40 GeV (or higher) from approved projects.

It is remarkable to witness the rapid expansion of SRF technology and accelerator applications from the Europe/Americas regions to Asia, and especially to China where the SRF community is expanding rapidly. In China, 3 facilities are in operation, 4 projects are running, and 3 projects are in proposal stages. Numerous institutions are engaged in SRF research and numerous SRF facilities are well underway, as reviewed in [469, 470].

The accounts in this chapter are organized into Section 11.2 Low-beta accelerators for heavy ions for nuclear physics, nuclear astrophysics, rare isotope beams (RIB), and medical isotope production; Section 11.3: High-intensity proton accelerators for neutron sources, high-intensity neutrino sources, transmutation of nuclear waste, generation of nuclear power, Section 11.4: Electron linacs for light sources for materials science, and for biological applications, Section 11.5 Electron storage rings for light sources, Section 11.6 Electrons for energy recovery linacs for light sources and for electron–ion colliders for high-energy physics; Section 11.7 Electrons for nuclear physics, nuclear astrophysics, and radio-isotope production; Section 11.8 Crab cavities for high luminosity. The chapter concludes with a summary of ongoing SRF applications with the expected number of cavities and cryomodules.

Superconducting Radiofrequency Technology for Accelerators: State of the Art and Emerging Trends, First Edition. Hasan Padamsee.

11.2 Low-Beta Accelerators for Nuclear Science and Nuclear Astrophysics

Superconducting linacs providing precision beams of heavy ions have consistently been one of the most successful SRF applications. Heavy ions, from helium to uranium, are accelerated to energies from a few to 20 MeV/nucleon and used to bombard other nuclei. Above 5 MeV/nucleon, ions have sufficient energy to overcome the Coulomb barrier and penetrate the nucleus. The collisions cause energy, mass, and angular momentum to be transferred between the projectile and target nuclei, enabling structure research on the evolution of nuclear shape as a function of excitation energy and other aspects, such as spin.

Heavy-ion and proton accelerators must efficiently accelerate particles whose velocity changes along the accelerator. They must also be able to accelerate a variety of ions with different velocity profiles. Several structure geometries are therefore needed, each of which must be optimized for a particular velocity range. A major advantage of superconducting resonators is that a high cw voltage can be obtained in a short structure. The LINAC to boost ion energies can therefore be formed as an array of independently-phased resonators, making it possible to vary the velocity profile of the machine. The superconducting booster is capable of accelerating a variety of ion species and charge states.

The last 20 years have seen a renaissance in low and medium-β SRF cavity development for RIB post-accelerators (e.g. ISAC-II, ReA [NSCL], HIE-Isolde), RIB drivers (SPIRAL-II, FRIB, and RAON). Impressive progress in performance of TEM cavities continues to reach higher values of E_{pk} and H_{pk}. The best 72 MHz QWR cavity performance reported [471] showed $E_{pk} = 117$ MV/m and $H_{pk} = 165$ mT, beating the best *Epk* values reached for 9-cell TESLA cavities $E_{pk} = 90$ MV/m, and single cells of $E_{pk} = 100$ MV/m. The record QWR would be able to provide an accelerating voltage gain of 7.4 MV.

11.2.1 ATLAS at Argonne

At Argonne, ATLAS has been operating for more than 35 years as a national user facility for heavy-ion nuclear and atomic physics research [6]. A series of 64 superconducting resonators span β values between 0.009 and 0.037. ECR ion sources have provided beams of virtually all stable isotopes at ever-increasing intensities.

More recently, ATLAS upgraded its energy with a single 5-m-long cryomodule containing seven 109 MHz $\beta = 0.15$ QWR resonators at 7.6 MV/m [472] providing nearly 3 MV per cavity (Figure 11.1). Improvements continue with 72 MHz $\beta = 0.077$ quarter-wave cavities [473]. The maximum cw accelerating gradient is 13.7 MV/m at 4 K and 15.5 MV/m at 2 K (limited by available X-ray shielding). High performance resulted from cleanroom assembly, separate cavity and insulating vacuum spaces, low peak surface fields design optimized geometry and electropolishing cavity subcomponents.

Figure 11.1 Completed ATLAS upgrade CM string with 72 MHz QWR resonators. Source: S-H Kim, Argonne National Laboratory [474]/Courtesy of S-H Kim.

Besides the pioneer heavy-ion accelerators at Argonne, many facilities have operated, or are still operating. See the review of low-β accelerators [6]. In total, these utilize over 300 resonators. The largest is ALPI at INFN Legnaro which has been in operation since 1994 [475]. In all, 64 QWRs at 160 MHz are in the main beam line delivering a total of 49 MV. Both Nb and Nb–Cu resonators reach 6–8 MV/m at 7 W in vertical tests.

11.2.2 ISAC and ISAC-II at TRIUMF

TRIUMF in Canada is operating an isotope separation-on-line (ISOL) facility-based radioactive beam facility, ISAC [476]. The accelerator consists of 40 QWRs with two β values, 0.057 and 0.071, where each cavity provides approximately 1 MV. To extend the final energy, TRIUMF installed ISAC-II using $\beta = 0.11$ QWRs, which added 20 MV with 20 cavities. These have achieved surface E_{pk} up to 35 MV/m in operation (Figure 11.2).

11.2.3 SPIRAL II at GANIL

SPIRAL-II at the Grand Accélérateur National d'Ions Lourds (GANIL) in Grenoble, France is a high power, superconducting linac [477, 478], which delivers a high-intensity, 40 MeV deuteron beam as well as a variety of heavy-ion beams with mass-to-charge ratio of 3 and energy up to 14.5 MeV/nucleon. Acceleration is based on twenty-six 88 MHz, QWRs, divided into two families ($\beta = 0.07$ and $\beta = 0.12$) to operate at 6.5 MV/m and 4.2 K. The low-β cavities and cryomodules (with single resonators) were developed at CEA-Saclay, while the high-β ones (two resonators per module) at IPN-Orsay. During vertical tests, cavities achieved gradients above 10 MV/m with $Q_0 \sim 1 \times 10^9$ at 4.2 K, well above design. The LINAC has been operating since November 2017 (Figure 11.3).

(a) (b)

Figure 11.2 The ISAC-II (a) QWR cavity and (b) Cryomodule. Source: R. Laxdal, TRIUMF [8]/JACoW/CC BY 3.0.

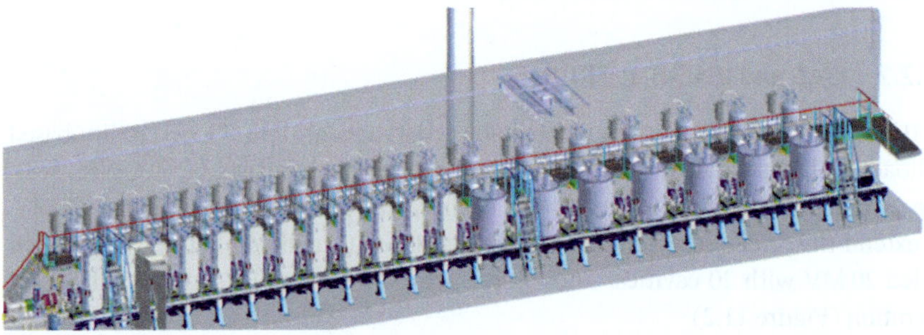

Figure 11.3 The SPIRAL-II cryomodule with QWRs. from TTC 2021. Source: A. Ghribi, GANIL [478]/Courtesy of A. Ghribi.

11.2.4 HIE ISOLDE

The HIE-ISOLDE linac serves a large physics community at CERN [479, 480]. It uses Nb/Cu technology for QWRs prepared by dc magnetron sputtering at about 500 °C (Section 9.1.2). More than 20 cavities were coated and tested. Good substrate quality is essential for good performance. 4 cryomodules with 5 cavities each have been assembled, installed, and commissioned in the linac.

Five cavities are running at $V_{acc} = 1.8$ MV each (5–6 MV/m) at 4.5 K requiring ~100 W RF power for ~1.8 MV V_{acc}. The cavity dissipation in the linac is less than 10 W at 6 MV/m. Field emission can be processed online. There is no interference from the magnetic fringe fields of the neighboring superconducting solenoids.

For good Q_0, it is important to have homogeneous cooldown across T_c with minimum temperature differentials to avoid thermocurrent-induced magnetic flux trapping arising from composite material junctions (Section 9.1.2). Better results were obtained with sputtering onto a seamless copper substrate cavity (Section 9.1.3).

11.2.5 RILAC at RIKEN

The RIKEN heavy ion linac is upgraded to RILAC [481–483] to provide intense heavy-ion beams to continue super heavy element search beyond Nh-Nihonium – atomic number 113). Nihonium was discovered at RILAC. Ten QWRs (73 MHz) in 3 CM are operating at 4.2 K. The design gradient of 6.8 MV/m has been exceeded. Ions with A/Q = 6 are accelerated to 6.5 MeV/U. Operations began in early 2020. An argon beam was accelerated to 6.2 MeV/U.

11.2.6 SPES Upgrade of ALPI at INFN

An upgrade of the long-operating ALPI facility, the selective production of exotic species (SPES) at ALPI (LNL-INFN) will be a second generation ISOL facility for nuclear physics [484]. Exotic beams of rare isotopes will be produced and reaccelerated with the ALPI linac along with research and production of radioisotopes for nuclear medicine: Two new cryostats with 8 QWRs of NbCu are planned to be installed. The best two performing cavities reached 5.5 and 6.6 MV/m with Q values $3–4 \times 10^8$ (Figure 11.4).

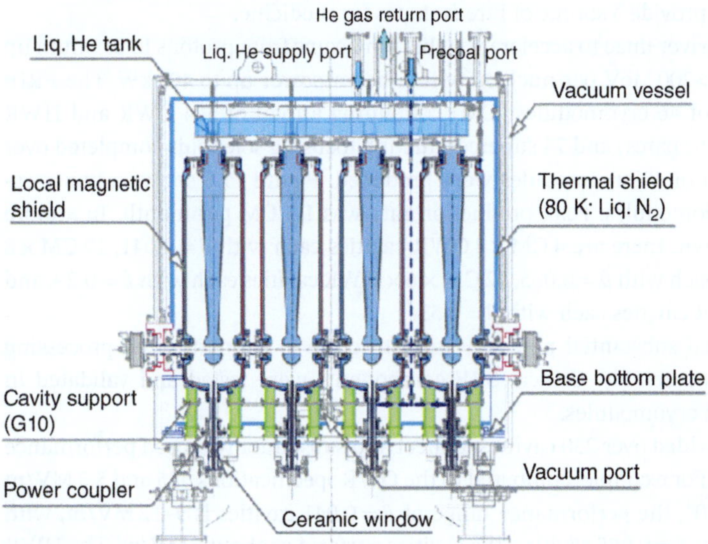

Figure 11.4 (a) Schematic for the four-cavity CM for RILAC at RIKEN. Source: N. Sakamoto, RIKEN [482, 485]/JACoW/CC BY 3.0.

Figure 11.5 FRIB linac installation segment with 15 cryomodules. Source: J. Wei et al., FRIB, MSU [489]/JACoW/CC BY 3.0.

11.2.7 FRIB at MSU

Based on the wide success of low-β SRF cavities for heavy-ion accelerators, a major new Facility for Rare Isotope Beams (FRIB) is now complete at Michigan State University (MSU) for the study of nuclear structure, nuclear reactions, and nuclear astrophysics. FRIB is the world's largest low-β SRF linac with user experiments starting in 2022 [486–490].

Experiments with the new isotopes will lead to a comprehensive description of nuclei, elucidate the origin of the elements in the cosmos, and provide an understanding of matter in the crust of neutron stars, among other salient physics topics. FRIB will also provide a source of rare isotopes for medicine.

FRIB has a driver linac to accelerate stable ion beams from protons to uranium up to energies of >200 MeV per nucleon with a beam power up to 400 kW. The FRIB linac consists of 46 cryomodules (CM) containing a total of 324 QWR and HWR cavities (plus 16 spares) and 74 superconducting focusing solenoids completed over 6 years. A total of 49 cryomodules were produced. Figure 11.5 shows a linac segment of 15 cryomodules. Peak production rate was 1.5 CM per month. In a more detail breakdown, there are 4 CM × 4 QWR cavities each with $\beta = 0.041$, 12 CM × 8 QWR cavities each with $\beta = 0.085$, 12 CM × 6 × HWR cavities each with $\beta = 0.29$, and 18 CM × 8 HWR cavities each with $\beta = 0.53$.

FRIB-installed substantial new SRF infrastructure to support cavity processing and cold mass assembly. Critical SRF components were tested and validated in pre-production cryomodules.

Industry provided over 230 cavities to meet and substantially exceed performance specifications. For example, compared to the QWR specifications of 5 and 5.5 MV/m at $Q_0 = 1–2 \times 10^9$, the performance range of $\beta = 0.041$ cavities is 9–12 MV/m, with Q_0 near 10^9. For $\beta = 0.085$ cavities, the cavities perform at about 7 MV/m. The HWR specification is 7.5 MV/m and the performance range is 12–15 MV/m with Q_0 near 10^{10} (Figure 11.6).

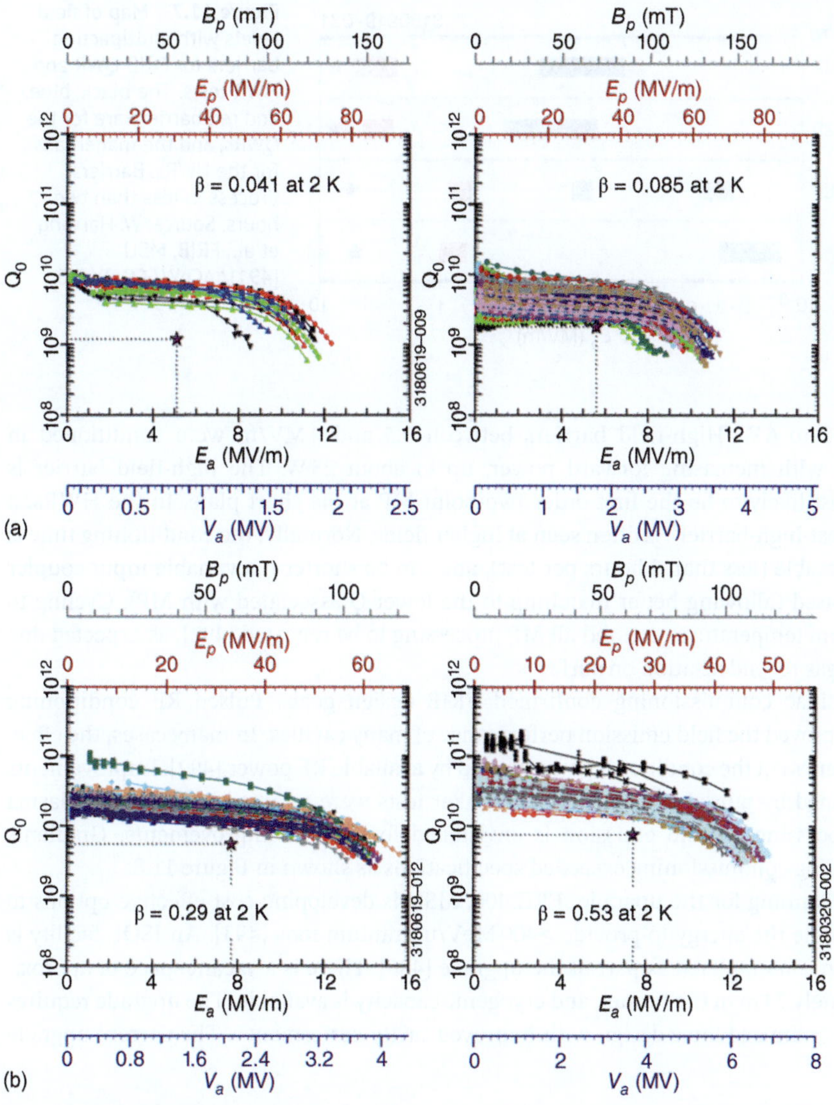

Figure 11.6 Performance of (a) QWR and (b) HWR cavities for FRIB exceed design goals. Ultimate performance is limited by HFQS as expected for BCP-treated cavities. Source: K. Saito, FRIB, MSU [252].

Most of the cavities were limited by the HFQS (Figure 11.6) due to BCP treatment adopted (Sections 6.1 and 6.2). The onset field for HFQS is 70–100 mT, similar to elliptical cavities. As expected, 120 °C bake does not help BCP-treated surfaces to suppress HFQS and reach higher gradients [252]. Most structures exhibited multipacting (MP) barriers [491], which could be processed. QWRs show low-field MP barriers (Figure 11.7), easily processed during filling the cavity with RF power. Mid-field barriers below 0.5 MV/m were conditioned with constant forward power

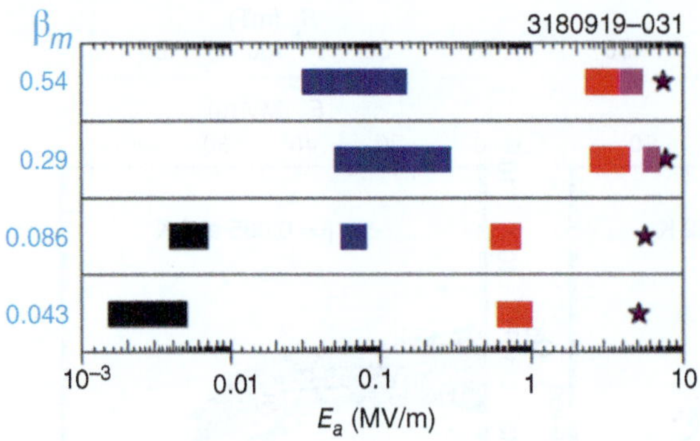

Figure 11.7 Map of field levels with multipacting barriers for FRIB QWR and HWR tests. The black, blue, and red barriers are for the QWRs, and the magenta is for the HWRs. Barriers process in less than two hours. Source: W. Hartung et al., FRIB, MSU [491]/JACoW/ CC BY 3.0.

of 1 to 6 W. High-field barriers between 0.5 and 4 MV/m were conditioned in cw with increasing forward power, up to about 25 W. The high-field barrier is most likely to be the first-order two-point MP at the short plate. In the HWRs, a "post-high-barrier" is often seen at higher fields. Normally, the conditioning time is tolerable (less than 2 hours per test), and can be shorter if a variable input coupler is used (allowing better matching to the lower Q associated with MP). Cycling to room temperature required all MP processing to be repeated [492], as expected due to gas recondensation on surfaces.

Linac commissioning confirmed FRIB design goals. Pulsed RF conditioning improved the field emission performance of many cavities. In many cases, the effectiveness of the conditioning was limited by available RF power [492]. Improvements gained by pulsed conditioning in bunker tests were retained in the linac. Plasma processing of field emission is envisioned for further improvements. Gradients during commissioning exceeded specifications as shown in Figure 11.8.

Planning for the upgrade, FRIB400, MSU is developing cost-effective options to double the energy to provide ≥400 MeV/u uranium ions [493]. An ISOL facility is also contemplated as part of the upgrade [486]. There is a vacant space of approximately 74 m in the tunnel, and cryogenic capacity is available. The upgrade requires an optimized linac design with improved cavity performance. The current upgrade

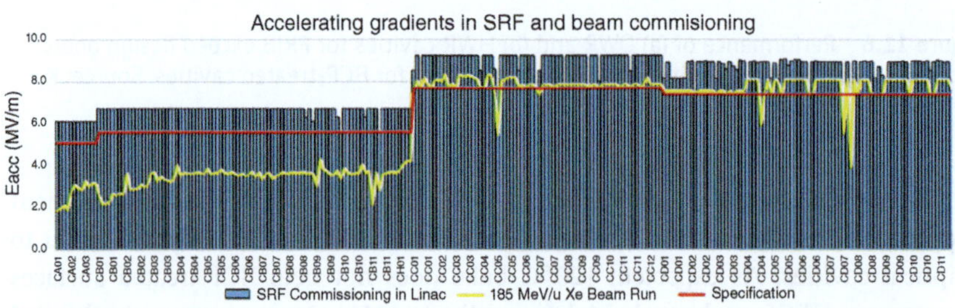

Figure 11.8 During beam commissioning, accelerating gradients (blue bars) exceed specifications (red line). Source: J. Popielarski, FRIB, MSU [492]/Courtesy of J. Popielarski.

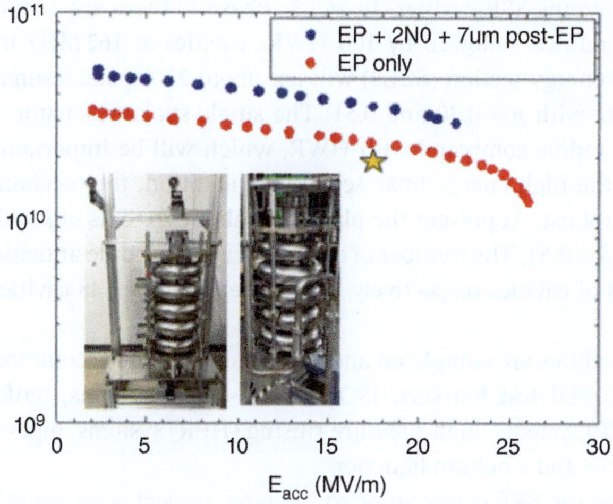

Figure 11.9 Performance of prototype cavities for the FRIB400 upgrade. Source: T. Xu, FRIB, MSU [486]/Courtesy of T. Xu.

scenario would need 11 cryomodules with 5 elliptical cavities each at 644 MHz at a gradient of 17.5 MV/m. The cavities have a design optimum $\beta = 0.64$. Prototype cavities have been built and tested in collaboration with ANL and FNAL to exceed design specifications as shown in Figure 11.9 [486]. In a collaboration with FNAL, N-doping followed by cold EP resulted in 30 MV/m gradient [494, 495]. This is double the SNS medium-β elliptical cavity gradient.

The National Superconducting Cyclotron Laboratory (NSCL) at MSU has reaccelerated beam facilities, ReA3 and ReA6 to provide low-energy rare isotope beams [490]. In ReA 3 the first cryomodule houses two solenoids and one $\beta = 0.041$ cavity. The second cryomodule incorporates six $\beta = 0.041$ cavities and three solenoids and the third cryomodule contains eight $\beta = 0.085$ cavities and three solenoids. They recently completed the ReA6 linac upgrade provides ion beam energies up to 12 MeV/u for $M/q = 2$ ions. The cryomodule contains eight $\beta = 0.085$ 80.5 MHz QWRs and three SC solenoids. All eight cavities operated at 4.3 K, and average 6 MV/m.

11.2.8 RAON

The Institute of Basic Sciences (IBS) in S. Korea has launched RAON [496–500], a large facility for rare isotope beams (Rare Isotope Science Project, RISP) anticipated with a 400 kW superconducting linac. As the ultimate goal, proton and uranium ions will be accelerated up to 600 MeV and 200 MeV/u respectively. The accelerator will operate with a 4.2 kW cryo-system.

The first phase (SCL3) is based on QWRs and HWRs for the low energy sections, and the second phase (SCL2) will be based on SSR (spoke) resonators for the high-energy sections. SCL2 will be installed after success of the first phase and the completion of prototyping SSR cavities. In SCL3 (Phase 1) there are 22 QWRs at 81 MHz in 22 cryomodules, followed by 106 HWRs cavities at 162 MHz in 34 cryomodules. The high-energy section (SCL2) will use about 220 spoke resonators of two types at 325 MHz with $\beta = 0.30$ and 0.51. The single-spoke-resonator type will have a larger bore radius compared with HWR, which will be important for reducing beam loss in the high-energy linac section. In addition, the mechanical design for SSR is more robust. At present the plan is for about 70 SSRs of $\beta = 0.30$, and about 150 SSRs for $\beta = 0.51$. The number of cavities per cryomodule anticipated is 3 and 6 for two types of cavities respectively. In all, there will be 348 cavities in 104 cryomodules.

Two large SRF test facilities are completed and in operation. They accommodate vertical test pits, horizontal test bunkers, ISO class 4 ~ 7 cleanrooms, buffered chemical polishing (BCP) systems, high-pressure rinsing (HPR) systems, high- and low-temperature furnaces and a helium liquefier.

RISP is a newcomer to the SRF community, so the large project is an ambitious undertaking. Partnering with experienced laboratories will be important. RAON is collaborating with Cornell, CEA, GANIL, IHEP, IMP, KEK, TRIUMF, and RIKEN.

In the prototyping stage, the QWR prototype bare cavity demonstrated 11 MV/m (design 6 MV/m), and the dressed cavity QWR reached 8–9 MV/m. The HWR prototype demonstrated 13–14 MV/m in the vertical test (design 9 MV/m). After successful prototyping, all the QWR resonators and modules along with 23/34 HWR modules are complete and installed. Figure 11.10 shows vertical test results for QWR and HWR cavities [500]. The broad scatter in field emission for the HWR cavities has been found to arise from high organic content of the rinsing water.

Figure 11.11 shows HWR cavities on the vertical test stand. Figure 11.12 shows one stage of the QWR linac installation in the RAON linac. The first QWR cryomodule reached design performance of 6.1 MV/m with less than 20 W dissipation. The first HWR Type A cryomodule achieved 6.6 MV/m at 14 W. An oxygen beam was accelerated to 700 keV/nucleon through the QWR module.

The SSR1 design for phase 2 is developed with TRIUMF to be of the balloon shape (Figure 11.13) to avoid mulitpacting [501]. Cold tests on the prototype have demonstrated the principle of the balloon concept. There is no multipacting barrier around the operational gradient (8–10 MV/m). Several barriers, consistent with simulations, do exist in the low field regime (0.5–2.5 MV/m). Low-temperature baking mitigates their intensity.

11.2.9 Spoke Resonator Structure Developments to Avoid Multipacting

TRIUMF in Canada and RISP in S. Korea improved the design for the single-spoke resonator to avoid MP in the operating region. The various orders of MP in the original spoke design Figure 11.13 a are shown in Figure 11.13b. The new spherical SSR1 design is of the balloon shape (Figure 11.13c) and the calculated barrier levels are shown in Figure 11.13d. The prototype is fabricated and tested. The cold tests have demonstrated the principle of the balloon concept. There is no MP barrier around the

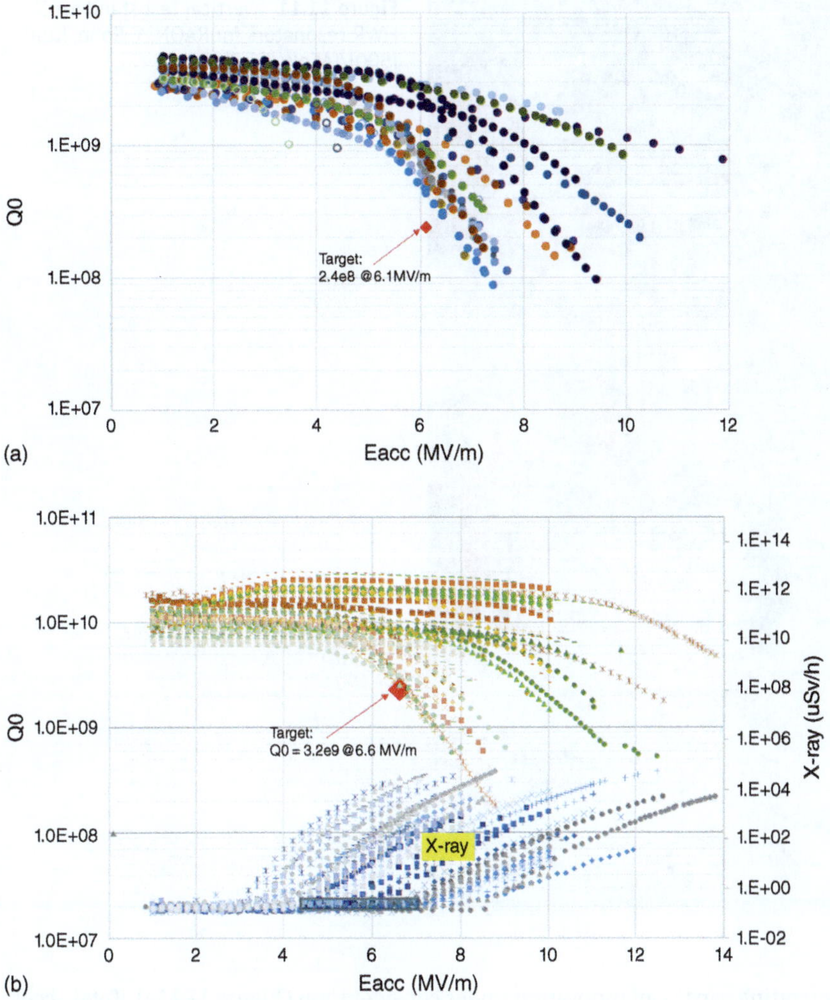

Figure 11.10 (a) Vertical test results for (a) QWR. Source: Y. Sohn, RISP [500]/ and (b) HWR cavities for RAON. Source: Y. Sohn, RISP [500]/JACoW/CC BY 3.0.

operational gradient (8–10 MV/m). Several barriers, consistent with simulations, do show up in the low field regime (0.5–2.5 MV/m). Low-temperature baking mitigates MP intensity. Figure 11.14 compares the MP intensity of the balloon shape with the standard spoke resonator. The cavity performance matches design with MP limited from 0.2 to 1.8 MV/m well away from the operational gradient at 9 MV/m [501].

11.2.10 JAEA Upgrade

The Japan Atomic Energy Agency JAEA is planning to upgrade its tandem booster using 40 130-MHz QWR resonators in 10 cryostats to accelerate up to Au ions to 10 Mev/nucleon, and up to U from 0.5 to 4.5 MeV. A twin drift tube QWR with three

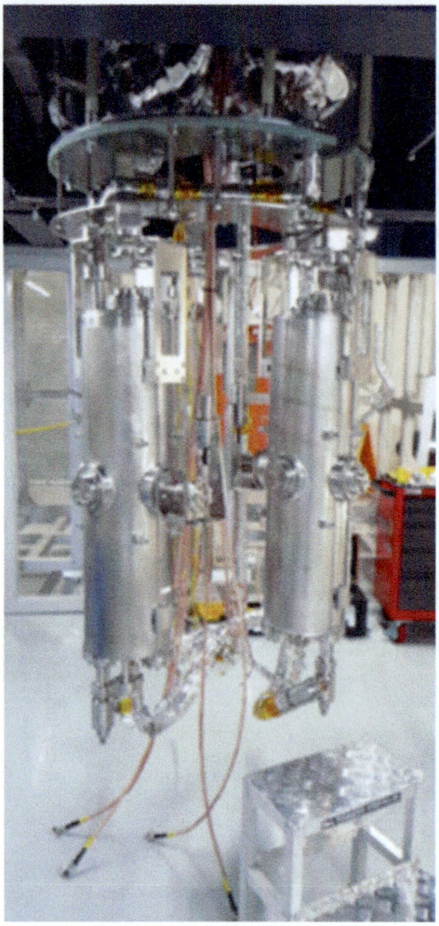

Figure 11.11 Vertical test stand for 3 HWR resonators for RAON. Y. Sohn, RISP [500]/JACoW/CC BY 3.0.

accelerating gaps is an innovation under consideration (Figure 11.15a). Total acceleration voltage will be 30 MV [502, 503].

11.2.11 HELIAC

The superconducting Helmholtz Linear Accelerator (HELIAC) is under construction at GSI in Darmstadt for the FAIR project [505–507]. The cw-mode linac with a final energy of 7.3 MeV/u at a mass-to-charge ratio of $A/q = 6$ and a frequency of 217 MHz. 11 CH (cross-bar, H mode) cavities will be constructed. The CH cavity is a multi-gap cavity for ion acceleration to reduce the number of drift spaces between resonators. The β value of the accelerated ions will change from 0.076 to 0.083 as they go through the structure. Figure 11.15b is one example. Four cryostats will contain three (CH) type cavities.

Figure 11.12 QWR cryomodule installation in the RAON linac. Source: Y. Sohn [500]/JACoW/CC BY 3.0.

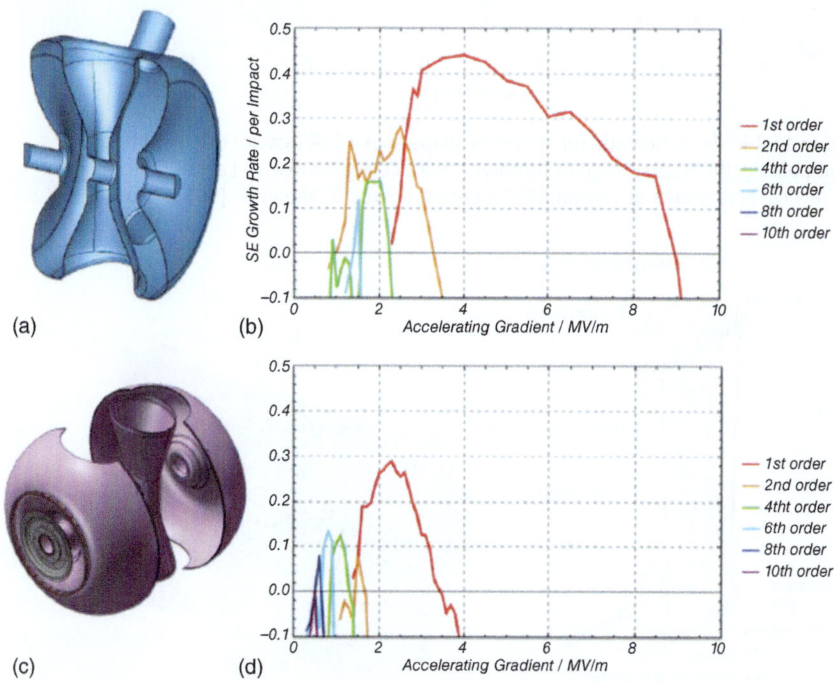

Figure 11.13 (a) The original single spoke resonator [8] with (b) calculated MP barriers [8] (c) The new spherical shape spoke resonator with (d) calculated MP barriers. Source: R. Laxdal, TRIUMF [8]/Courtesy of R. Laxdal.

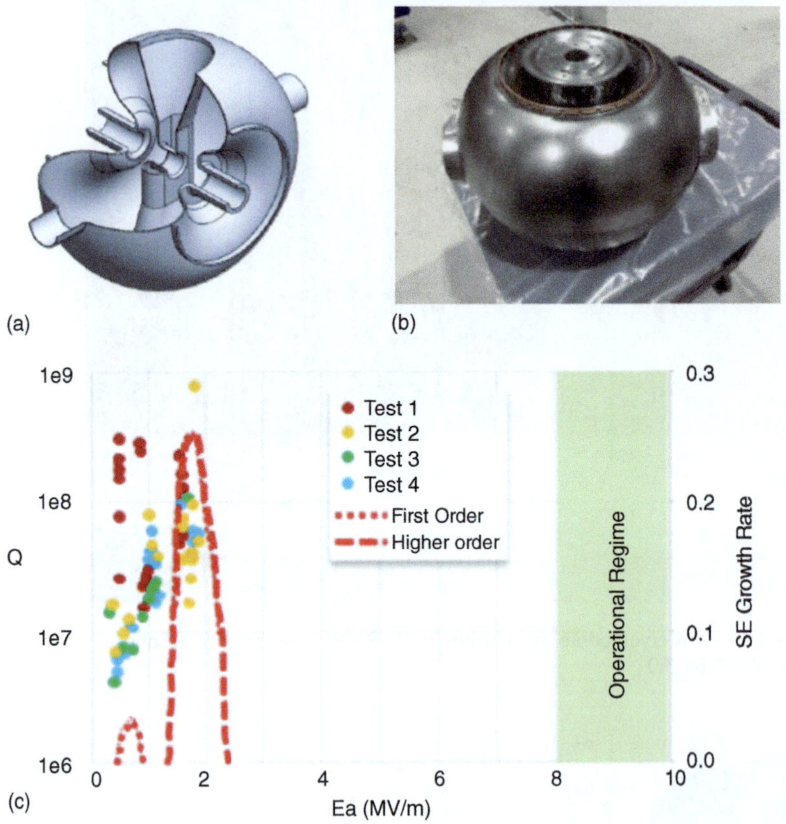

Figure 11.14 (a) Schematic for spherical shape spoke SSR1 cavity designed to avoid multipacting. (b) Photograpgh of prototype SSR1 cavity. Source: R. Laxdal, TRIUMF [8] (c) Vertical test results of several SSR1 cavities. Source: Z. Yao et al. [501]/JACoW/CC BY 3.0.

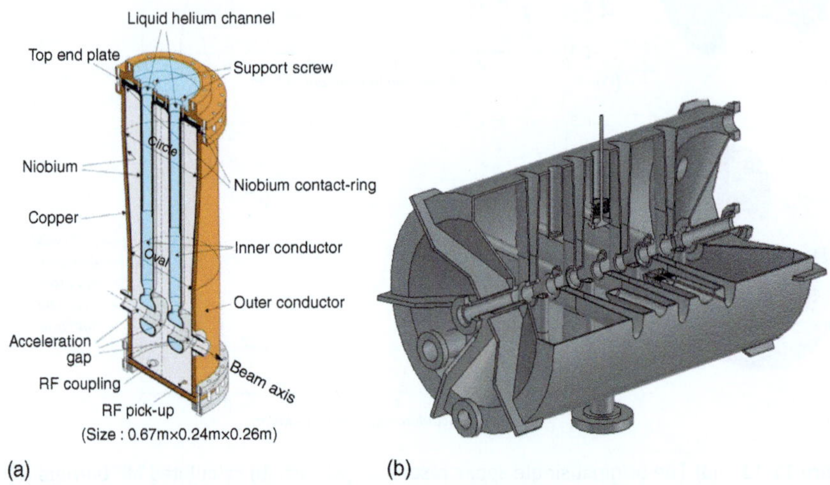

Figure 11.15 (a) QWR resonator with two-center conductors and 3 gaps for upgrading the JAEA booster. Source: [502, 504] with permission of Elsevier. (b) The cross-bar H-Mode (CH) cavity with 8 gaps for HELIAC. Source: T. Conrad et al. [505]/JACoW/CC BY 3.0.

11.2.12 SARAF

SARAF's goal is to accelerate 5 mA beam of protons from 1.3 to 35 MeV, or deuterons from 2.6 to 40 MeV. The mission is nuclear physics and astrophysics as well as medical isotope production. The first two identical cryomodules host 6 half-wave resonators (HWR) with $\beta = 0.09$ at 176 MHz. The last two identical cryomodules will host 7 HWR high-β cavities with $\beta = 0.18$ at 176 MHz. SARAF has contracted CEA saclay to deliver the superconducting linac [508]. The design gradients are 7 and 8 MV/m for the low-β and high-β cavities, with Q values near 10^9 at 4.45 K. Gradients up to 10 MV/m were achieved for the low-β, and 11.5 MV/m for the high-β. All cavities show a significant Q-slope. The low-β cavities are complete and qualified, and the CM is under assembly [509]. High-β cavities are under construction.

11.2.13 HIAF at IMP

The high-intensity heavy ion accelerator facility (HIAF) [510] for nuclear structure research at the Institute of Modern Physics (IMP-Academy of Sciences) in China aims to provide 17 ~ 22 MeV/u ($U^{35+\sim46+}$). It will consist of 30 QWRs at 81 MHz in 6 CM, ($\beta = 0.0076$) followed by 66 HWR cavities at 162.5 MHz in 11 CM ($\beta = 0.015$).

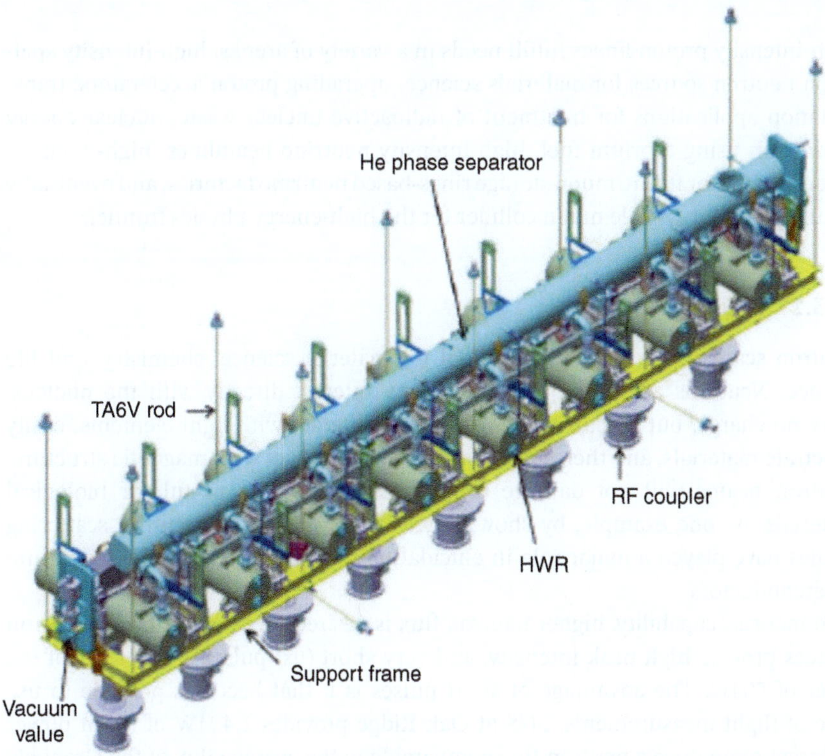

Figure 11.16 HWR Cavity String for IFMIF cryomodule. Source: N. Bazin & S. Chel [512]/JACoW/CC BY 3.0.

Operation is planned at 2 K for stability. Gradient design is for 6 MV/m. Anticipated completion is 2025.

11.2.14 IFMIF

The international fusion material irradiation facility (IFMIF), a low-β deuteron accelerator will be installed and commissioned in Rokkasho, Japan [511]. Understanding the degradation of physical properties of the materials exposed to critically intense, high-energy (14.1 MeV) neutron flux is key to the successful design of a fusion power plant, such as the international thermonuclear experimental reactor, ITER. IFMIF plans to accelerate two 125 mA cw deuteron beams at 40 MeV to generate an intense flux of neutrons leading to 5 MW average beam power. The SRF-based linac will have 46 HWRs at 175 MHz housed in five cryomodules including the injector CM. The design gradient is 4.5 MV/m. A prototype injector has delivered a cw 125 mA deuteron beam at 9 MV [512] (Figure 11.16).

11.3 High-Intensity Proton Accelerators

High-intensity proton linacs fulfill needs in a variety of arenas: high-intensity spallation neutron sources for materials science, upgrading proton accelerators, transmutation applications for treatment of radioactive nuclear waste, nuclear energy production using thorium fuel, high-intensity neutrino beamlines, high-intensity muon sources for future muon storage rings-based neutrino factories, and eventually a multi-TeV energy scale muon collider for the high-energy physics frontier.

11.3.1 SNS

Neutron scattering is an important tool for material science, chemistry, and life science. Neutrons are a unique in that they interact directly with the nucleus, carry no charge, but do carry spin. Thus they interact with light elements, easily penetrate materials, and their magnetic moments interact with magnetic structure. Neutron beams will not damage cells, making neutrons useful for biological materials. As one example, by showing where the atoms are, neutron scattering studies have played a major role in elucidating the structure of high-temperature superconductors.

To increase capability, higher neutron flux is desired. Accelerator-based neutron sources provide high peak intensity, and very short (μs) pulses at rep rates of the order of 50 Hz. The advantage of short pulses is it that becomes possible to use time-of-flight measurements. SNS at Oak Ridge provides 1.4 MW of beam power on target to produce a neutron flux comparable to the average flux of the Grenoble reactor, the largest neutron science reactor-based facility.

SNS has been operating for many years, with an upgrade on the way [513–515]. SNS has 81 × 6-cell cavities in 23 cryomodules. About half the cavities are medium-β (0.62), and half are high-β (0.81). The CMs and vendor-built cavities were supplied by JLab.

During the last 5 years, the availability for the SRF system alone is 99.5%, and for the linac (including RF, HVCM, control, vacuum, etc.) the availability is 98%. The average trip-related downtime is <1 trip/day (<5 min/day).

The plan for the proton power upgrade (PPU) [516, 517] is to increase total beam power from 1.4 to 2.8 MW and beam energy to 1.3 GeV. This will be accomplished by increasing the beam current and adding 28 new high-β (0.81) cavities in 7 new cryomodules. The cavities received cold EP instead of BCP used for the original SNS cavities. To date, 30 cavities have been delivered by vendors to JLab and 16 are vertically tested for qualification for 18 MV/m as compared to the original SNS specification of 15 MV/m. The rejection rate for these cavities was 56% due to early onset of field emission, these cavities required a second HPR at JLab to meet the performance goal (Figure 11.17).

Plasma processing [342, 343, 518, 519] has been successfully applied (Section 8.5) to reduce field emission and to raise gradients for cavities installed in SNS. The procedure was first proved with off-line cryomodules and then applied to several cryomodules in the tunnel [515]. The average improvement in E_{acc} obtained is 2.5 MV/m (21% increase). The individual improvements range from 0.2 to 5.5 MV/m. No cavity performance degradation from plasma processing was observed.

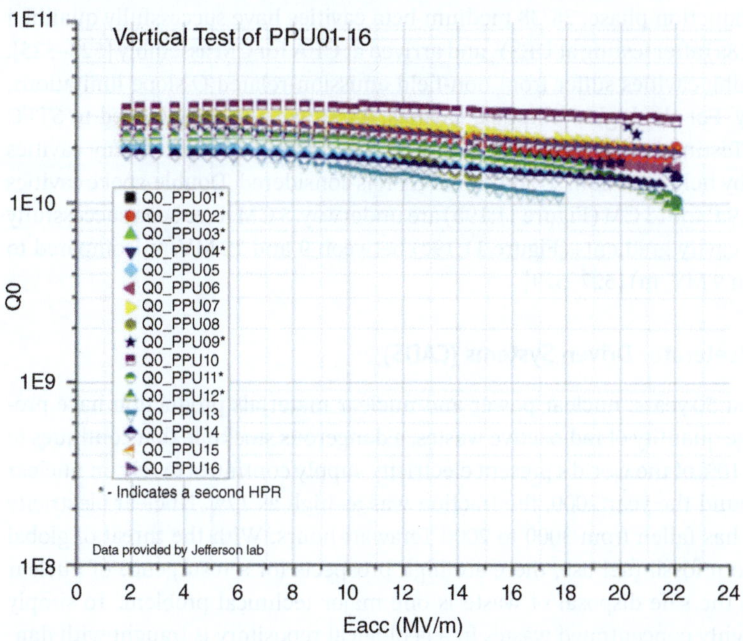

Figure 11.17 Vertical test results of the first 16 SNS upgrade cavities. Acceptance gradient is 18 MV/m. Source: J. Mammosser et al., SNS, ORNL [516]/JACoW/CC BY 3.0.

11.3.2 ESS

The European spallation neutron source (ESS) is under construction in the city of Lund, in southern Sweden [520–522]. 18 European countries are partners who will provide all components of the linac as in-kind contributions. The linac will have 2 GeV energy with 2 MW beam power rising to 5 MW in the final stage. Construction of the entire facility will be complete by 2025. The prototype phase has been successfully completed and the production phase is underway with much progress.

The linac will be built from 26 double-spoke resonators (13 CM), 36 medium-β (0.67) elliptical cavities (9 CM) at 704 MHz and 84 high-β (0.86) elliptical cavities (21 CM). There will be a total of 43 CM and 150 cavities.

ESS is partnering with CEA Saclay for the elliptical cavity prototype and CM, with IPN Orsay and FREIA (Sweden) for the spokes and CM, with INFN Milan (LASA) and DESY for the medium-β cavities, and with Science and Technology Facilities Council (STFC) in Daresbury, UK for the high-β elliptical cavities production. Prototype double-spoke CM exceeded specifications in the acceptance test to reach $E_{acc} = 15\text{–}16$ MV/m ($H_{pk} = 102$ mT, $E_{pk} = 65$ MV/m) compared to the design of 9 MV/m. The CEA saclay elliptical cavity prototypes reached 22 and 25 MV/m compared to the design of $E_{acc} = 19.9$ MV/m. A medium-β prototype at LASA (INFN) treated by BCP reached 22 MV/m ($H_{pk} = 110$ mT, $E_{pk} = 56$ MV/m) compared to design value of 16.7 MV/m, at $Q = 5 \times 10^9$. Initial batch of medium-β cavities reached 22 MV/m at Q of 10^{10}. All cavities are treated with BCP.

In the production phase, 28/38 medium-beta cavities have successfully qualified (Figure 11.18a) after testing at DESY, and arrived at CEA for CM assembly [523–525]. The remaining cavities suffer from non-field emission-related Q-slope limitations, under study. For the high-β elliptical cavities, 43/88 have been delivered to STFC [525, 526]. Testing is underway; see Figure 11.18b for initial results. Many cavities are limited by field emission, so additional HPR is considered. Double-spoke cavities (Figure 11.19a) for 13 CM (Figure 11.19b) are underway. 5 CM have been successfully tested with cavity gradients (Figure 11.19c) between 9 and 12 MV/m, compared to specification 9 MV/m) [527–529].

11.3.3 Accelerator Driven Systems (CADS)

Over the last 50 years, nuclear power and nuclear materials production have produced a large quantity of radioactive wastes, a dangerous stockpile that continues to grow while 10% of the world's present electricity supply continues to rely on nuclear energy. Around the year 2000, the fraction was as high as 17%. Nuclear electricity production has fallen from 3000 to 2000 Terawatt-hours. With the threat of global warming from fossil fuel use, there are high prospects for a resurgence of nuclear energy. But the safe disposal of waste is one major technical problem. To simply store the highly concentrated wastes in a geological repository is fraught with dangers because of the 10 000-year lifetime of some of the radioactive waste products. This danger may be reduced if the long-lived species can be transmuted to isotopes with shorter life.

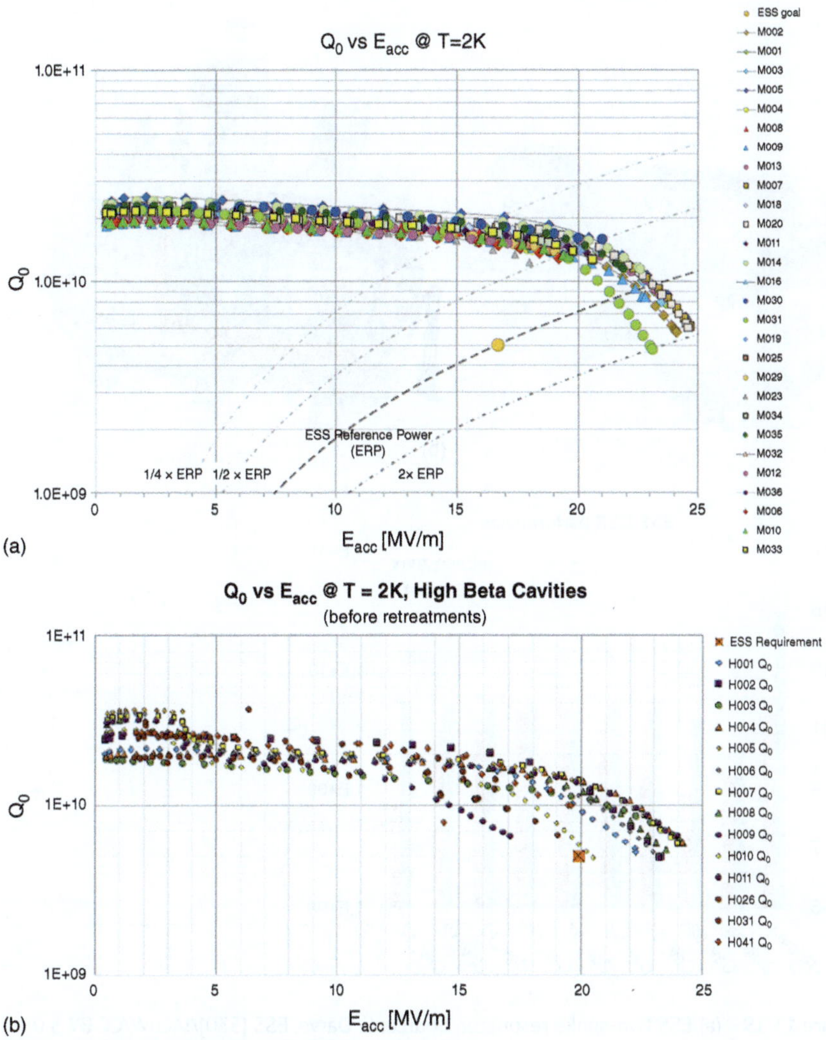

Figure 11.18 (a) Performance of medium-β elliptical cavities for ESS. 28 cavities have been successfully qualified. Source: [522] Courtesy of C. Maiano. (b) Performance of high-β cavities. Out of 16 tested, 9 exceeded ESS specifications for gradient and Q. Source: C. Maiano [522]/Courtesy of C. Maiano.

In accelerator-based transmutation of waste (ATW), spallation neutrons transmute long-lived actinide isotopes and fission products to stable isotopes, or to isotopes that decay to stable products over 100 years instead of 10 000 years. No additional transuranic waste is produced. This approach can lessen the technical problems of storing long-lived high-level radioactive waste. In an optimistic design, a single accelerator driven subcritical system (ADS) accelerator can burn the waste from ten 1 GW reactors, while providing enough power to run itself. Unlike a reactor, the ADS can never become critical because the linac can be turned off.

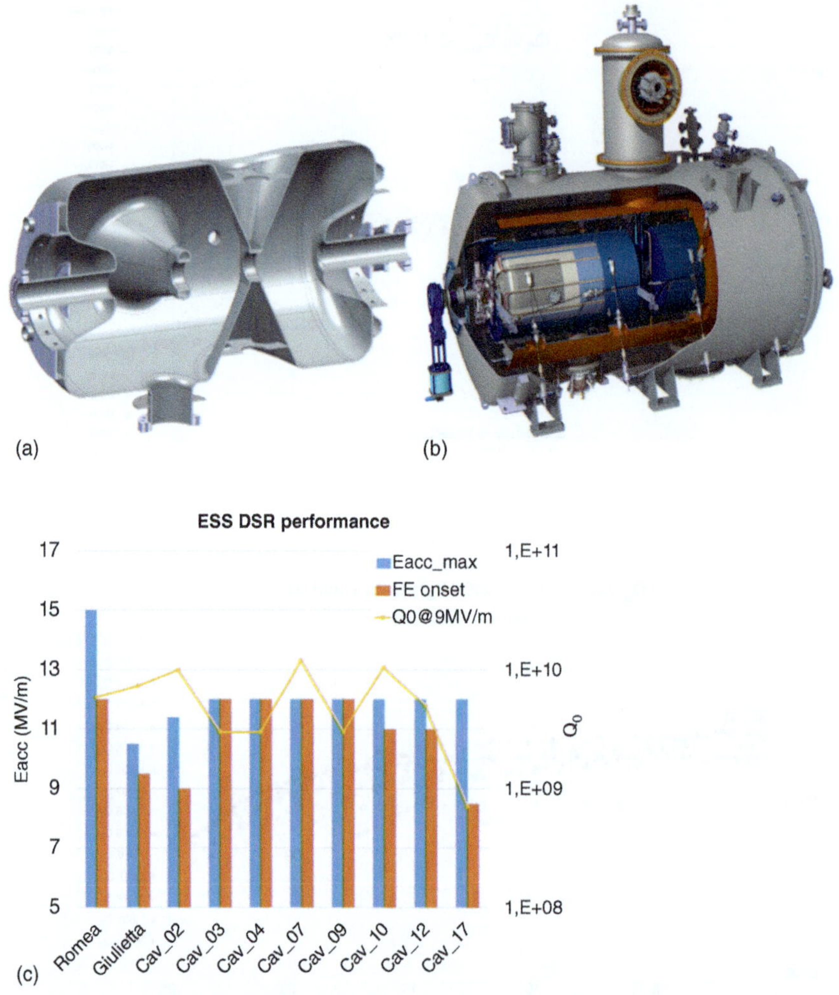

Figure 11.19 (a) ESS two-spoke resonator. Source: C. Darve, ESS [530]/JACoW/CC BY 3.0. (b) Two-cavity spoke cryomodule. F. Schlander et al. [520]/JACoW/CC BY 3.0. (c) Gradient and Q performance of 10 two-spoke resonators for the first five cryomodules. Source: Han Li, Uppsala University [529]/JACoW/CC BY 3.0.

China has 18 nuclear reactors in operation supplying 14 GW of power, and 28 reactors under construction to provide 28 GW in the near future. By 2020, China plans to expand nuclear power capability to 59 GW, and from there to 400 GW by 2050 [531]. Therefore, nuclear waste management is a critical issue for sustainable nuclear power development plans. Key technologies for a superconducting linac are under development for nuclear transmutation of waste.

Two SRF-based injector linacs have been built for 10 MeV to satisfy the strict reliability and stability demands of the system [532]. A final injector scheme will be chosen, and two identical injectors will be built, one for operation, and the other as a hot spare standby. Injector I at IHEP is based on 14 $\beta = 0.12$ spoke resonators at 325 MHz in two cryomodules, and Injector II at 162.5 MHz is based on half-wave resonators by IMP. Injector II has two HWR cryomodules with 6 cavities each with $\beta = 0.1$. Both injectors were commissioned successfully. Injector I reached 10.7 MeV with 10.6 mA in pulsed mode, and 10 MeV with 2 mA in cw mode. Spoke cavities reached E_{pk} between 17 and 34 MV/m in pulsed mode and from 17 to 25 MV/m in cw mode. Injector II reached a milestone with cw acceleration of 2.7 mA acceleration to 9.55 MeV. Pulse beam to 10 mA was delivered. Average E_{pk} was between 23.5–31 MV/m.

IHEP is also contemplating a 500 kW, 300 MeV upgrade of the Chinese spallation neutron source (CSNS) from the present capability of 100 kW with 80 MeV linac plus rapid cycling synchrotron (RCS). They are considering either all spoke cavities or a combination of spoke and elliptical cavities. The spokes will be double spokes with a record high $\beta = 0.5$ at 324 MHz. There will be 19 cryomodules. If elliptical cavities are selected, the linac could be shorter leaving room for higher energy. IHEP has also developed the medium β elliptical cavity [533].

CAFe (China accelerator facility for new elements) is the CADS front end demo linac that has achieved 17.3 MeV, 10 mA, 127 kW for >100 hours in 2021. 200 kW was achieved for a short period of 12 minutes. Commissioning started in 2017. Original spoke cavity resonators were replaced by better-performing HWR resonators. 3 CM are populated by 6 HWR cavities each ($\beta = 0.010$) and one CM with 5 HWR cavities ($\beta = 0.015$). The total number of cavities is 23. The average gradient of the first 3 CM was 3 MV/m and for the higher β CM is 1.2 MV/m. Best surface E_{pk} achieved was 30 MV/m. The facility plans to evolve to CAFE2, for material irradiation studies. It will provide 4.5–7 MeV ions from Ca to Zn. Most trips were rapidly recovered in less than 1 minutes. For the future, recovery times of 10 seconds or less are needed for successful ADS applications.

11.3.4 CiADS (China Initiative Accelerator Driven System)

Phase I of CiADS at IMP (Institute of Modern Physics) in China, will be a 500 MeV 5–10 mA cw proton linac with up to 2.5 MW beam power for completion by 2025 to become the world's first MW level ADS prototype facility, aiming to demonstrate nuclear waste transmutation. The specification of beam trip on the reactor is defined in the time scale of seconds, which is the most challenging issue for the ADS linac. Reliability, availability, maintainability, and inspectability (RAMI) analysis will be of prime importance in driving the design and evaluating the performance.

The SRF section accelerates proton from 2.1 to 500 MeV and includes three families of cavities with different frequencies and structures: HWR010 & HWR019 at 162.5 MHz, double spokes 042 at 325 MHz, elliptical 062 & 082 at 650 MHz [510] (Note HWR010 stands for HWR with $\beta = 0.010$.). IMP also plans to construct and operate the High Intensity heavy-ion Accelerator Facility (HIAF) with the capability to accelerate uranium ion beams to energies up to 17MeV/u for nuclear physics. There will be 6 CM each with 5 QWR resonators (beta 0.007) followed by 11 CM with 6 HWR (beta 0.15). The QWRs have already been developed for CAFÉ. Figure 11.20 shows 3D models for the long succession of cavities for CiADS and HIAF. In all there will be about 50 CM with over 230 cavities. Successful commissioning of the linac up to 25 MeV has been achieved. Field emission has been a limiting factor.

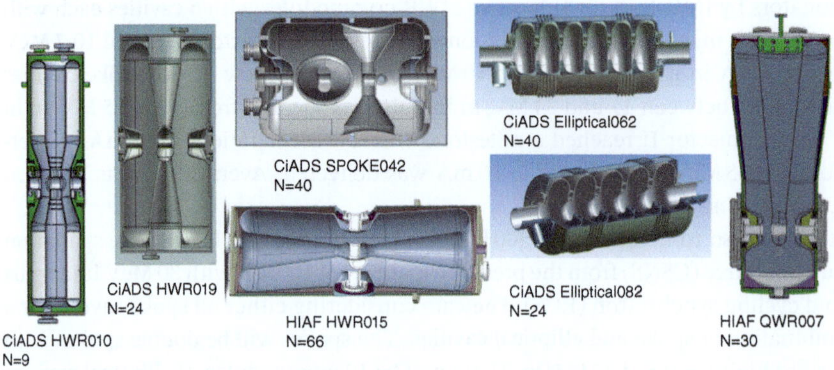

Figure 11.20 3D models of 7 types of cavities under development for CIADS and HIAF. A total of 233 structures will be constructed, 137 for CIADS and 96 for HIAF. Source: M. Xu and Y. He et al. [510]/JACoW/CC BY 3.0.

Plasma and helium processing are being explored as complementary techniques to mitigate field emission. Phase II will be 1.5 GeV targeted for 2032.

IMP and IHEP (Institute of High-Energy Physics) have developed and operated preliminary processing facilities for BCP, HPR, high-temperature heat treatment, and bake.

To realize CiADS, as well as other future large superconducting accelerator projects in China (discussed in Sections 11.2.12, 11.3.5, 11.4.3), it is planned to build several large-scale production and SRF R&D facilities [470]. The Beijing local government funded IHEP to build PAPS (Platform of Advanced Photon Source) as a large SRF facility (4500 m^2) as part of the integrated facility to support SRF Projects in China. The PAPS SRF facility is aimed at processing and testing several hundreds of SRF cavities and couplers, and assembly and testing of about 20 cryomodules per year for different users. It is also planned to explore new SRF material and technology development. Another large SRF facility for SHINE is in Wuxi.

11.3.5 Japan Atomic Energy Agency (JAEA) – ADS

JAEA is designing a reliability-oriented ADS to deal with ATW [503]. A 800 MW subcritical reactor is driven by a 30 MW cw proton linear accelerator at 1.5 GeV with beam current of 20 mA. The low-β cavities are designed to operate with the accelerating gradients of 9.7, 10, and 11.4 MV/m for HWR, SSR1, and SSR2, respectively. These values are chosen to keep the peak surface electric field lower than 60 MV/m, and the peak surface magnetic field lower than 120 mT. More than 290 cavities of five types are considered, from HWR to SSRs to elliptical.

11.3.6 High-Intensity Proton Accelerator Development in India

Several laboratories in India are collaborating with PIP-II [104] (Section 11.3.7) at Fermilab with the overall goal of building a 1 GeV spallation neutron source at

Raja Ramanna Center for advanced technology (RRCAT) in Indore to be followed by an ADS accelerator at the Bhabha Atomic Research Centre (BARC) [534, 535]. The layout of the spallation source would be similar to PIP-II except for the addition of 200 MeV, with about twelve 650 MHz cavities operating at 20 MV/m housed in two additional cryomodules (author's preliminary estimate).

There is substantial SRF Infrastructure development at RRCAT together with SRF cavity and cryomodule R&D at RRCAT, Inter-University Accelerator Center IUAC (Delhi), Variable Energy Cyclotron Center VECC (Kolkatta), and *BARC* (*Mumbai*).

As part of the experience of mastering SRF technology, RRCAT made significant progress in development and testing high-β 650 MHz cavities for PIP-II in collaboration with Fermilab. All the necessary SRF infrastructure (EP, HPR, and clean rooms) along with vertical and horizontal test stands have been created at RRCAT. BARC/IUAC has made significant progress in the development of SSR1 cavities, which have performed very well. VECC is developing five-cell low-β 650 MHz cavities, starting with single cells, in collaboration with IUAC (New Delhi).

11.3.7 PIP-II and Beyond

The Proton Improvement Plan-II (PIP-II) at Fermilab will be a 800 MeV SRF linac to be part of an upgrade to the existing accelerator complex [104, 536] at Fermilab. It will provide record intensity neutrino beams to two detectors long baseline neutrino facility (LBNF) and deep underground neutrino experiment (DUNE).

Neutrinos are elementary particles with smallest mass (<one millionth that of the electron). They are the most ubiquitous particle in the universe, but the least understood. Being electrically neutral, and ultralight they pass through matter unimpeded and undetected. About 100 trillion pass through our bodies every second without any interaction. Neutrinos come in three flavors: electron, muon, and tau. Unlike other elementary particles, they oscillate in flight between the different flavors. The parameters of these oscillations are very important for high energy physics theories that hope to unlock our understanding of the universe, but the values are still not known precisely. The existing neutrino detectors focus on measuring the neutrino oscillations.

PIP-II will be built as a collaboration with ANL, INFN-Milan (LASA), and Indian labs (BARC-Mumbai, VECC-Kolkata, RRCAT-Indore, IUAC-Delhi), UK (STFC), France (CEA, CNRS, IN2P3, IJC), Poland (WUST, WUT, and TUL), and LBNL.

The linac is planned to be a cw-capable SRF accelerator to accelerate H^- ions with 2 mA average beam current. It will be operated initially in the pulsed mode. The beam will be transported to the existing booster and main injector accelerators at Fermilab to deliver a 120 GeV beam with 1.2 MW beam power to the long-baseline neutrino facility (LBNF) target. The future upgrade goal is for beam power > 2.4 MW. The cavities and CMs are designed to operate both in a low-duty factor, pulsed mode, as well as in the cw mode.

The PIP-II SRF linac consists of five cavity/cryomodule types (see Figure 11.21) starting with HWRs at 162.5 MHz, followed by single spokes at 325 MHz, concluding with 650 MHz elliptical cavities of medium-β and high-β. There will be a total of 119

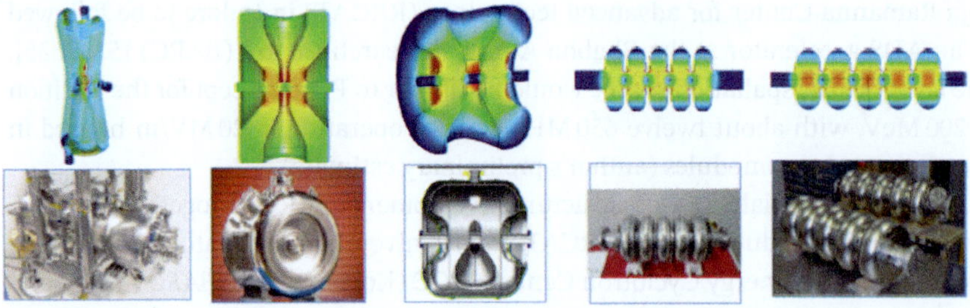

Figure 11.21 Five varieties of cavities for PIP-II, starting from low-beta HWR, followed by two types of spoke resonators and finally two types of elliptical cavities. (Upper set) 3D models (lower set) photographs of structures built. Source: A. Klebaner et al., Fermilab [104]/JACoW/CC BY 3.0.

cavities distributed in 23 cryomodules. The first SRF structure for the beam will be a single HWR cryomodule delivered by ANL. It has eight $\beta = 0.11$ HWR cavities with a design performance of 9.7 MV/m with a Q_0 of 8.5×10^9. The qualification tests of these cavities gave $Q > 10^{10}$, and accelerating gradients well above specifications.

After the HWR module, there will be two modules of SSR1s, each with eight $\beta = 0.22$ cavities to perform at 10 MV/m at a Q of 8.2×10^9. Seven cryomodules of five SSR2 cavities each with $\beta = 0.47$ will follow with a gradient performance specification of 11.5 MV/m. Finally, there will be two sets of elliptical cavity cryomodules, nine at medium-β (0.61) carrying four cavities each, and four CM at high-β (0.92) carrying 6 cavities each. The performance specification for the low-β elliptical cavities is 16.8 MV/m at Q of 2.2×10^{10}, and for the high-β cavities is 18.7 MV/m at Q_0 of 3.3×10^{10} using N-doping. Doping studies of 5-cell high-β cavities are yielding high Q at high gradients near 25 MV/m.

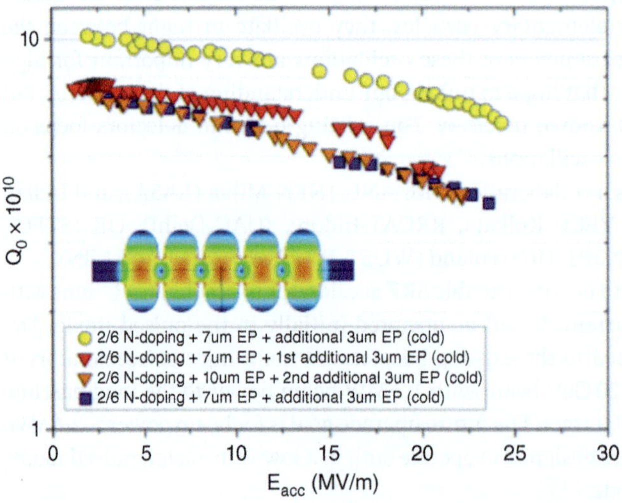

Figure 11.22 Performance of HB 650 MHz elliptical cavities with N-doping. Source: Ref. M. Martinello et al., Fermilab [538]/AIP Publishing/with permission from Journal of Applied Physics.

A beam test provides successful validation of critical technologies. The prototype SSR1 CM performed well above specifications, and with beam. The HWR and SSR1 CM accelerated beam to 17 MeV in the Fermilab beam test facility. The beam passed through both CMs on first shot with excellent transmission [104].

As one scenario for the future upgrade, the PIP-II LINAC will be the front end of PIP-III that will deliver multimegawatt beam power to DUNE via the main injector over the energy range 60–120 GeV [537]. The SRF-based alternative for PIP-III extends from 800 MeV to 8 GeV to inject beam into the main injector. The first stage will be from 800 MeV (PIP-II) to a new 2.4–3 GeV linac using 650 MHz SRF, to be followed by a 3–8 GeV linac with 1.3 GHz TESLA-type technology based upon the LCLS-II-HE cryomodules developed at Fermilab. The 8 GeV linac plans to incorporate many recent improvements in SRF technology.

N-doping is under exploration for 650 MHz low-β and medium-β cavities for PIP-II. The MSU400 upgrade, based on similar 644 MHz cavities would also adopt similar enhancements [538]. There are now excellent results (Figure 11.22) with 2/6 doping and cold EP, and also with later developed LCLS-HE recipes of 3/60 doping and 2/0 doping [157].

11.4 Electrons for Light Sources – Linacs

11.4.1 European X-ray Free Electron Laser (EXFEL)

Free electron lasers (FELs) are sources of tunable, coherent radiation at wavelengths covering a wide range from mm to UV to X-rays [539]. At ultrashort wavelengths, less than 100 nm, mirrors are not available for FELs. The alternative is to use coherent bunching of the electron beam in a single-pass through a long undulator (a magnet designed to force the beam to follow a sinusoidal path). As the bunch interacts with the undulator field, micro-bunches develop and emit radiation coherently. The radiation field amplitude grows exponentially with distance along the undulator. The power increases as the square of the number of particles per bunch. This process is called *self-amplified-spontaneous-emission* (SASE) [540]. SASE FELs are the most attractive candidates for extremely high brilliance coherent light with wavelength in the angstrom regime. Linear accelerator-driven FELs using the SASE principle provide a promising approach to produce X-rays with unprecedented quality.

With the ultrahigh brilliance, coherence, and ultrashort (10–100 fs) pulse length, research with SASE sources have launched a new era, impacting the full span of materials and biological sciences. The femtosecond (fs) timescale opens the possibility for novel time exposure experiments in biological, chemical, and physical processes to investigate structural changes, or to capture bond formation processes. It becomes possible to acquire holographic snapshots with atomic resolution in space and time on the scale of chemical bond formation and bond breaking. Modern X-ray sources, like the EXFEL, open the ability to focus on individual atoms, even when embedded in a complex molecule, and view electronic and nuclear motion on

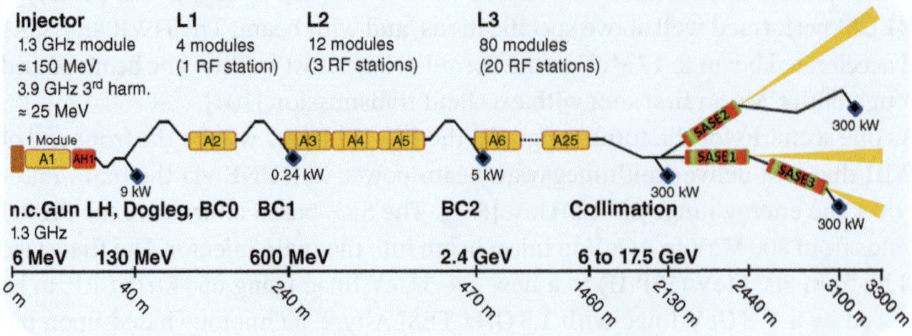

Figure 11.23 Schematic overview of the European XFEL accelerator. Source: D. Reschke et al. & W. Decking et al., DESY [105, 542]/JACoW/CC BY 3.0.

their intrinsic scales (attoseconds and Ångstroms). For comparison, such timescales can be compared to the 150 attoseconds Bohr orbital period in the hydrogen atom.

EXFEL is the largest SRF application to date [105]. The accelerator has been operating successfully for more than four years with 89 modules (712 cavities). During commissioning, the EXFEL has operated at 14 GeV to lase at 1.5 Angstrom. The 17.5 GeV design energy has been successfully demonstrated [541].

Originally EXFEL (Figure 11.23) was conceived with a dual mission, an X-ray laser, and, in parallel, a prototype for the international linear collider (ILC) (Section 12.1). Therefore, some comparisons between EXFEL SRF performance, and the needs for ILC are appropriate.

The entire LINAC for EXFEL (Figure 11.24) consists of 97 accelerator modules, each housing eight 1.3 GHz TESLA-type 9-cell cavities. Fabrication of the cavities required approximately 20 tons of Nb. Cavities operate with an average gradient of 23.6 MV/m at 1% duty factor, 10 Hz repetition rate, and 1.4 ms RF pulse length. There is one-third harmonic cryomodule with eight 3.9 GHz cavities for bunch shaping. There are a total of 24 rf stations with 4 cryomodules per station.

Modules were assembled at CEA Saclay [543, 544] at an average rate of one module per week, after an initial learning curve at a slower rate. The module performance achieved was 27.5 MV/m (±4.8) [105] as compared to the starting cavity "usable" performance average in vertical tests of 28.3 MV/m (±3.5) for the cavities used in the modules. The "usable" performance gradient is defined as when the cavity Q falls below 10^{10}, or when the X-ray output from field emission exceeds a set limit. The gradient during the module test was held at an administrative limit of 30 MV/m due to maximum allowed RF power (to protect components). Many cryomodules reached the administrative limit of 30 MV/m, close to the ILC specification of 31.5 MV/m. In five of the best modules, all eight cavities reached the 30 MV/m set limit. The true maximum gradient is likely to be higher. Figure 11.25 shows the distribution of cavity performance in the cryomodules.

Figure 11.24 View of the 1.5 km EXFEL linac populated with nearly 100 cryomodules. Source: D. Reschke et al., DESY [105]/JACoW/CC BY 3.0.

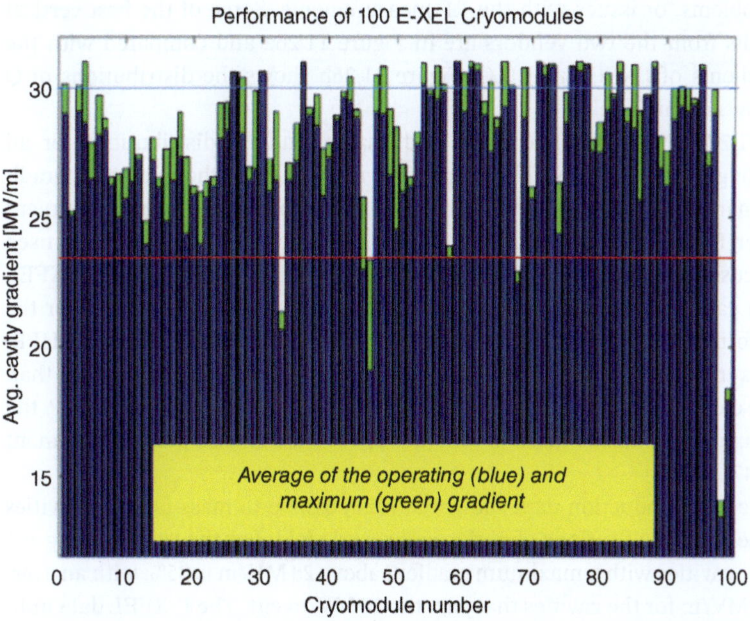

Figure 11.25 Performance of the cavities inside their cryomodules. Source: K. Kasprzac et al. [545]/JACoW/CC BY 3.0. Modules 98 and 99 show low performance as these were assembled from the lowest-performing cavities.

23 out of 776 cavities had to be detuned due to various conditions, such as high field emission, or coupler overheating. For all others, no significant degradation of individual cavities or modules has been observed. The average Q_0 value estimated from dynamic cryogenic losses is about 10^{10}. The cryogenic power for the facility is 2 kW at 2 K.

Plans for future upgrades include cw operation at 8 GeV, or long-pulse operation at 10 GeV with 35% duty factor along with the present capability of short pulse (duty factor 1.4%, 17.5 GeV) mode of operation. Performance of CMs under cw conditions is under study to evaluate heat loads, microphonics, and low-level rf (LLRF) control, with a view to upgrades. CW gradients up to 19 MV/m could be possible with Q values above 10^{10}. RF control and microphonic challenges for higher loaded Q values (10^7) have to be met [546].

During vertical tests to qualify the cavities from industry, the average maximum gradient for all 832 cavities was 33.1 MV/m (±4.9) [547, 548]. 29% of all the cavities were re-treated by HPR to establish this best performance level. Before re-treatment, the average max gradient for "as received" cavities was 31.4 MV/m (±6.7). Max gradient is defined as the highest gradient possible in the vertical test. About half of the "as received" cavities were limited by a quench. The other half were limited by the maximum RF power (35%) and the remaining 15% were due to HOM coupler heating problems, or issues with the RF measurements. Some of the best vertical cavity results from the two vendors are in Figure 11.26a and compared with the design gradients of EXFEL and ILC. Figure 11.26b shows the distributions of Q values at low and high fields.

Figure 11.27a shows the maximum and usable gradient distributions for all cavities, along with gradient yield. It is possible to distinguish the maximum gradient performance between two cavity suppliers. One vendor used a final treatment of the older BCP recipe after removal from the furnace. The other vendor used EP. As discussed in Chapter 6, EP final treatment gives superior results. E-XFEL production data from the vendor who used EP provides excellent statistics for the cavity performance "as received" from the vendors. After retreatment by HPR, 417 cavities from this vendor reached E_{max} = 34.7 (± 4.4) MV/m [547]. More than 10% of these cavities exceeded 40 MV/m. For the other vendor who used BCP, the average maximum gradient was 3 MV/m (about 10%) lower, also after re-treatment by HPR [547].

E-XFEL cavity production data show that it is possible to mass-produce cavities meeting the ILC specifications of performance and yield. For the vendor using EP, the yield for cavities with a maximum gradient above 28 MV/m is 85%, with an average of 35.2 MV/m for the cavities that pass the 28 MV/m cut. The E-XFEL data indicate that after re-treating cavities below 28 MV/m with HPR, a yield of 94% for a maximum gradient above 28 MV/m can be achieved, with an average value of 35 MV/m, meeting the ILC specification. E-XFEL experience shows that re-treatment for most cavities can be limited to a simple HPR.

Taking into account limitations from Q and the onset of field emission, the "usable gradient" is lower. From EXFEL results we can expect (by extrapolation) an 82% (91%) yield after one (two) HPR re-treatments to reach an average usable gradient of 33.4 MV/m.

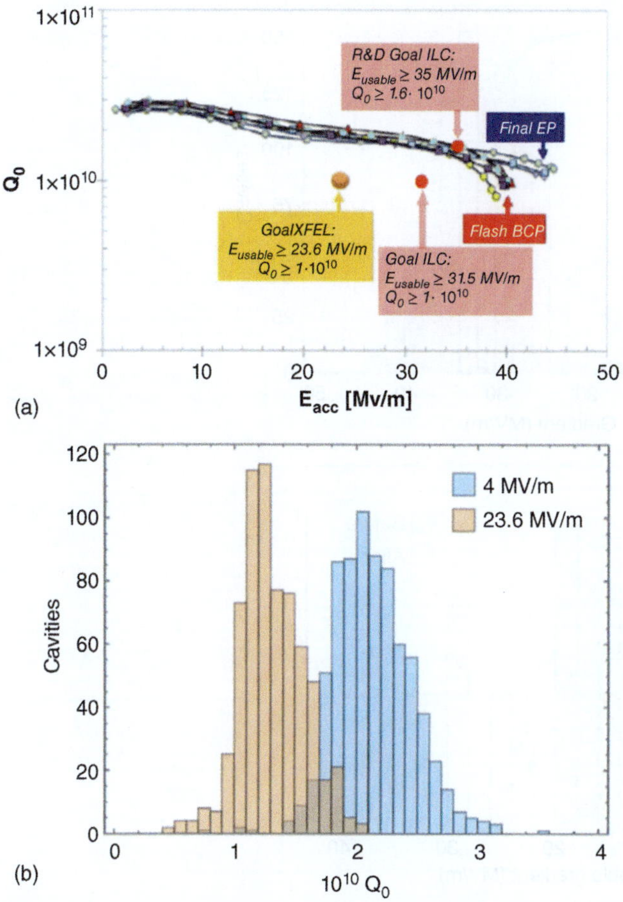

Figure 11.26 (a) Examples of the cavity performance for the best cavities, treated at one vendor using EP, or at another vendor, using BCP as the final treatment. Source: W. Singer et al., DESY [549]/American Physical Society/CC BY 4.0. The ILC gradient/Q goals are included. (b) Q distributions for the "as received" tests, measured at 4 MV/m and 23.6 MV/m. (Note that 15% of the cavities did not reach 23.6 MV/m in the "as received" and are omitted.) Source: D. Reschke et al., DESY [550]/American Physical Society/CC BY 4.0.

11.4.2 Linac Coherent Light Source LCLS-II and LCLS-HE (LCLS-High Energy)

The LCLS at SLAC at 15 GeV is the world's first X-ray free electron laser based on the original pulsed SLAC copper linac. The upgrade, LCLS-II [551] will be a 4 GeV cw SRF-based electron accelerator to produce both hard and soft X-rays. The SRF segment will use about 700 m out of the 3200 m SLAC tunnel. It will become operational in 2023. LCLS-II is the first large cw linac based on the TESLA/XFEL/ILC technology.

LCLS-II will be a cw electron accelerator operating at 1.3 GHz using 280 TESLA type 9-cell cavities operating at a medium gradient of 16 MV/m. The 16 MV/m cw gradient was selected to keep the 2 K refrigerator load manageable, or similar to the

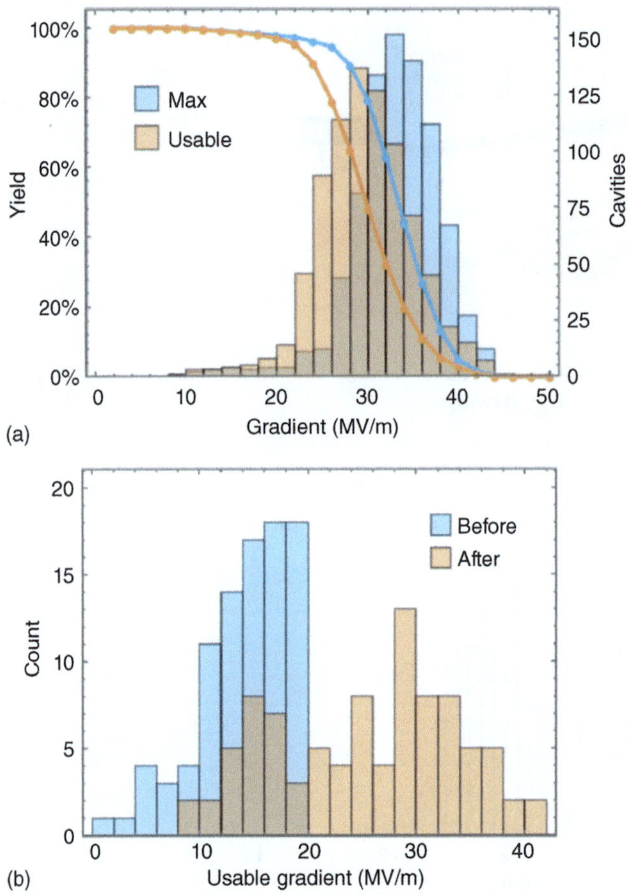

Figure 11.27 (a) Maximum and "usable" gradient distributions of cavities accepted for string assembly (including retreatments). Source: D. Reschke et al. [550]/American Physical Society/CC BY 4.0. (b) Improvement in the usable gradient distribution after the application of HPR. Source: D. Reschke et al., DESY [550]/American Physical Society/CC BY 4.0.

large refrigerator at JLab – at or below 4 kW. Each cavity will be powered by a 4.8 kW solid-state amplifier with nominal loaded Q_{ext} of 4×10^7.

To generate the 4 GeV electron beam, the project requires 35 cryomodules at 1.3 GHz with eight 9-cell cavities each, and two modules with six 9-cell higher-harmonic cavities at 3.9 GHz for bunch shaping. 18 cryomodules are provided by JLab and 19 by Fermilab. The higher harmonic cavities and CM are provided by Fermilab. Two cryoplants based on the JLab design have been installed to provide 4.5 and 4 kW each at 2 K. The expectation is to operate at 4 GeV with one cryoplant so that the other plant will be available in the near future for the 8 GeV upgrade, called LCLS-II-HE [101, 552].

The SRF technology closely follows that developed for the European XFEL. Niobium, cavity, coupler, cryomodule, and auxiliary components are almost identical to those used for the European XFEL. Appropriate changes to the cavity higher-order-mode extractor, fundamental power coupler (FPC), cavity tuner, and magnetic shielding were incorporated to manage the (i) higher dynamic heat load for cw operation, (ii) higher Q_{ext}, (iii) higher average beam current, and (iv) higher sensitivity to magnetic field of the N-doped cavities (Section 3.7.1). The CM hardware was fabricated by many of the same companies as for EXFEL.

The cavities utilize the novel N-doping process to achieve an average Q_0 of $>2.7\cdot 10^{10}$ at 16 MV/m (Chapter 3). It will be the first application of N-doping to a large-scale accelerator. As discussed earlier, N-doping reduces the niobium "BCS" resistance by a factor of 3–4 enabling a large cw accelerator facility such as LCLS-II to operate at 15–20 MV/m. The qualifying cavity gradient in the vertical test is 19 MV/m at a $Q_0 > 2.5\times10^{10}$ at 2 K. For the 12 higher harmonic cavities at 3.9 GHz, the operating specifications are a gradient of 13.4 MV/m and $Q_0 \sim 2.0\times10^9$.

The cavities meet the performance specifications. Figure 11.28 summarizes the Q and gradient performance statistics of 9-cell, LCLS-II cavities (~373 cavities) from

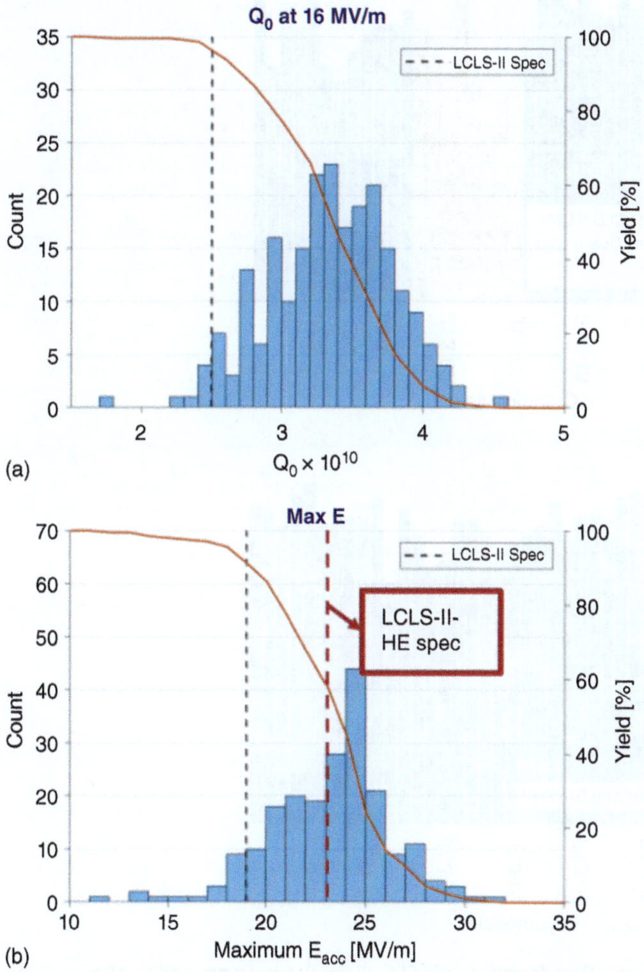

Figure 11.28 Performance of LCLS-II cavities for (a) Q and (b) gradient. Source: D. Gonnella, SLAC [167, 168] Courtesy of D. Gonnella. Cavities meet and exceed specifications.

vertical tests. The cavities were commercially produced at two different vendor sites, from raw Nb materials, also from two different vendors. The cavities were doped with 2 N6 recipe followed by 5–7 μm EP at vendor sites. The average gradient performance of 22 MV/m exceeded LCLS-II quench specifications of 19 MV/m, to allow reliable operation at an average gradient of 16 MV/m. The average Q values is 3.3×10^{10}, above the specification of 2.7×10^{10}. Increasing the heat treatment temperature to 900 °C improved the flux expulsion properties and consequently the Q of some of the Nb batches [167].

The cryomodule performance meets specifications (Figure 11.29) with sufficiently fast cooldown rate (using 32 g/s He flow rate) as needed to expel the ambient dc magnetic flux (Section 5.6.1). Three 3.9 GHz Cryomodules reached 18 MV/m at Q_0

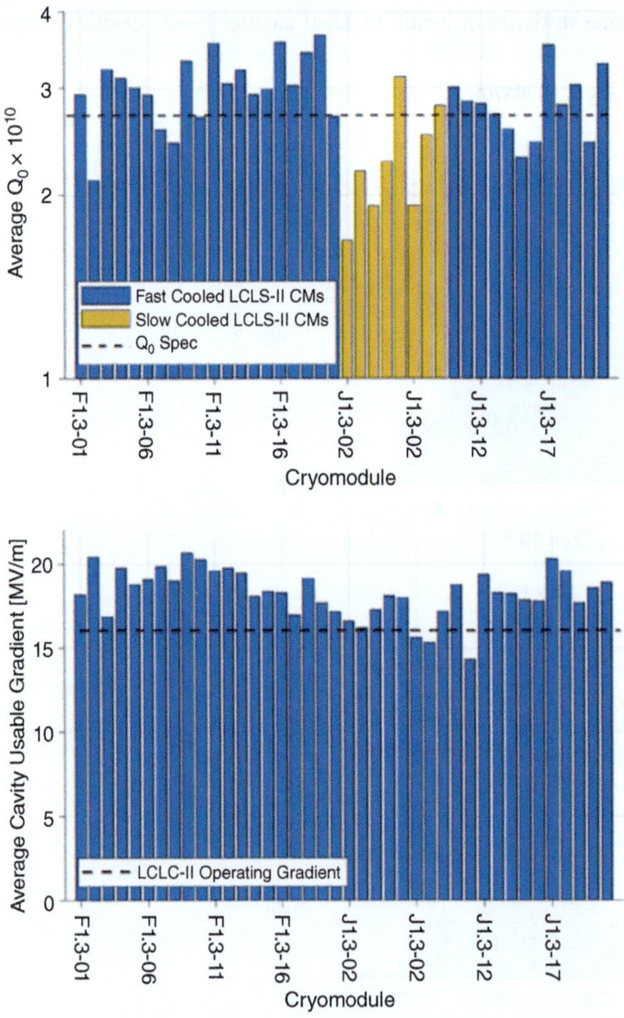

Figure 11.29 Gradient and Q performance of LCLS-II cavities in cryomodules, after N-doping. Source: [553] Courtesy of A. Burrill.

Figure 11.30 SLAC tunnel populated with LCLS-II cryomodules. Source: M. Ross, SLAC [554]/JACoW/CC BY 3.0.

of 3×10^9 in the qualifying test to meet specifications. 38 CMs with 304 cavities have been received at SLAC and installed in the tunnel (Figure 11.30). Figure 11.29 shows the cavity performance for Q and gradient in CM tests [553]. Cavities with slow cooldown failed to meet the Q specification, due to insufficient flux expulsion, as expected from the discussions of Chapter 5.

To achieve adequate flux expulsion (Section 5.11), the starting material quality was screened for a low level of mechanical imperfections that could become flux pinning sites. One batch of Nb expels flux well with the 800 °C treatment, but another needed 900 °C.

At the start of CM shipments to SLAC, two CMs lost beamline vacuum due to cracking at the FPC cold bellows. One cause identified was the stiffness of the shipping frame which induced vibrations at the natural resonance frequency of the FPC assembly, which was also found to be insufficiently restrained [555, 557]. Adjusting the suspension springs in the transport fixture reduced the resonance frequency to decouple the transport vibrations from coupler motion. A small neoprene spacer clamped to the one of the flanges of the coupler assembly further restrained the coupler motion. The spacer is difficult to install due to congestion of the components in the cryomodule, and must also be removed very carefully on arrival without damaging other components. This could be accomplished in all but one case that did cause beamline damage.

Looking to the future, LCLS-HE will upgrade the energy to 8 GeV. The spectral reach will increase to more than 11.8 from 5 KeV, and potentially to 20 keV at 1 MHz rep rate. The goal is to add 20 additional cryomodules using the same laboratory and industrial partnership scheme as for LCLS-II. The new CMs will be installed

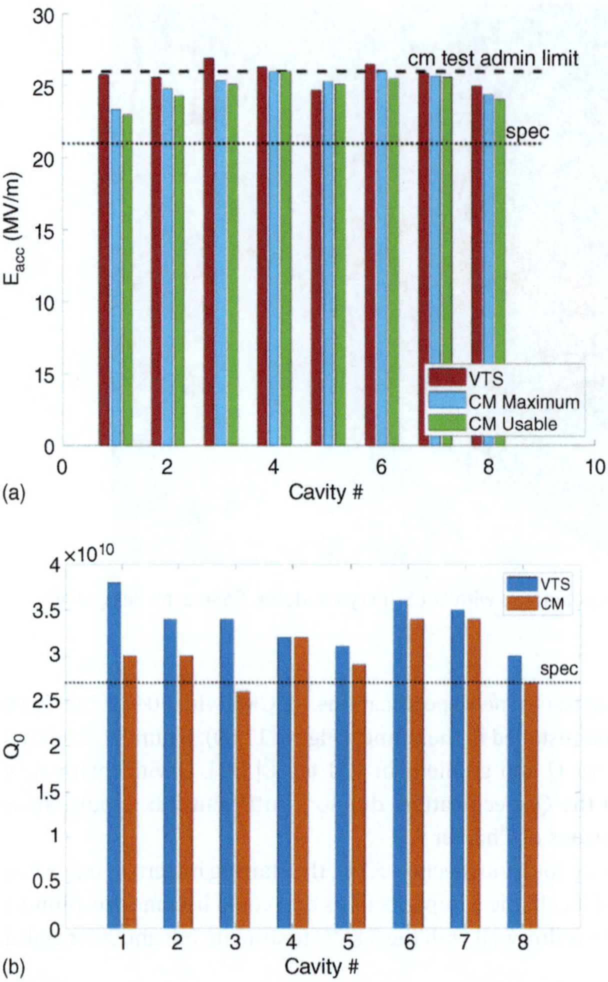

Figure 11.31 (a) Gradient distribution for the verification cryomodule for LCLS-HE. Total usable voltage >200 MV, with average gradient 25 MV/m, exceeding specifications. Source: S. Posen et al., Fermilab Ref. [556]/arXiv. (b) Q distribution for the verification CM. Average Q at 21 MV/m = 3×10^{10}. Source: S. Posen et al., Fermilab Ref. [556]/arXiv.

in the next 300 m of the refurbished SLAC tunnel. The gradient specification for the upgrade cryomodules is 21–25 MV/m at the same Q_0 value of 2.7×10^{10} as for LCLS-II. Continuing R&D on N-doping has been essential for the success. New recipes 2/0 and 3/60 were validated, along with cold EP [327] as well as chemical soaking [334], all essential for success. A verification CM for LCLS-II-HE with 8 cavities has been qualified to reach the desired specifications [168] Figure 11.31 shows Q and E_{acc} results for the verification module.

11.4.3 Shanghai Coherent Light Facility (SCLF) SHINE

Similar to LCLS-II and LCLS-HE, SHINE in China is a newly proposed XFEL project, to be carried out jointly by Shanghai-Tech University, and various laboratories in

China. It will be an 8 GeV cw SRF linac [558–562]. The facility will be built in a 3.2 km long tunnel at Zhang-Jiang High Tech Park, across the Shanghai Synchrotron Radiation Facility (SSRF) campus in Shanghai to provide radiation in the photon energy range of 0.2–25 keV (60–0.5 Angstroms). The proposal was approved by the central government in 2017, and is in the feasibility study phase, aiming at completion around 2025. New infrastructure including caves for vertical-cavity tests and bunkers for horizontal cavity tests have been installed along with clean rooms for cavity assembly and facilities for cryo-module assembly. Processing facilities for EP, N-doping, and baking are included [470]. The SRF platform (102 m × 18 m) for cavity preparation and treatment at Wuxi, about 150 km west of Shangai has been commissioned and put into operation [563].

The SRF linac design calls for 75 EXFEL type cryomodules with 616 TESLA type cavities, operating cw at gradients between 12 to 16 MV/m, and two-third harmonic modules to include sixteen 3.9 GHz 9-cell niobium cavities for linearization. The design Q_0 is 2.7×10^{10}, as for LCLS-II. Nitrogen doping will be used to reach the high *Qs*. The refrigeration system required is 5 kW at 2 K. Mid-T baking is considered an option, as it mimics the benefits of N-doping (Chapter 4). Some of the encouraging IHEP results on mid-T baking are discussed in Section 4.3. With the progress of LCLS-HE and developments at Fermilab and China on mid-T baking, there is still time to modify the design to accommodate higher gradient near 20 MV/m or higher *Qs*.

Several studies on high *Q* and high gradient have been carried out [562]. N-doping on 9-cells with the 3–60 recipe achieved excellent results (Figure 11.32a). Similar success has been demonstrated with mid-T baking (Figure 11.32b) with a 9-cell cavity. In the high gradient spirit of the SRF community, a single cell has been prepared with the two-step (75/120 °C) baking procedure to reach >46 MV/m (Figure 11.33) and *Q* above 10^{10}. These good results testify to the high quality of the SRF facilities and practices.

Cavity studies are carried out in collaboration with IHEP and Peking University (PKU). Three types of CMs (prototype, standard, and short) are assembled and cooled down successfully. Five more standard CMs will be assembled/tested before series production starting 2023. A 3000 m^2 cryomodule assembly and test workshop with an 1 kW @ 2 K cryogenic system is under construction at the SSRF campus [561].

11.4.4 Institute of Advanced Science Facilities (IASF)

Along similar lines as SHINE, IASF in Shenzhen has proposed a 2.5 GeV XFEL cw SRF accelerator with 26 1.3 GHz cryomodules and two 3.9 GHz cryomodules to be completed by 2028 [470].

11.4.5 Polish Free-Electron Laser POLFEL

POLFEL [564] in Świerk, near Warsaw, will provide tunable coherent radiation in the range from several nanometers (soft X-rays) to several hundred micrometers (Terahertz). The injector is based on a superconducting electron gun with a Pb film deposited onto a head of Nb demountable plug. The linear accelerator will consist

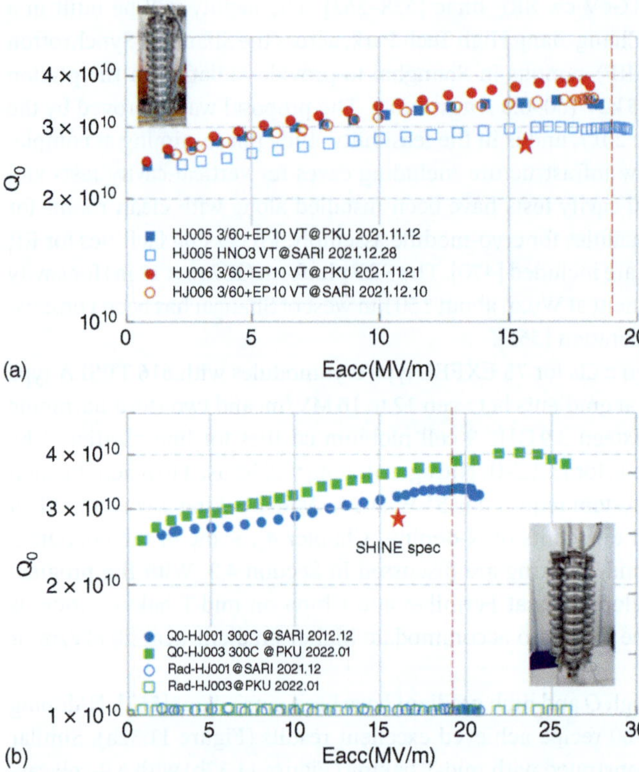

Figure 11.32 (a) Performance results for N-doping with 3–60 recipe on 9-cell cavities. Source: J. Chen, SARI (Shanghai Advanced Research Institute, China. [562]/Courtesy of J. Chen. (b) Performance results on mid-T baking with 9-cell cavities. The stars show the design goals. Source: [562] Courtesy of J. Chen.

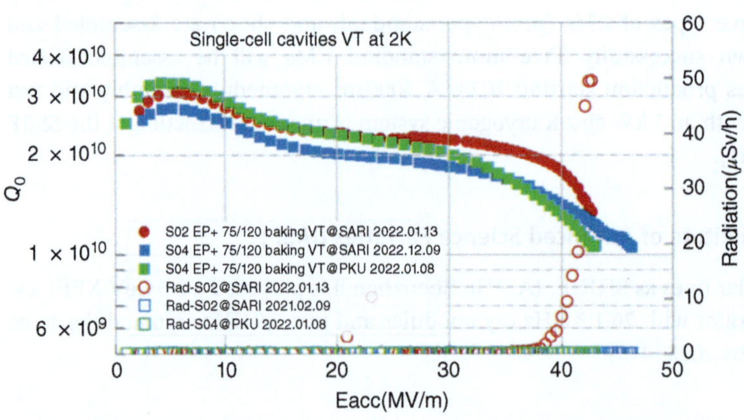

Figure 11.33 Performance results on single-cell cavities with two-step baking. Source: [562] Courtesy of J. Chen.

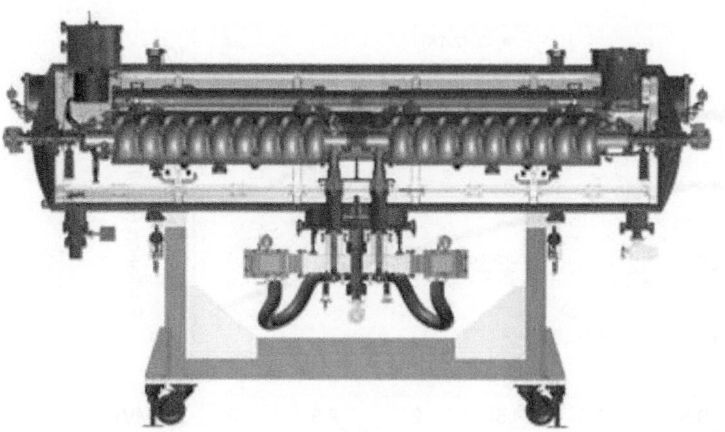

Figure 11.34 The Rossendorf cryomodule with two 9-cell cavities for POLFEL. Source: R. Nietubyć et al. [564].

of four, and later five, Rossendorf-RI-type cryomodules (Figure 11.34) to operate in cw mode. The design gradient for the TESLA 9-cell structures is 18 MV/m at a Q near 10^{10}.

A low-energy branch delivers 72 MeV to an IR undulator chain. The high-energy branch guides the electron beam through succeeding CMs, to boost their energy up to 187 MeV for supplying a VUV undulators chain. The FEL is approved and now under construction.

11.5 Electrons for Storage Ring Light Sources

11.5.1 High-Energy Photon Source (HEPS)

Beijing is developing HEPS [460], the next electron storage ring light source at 6 GeV, 200 mA with low emittance to be completed by 2025. HEPS plans to use four 166 MHz SRF cavities of the QWR design for acceleration (Section 10.3), and two 500 MHz SRF KEK type cavities for bunch lengthening (Section 11.5.2). Excellent performance (Figure 11.35) was shown in vertical tests for the innovative QWR accelerating cavity (Section 10.3). Some performance degradation was observed in the horizontal test for the dressed cavity due to coupler heating [565].

11.5.2 Taiwan Photon Source (TPS)

Powered by a single SRF module of the Cornell design [566], the Taiwan light source operated (from 2005) routinely at 1.5 GeV with 360 mA. A maximum beam current up to 400 mA was demonstrated. The input coupler power needed was 80 kW per cavity. The reliability of SRF operation grew better every year. In 2017, the mean time between RF failures rose to over 1400 hours. No Q degradation has been observed over operation.

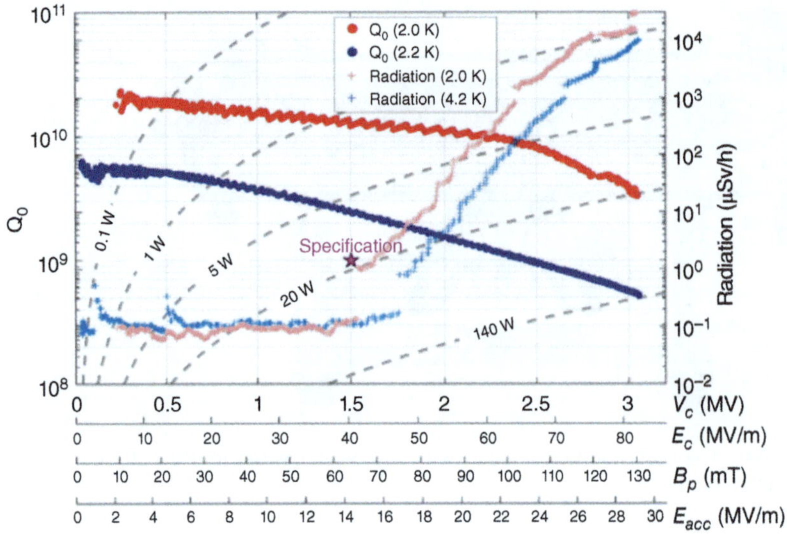

Figure 11.35 Performance of the BEPS QWR accelerating cavity in vertical tests. Design specification is for 3 MV at a Q of 2.4×10^9. Source: [565]/with permission P. Zhang.

In 2015, two 500 MHz modules of the KEK-B design [567] were installed for the upgrade, 3 GeV Taiwan Photon Source (TPS). TPS has been operating with two KEK-B cavities at 8 MV/m, and beam current 400 mA. A maximum beam current of 500 mA was achieved. The input coupler could deliver 220 kW per cavity. No Q degradation has been observed over operation. The mean time between RF failures was between 110 and 125 hours over operation (Figure 11.36).

11.5.3 Higher Harmonic Cavities for Storage Rings Chaoen WANG, NSRRC, Taiwan

Higher harmonic cavities are attractive for light source storage rings to increase bunch length, improve Touschek lifetime, and increase the single bunch charge limit. Landau damping is also enhanced by synchrotron frequency spreads. Superconducting harmonic cavities are a good choice due to their compact size and damping of higher-order modes.

A SUPER-3HC has been installed and operating [569] in ELETTRA (Sincrotrone Trieste) and at the Swiss light source (SLS) to provide a total 0.8 MV at 1.5 GHz with 2 cavities. The Nb/Cu cavity is scaled from the 500 MHz SOLEIL cavity. SOLEIL (*Source optimisée de lumière d'énergie intermédiaire du LURE*) – the synchrotron light facility near Paris installed SRF storage ring cavities [570].

A Superconducting 3rd harmonic cavity is under development for SSRF in China [571]. A bunch lengthening cryomodule using a single cell "higher-harmonic" superconducting cavity based on the TESLA shape will be used to improve the lifetime and increase the single bunch current limit for the APS at Argonne [572].

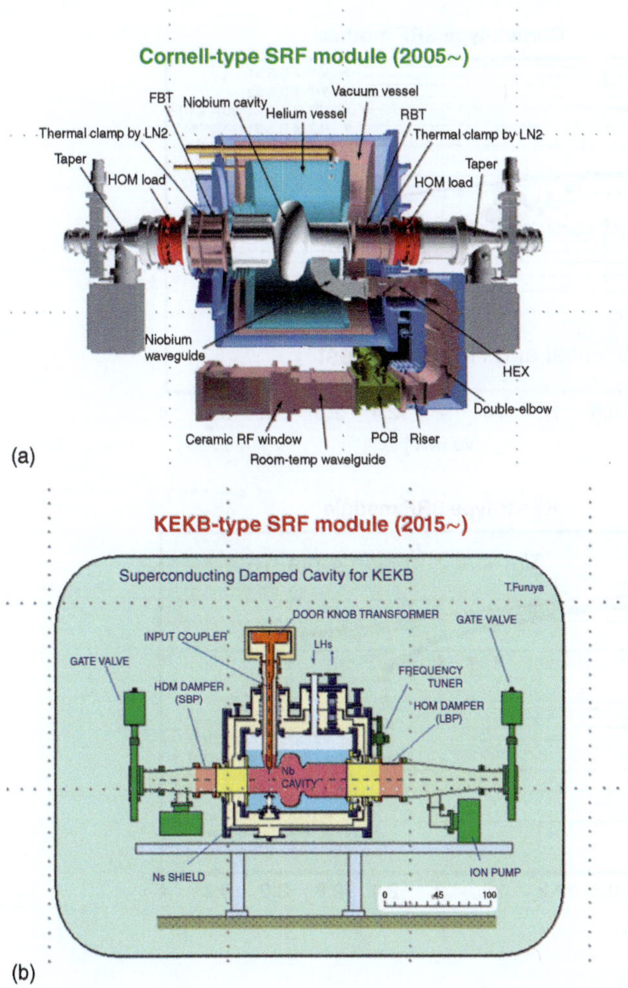

Figure 11.36 (a) The Cornell 500 MHz storage ring cavity (b) KEK-B storage ring cavity (c) Performance of the Cornell cavity. (d) Performance of the KEK-B cavity. Source: [568] Courtesy of Chaoen WANG.

Toward a next-generation light source, the Berlin Electron Storage Ring Society for Synchrotron Radiation (BESSY) is pursuing a variable pulse storage ring (VSR) project [573] to develop an alternative to diffraction-limited storage rings. VSR is a novel approach to create both long and short photon pulses simultaneously for all beam lines through a pair of superconducting bunch compression cavities. Such a system should be able to store "standard" (10 ps) and "short" (ps and sub-ps) pulses simultaneously in the storage ring, opening the door to picosecond dynamic and high-resolution experiments at the same facility. This feature can be created by introducing beating effects produced by two higher harmonic SRF cavity systems at 1.5 and 1.75 GHz. When combined with the existing 500 MHz, cavities will generate the required beating for bunch compression for half of the buckets while keeping the

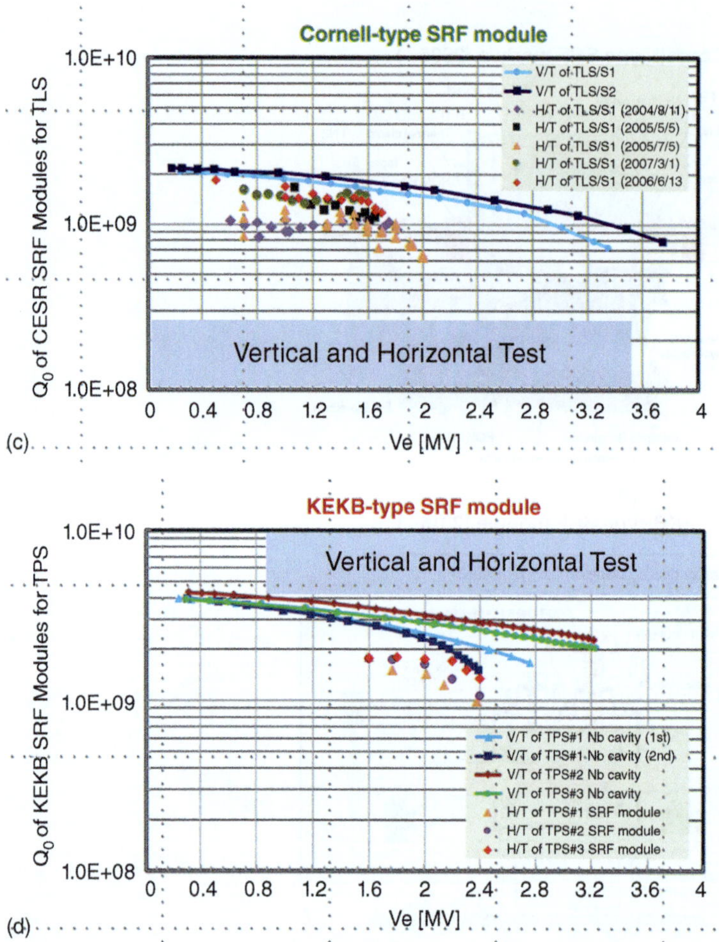

Figure 11.36 *(Continued)*

other half unaltered. By making use of pulse picking techniques each individual user would be able to freely switch between high average flux for X-ray spectroscopy, microscopy, and scattering or picosecond pulses up to 500 MHz repetition rate for dynamic studies.

The challenging design specifications call for strong HOM damping using waveguide-damped cavities under development to handle nearly 2 kW of HOM power with 300 mA beam current. Figure 11.37 shows the layout for a 2-cavity VSR cryomodule with two heavily damped harmonic SRF cavities.

11.5.4 BNL

At Brookhaven National Laboratory (BNL), a 56 MHz quarter-wave resonator cryomodule, used operationally for longitudinal bunch compression with up to 1 MV RF voltage, is under preparation to accommodate an expected beam current of 418 mA per ring in RHIC [575].

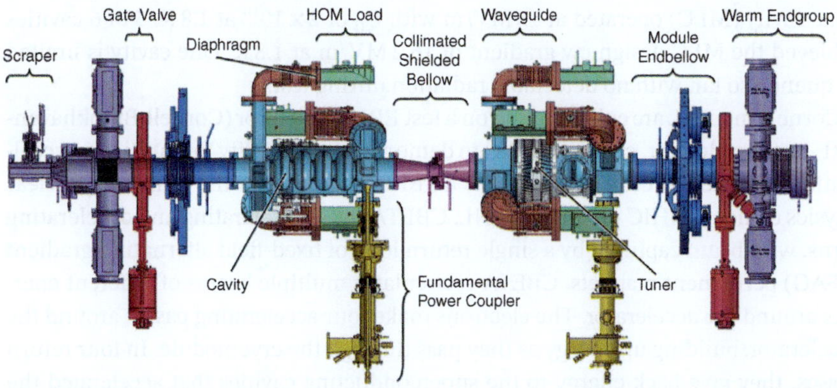

Figure 11.37 Two-cavity cryomodule in the SUPRALAB and bERLinPro facilities at HZB with HOM waveguide-damped cavities capable of carrying high current. Source: N. Wunderer et al., BESSY [574]/JACoW/CC BY 3.0.

11.6 Electrons in Energy Recovery Linacs (ERL) for Light Sources & Electron–Ion Colliders

An ERL-based light source is capable of producing smaller emittance, shorter bunches, and lower energy spread than proposed ultimate storage rings. ERLs are under development for X-ray light sources, FELs, electron–ion colliders and electron coolers. Some of the main challenges are cw operation of SRF installations at high Q_0 at medium gradients of 15–20 MV/m, as in LCLS-II. Precise phase and amplitude control of narrow-bandwidth SRF cavities are required. ERLs also require photo-injectors to produce high-brightness high current beams.

Although very active, the ERL field is still in the development and demonstration stage. Several ERL development projects [576] continue at JLab, Cornell, KEK, BNL, Berlin as well as ERLs for nuclear physics research, MESA at Mainz University, and Peking.

New ultimate storage ring designs come close to ERLs in terms of emittance, whereas shorter bunch length and lower energy spread are considered less critical at this stage. Rings that are proven in technology over ERLs can also store more current than ERLs. Hence light source facilities have favored the improved storage rings over ERLs for diffraction-limited hard X-ray beams.

Due to energy recovery the input power for ERLs is nearly independent of beam current, but small deviations can introduce strong effects so that the requirements for phase and amplitude stability become stringent. The RF power required is determined by the residual beam current, as well as phase and amplitude errors, and cavity resonant frequency fluctuation from microphonics.

11.6.1 Prototyping ERL Technology at Cornell

Cornell has developed injector and main linac prototypes for future ERLs [111]. The injector with five 2-cell cavities at 1300 MHz operated at 16 MV/m cw without beam, and with a world record beam current of 75 mA at 4 MV/m. A prototype main linac cavity (MLC) operated at 16 MV/m with Q_0 of 2×10^{10} at 1.8 K. 5 of 6 cavities achieved the MLC design cw gradient of 16.2 MV/m at 1.8 K. One cavity is limited by quench so far, with no detectable radiation during test.

Cornell and BNL are collaborating on a test ERL accelerator (Cornell-Brookhaven-ERL-test–accelerator, called CBETA) to demonstrate feasibility for multi-turn acceleration and energy recovery. CBETA is an ERL prototype for the 4 km long nuclear physics collider eRHIC planned at BNL. CBETA has 4 accelerating and decelerating turns, with beam captured by a single return loop of fixed-field alternating-gradient (FFAG) permanent magnets. CBETA recirculates multiple beams of different energies around the accelerator. The electrons make four accelerating passes around the accelerator, building up energy as they pass through the cryomodule. In four return passes, they give back energy to the superconducting cavities that accelerated the bunches. The high-brightness beam has applications in accelerator research, nuclear physics, and X-ray science. The CBETA requirement is 36 MeV per ERL turn. A beam with a total energy gain of 12 MeV was transported through the MLC, including active detuning compensation using the piezoelectric tuner, reaching the defined goal for the initial CBETA beam test.

Energy recovery has been achieved in the multipass accelerator (Figure 11.38). Energy delivered to the beam during the first four passes through the accelerating structure was recovered during four subsequent decelerating passes. The fixed-field alternating-gradient optical system used for the return loop successfully transported electron bunches of 42, 78, 114, and 150 MeV in a common vacuum chamber.

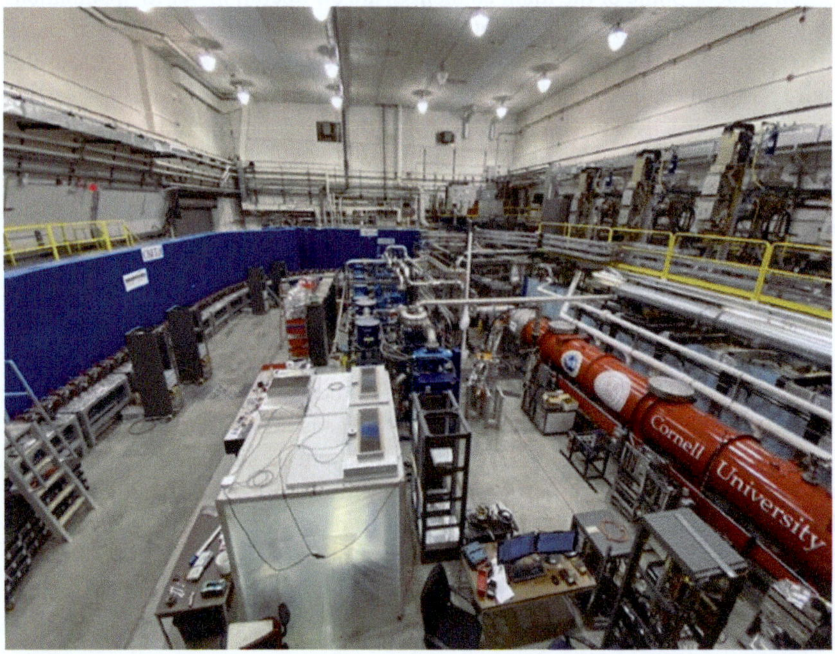

Figure 11.38 CBETA ERL demonstrator at Cornell with the SRF module in red. Source: G. Hoffstaetter, Cornell University [577]/CERN Talk, Courtesy of G. Hoffstaetter.

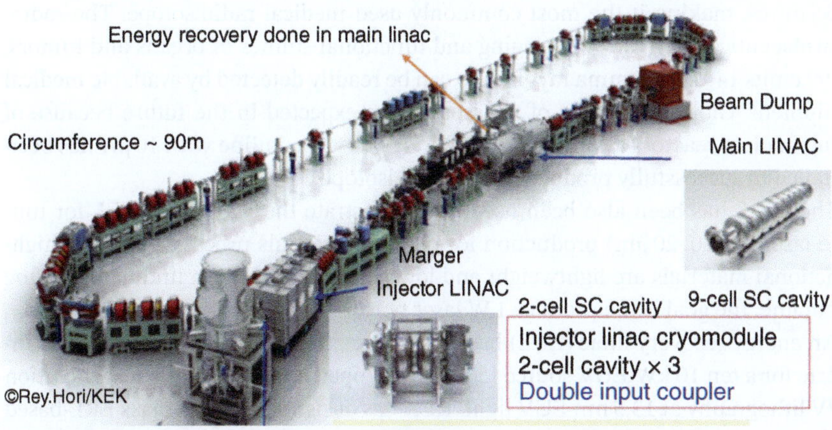

Figure 11.39 Compact ERL at KEK showing injector 2-cell and accelerator 9-cell cavities. Source: H. Sakai et al. & Umemori et al. [579]/JACoW/CC BY 3.0.

The multiple-pass energy recovery linac has the potential to accelerate much higher current than existing linear accelerators while maintaining small beam dimensions and consuming much less energy per electron.

With the successful demonstrations, CBETA is addressing important topics for other applications, such as medical isotope production, computer chip lithography, THz lasers, or high-energy X-rays from compton scattering. Inverse Compton scattering (ICS) can produce photon beams with properties and energies well beyond the limits of typical synchrotron radiation facilities. At 150 MeV, CBETA would be a significantly smaller accelerator system to produce X-rays with energies above 400 keV [578].

11.6.2 KEK ERLs

A Compact ERL (cERL) (Figure 11.39) has been constructed and operated at KEK as a 17.5 MeV prototype with ambitions for high current ERL-based light sources [580]. The prototype consists of two types of SRF cavities. Three injector 2-cell SRF cavities, and two main linac 9-cell SRF cavities. The beam operation started in 2013, with 100 nA (cw). Beam current increased to 1 mA (cw) without bream break up due to HOMs. Energy recovery was successfully achieved. The cERL injector cryomodule stably operated with beam for 3.5 years. The next goal is to increase beam current to 10 mA.

Heavy field emission in the main linac cavity that started during cryomodule operation was seen during beam operation, so that the performances of both the main linac and the injector cavities are sometimes degraded. One reason for degradation was sparking at beamline components due to charging. High-power pulsed RF processing helped to recover performance.

Various industrial applications are being explored at the prototype facility [581]. One example is non-fission production for the important medical isotope: Mo-99/Tc-99m. Tc-99m is used annually in tens of millions of medical diagnostic procedures, making it the most commonly used medical radioisotope. The radiopharmaceutical is in use for imaging and functional studies of organs and tumors. Tc-99 emits 140 keV gamma rays and so can be readily detected by available medical equipment. Global shortages of the isotope are expected in the future because of aging nuclear reactors that now produce it. A Mo-99 beamline with 10 μA has been installed to successfully produce the medical isotope.

The cERL has been also been used to demonstrate the ERL-SASE-FEL for tunable laser IR (10–20 μm) production for organic materials processing. These high-functional materials are lightweight and low-cost. With two 2-m undulators in the beam line, the goal is the produce 1 W laser power.

An energy recovery linac (ERL)-based free electron laser (FEL) is a possible candidate for a ten 10 kW EUV source [582]. It will open a new era for next-generation EUV lithography at 13.5 nm EUV light. KEK has designed a 10 mA class ERL-based EUV–FEL source to generate more than 10 kW power at 800 MeV beam energy. Key technologies to realize are acceleration gradient of 12 MV/m for reliable operation without field emission, and to strongly suppress HOMs. The cavity has been prototyped through the construction of the compact ERL.

11.6.3 Light-House Project for Radiopharmaceuticals

A Belgian company Institute for Radio Elements (IRE) has teamed up with advanced semiconductor materials lithography (ASML) and others to develop a new, non-fission production method for important medical isotopes such as Mo-99/Tc-99m in anticipation of near-future global shortages [583]. The consortium is planning to use high-energy, high-current electrons from an ERL injector (10 MeV, 75 mA) to generate the medical isotopes.

11.6.4 Peking ERL

Peking University is developing an ERL-FEL [558]. The system consists of a DC–SC photocathode injector, a superconducting linac composed of two nine-cell TESLA-type cavities. The objectives are to provide an infrared FEL, and to build a test bed for the study of beam dynamics and accelerator technology for energy recovery. The design accelerating gradient of the $3 + 1/2$ cell photocathode injector is 15 MV/m. Electrons will be accelerated to 30 from 5 MeV at a gradient of about 13 MV/m.

11.6.5 Berlin ERL

bERLinPro's is a single-pass ERL prototype [584, 585]. Its 6.5 MeV injector consists of a 1.3 GHz SRF 1.4-cell gun followed by a booster section with three 2-cell cavities. The beam is merged into the main linac and accelerated by three 7-cell cavities to 50 MeV. Beam is recirculated for energy recovery and finally is dumped in a 650 kW, 6.5 MeV beam dump. The Booster cavities are of Cornell ERL type with modified fundamental power coupling (FPC) section to accommodate modified KEK c-ERL coupler. 4 cavities were fabricated at JLab and tested well. Three cavities have helium vessels welded and tested well again.

11.6.6 MESA ERL

The Institute of Nuclear Physics at the Johannes Gutenberg University Mainz, Germany is constructing the mainz energy-recovering superconducting accelerator (MESA) [586]. The centerpiece consists of two superconducting cryomodules of the ELBE/Rossendorf type (Figure 11.34), which are operated at 1.8 K.

11.6.7 SRF Photo-injectors for ERLs

Several ERL projects have chosen SRF guns as the injector [587]. The SRF gun is a demanding component of a cw low emittance electron injector. RF guns have to operate at high accelerating gradients to suppress space charge effects which dilute emittance. SRF guns are attractive because they dissipate orders of magnitude less power at high gradient, cw operation. The challenge is the integration of a nonmetallic cathode of high quantum efficiency into a clean superconducting cavity, and to simultaneously achieving high Q_0, and high E_{pk} in the cavity, as well as high-quantum efficiency in the cathode. The material of the cathode can also contaminate the cavity.

Several laboratories are working on the development of SRF Guns: HZDR at Rossendorf, DESY, HZB at Berlin, BNL, KEK, and PKU Beijing. Many laboratories have achieved high E_{pk} at high Q_0 in the naked cavity, but it has been a challenge to maintain those results when a cathode stalk or cathode is inserted. First beams were obtained from several SRF guns. We give only two examples of detail results from an extensive collection of work.

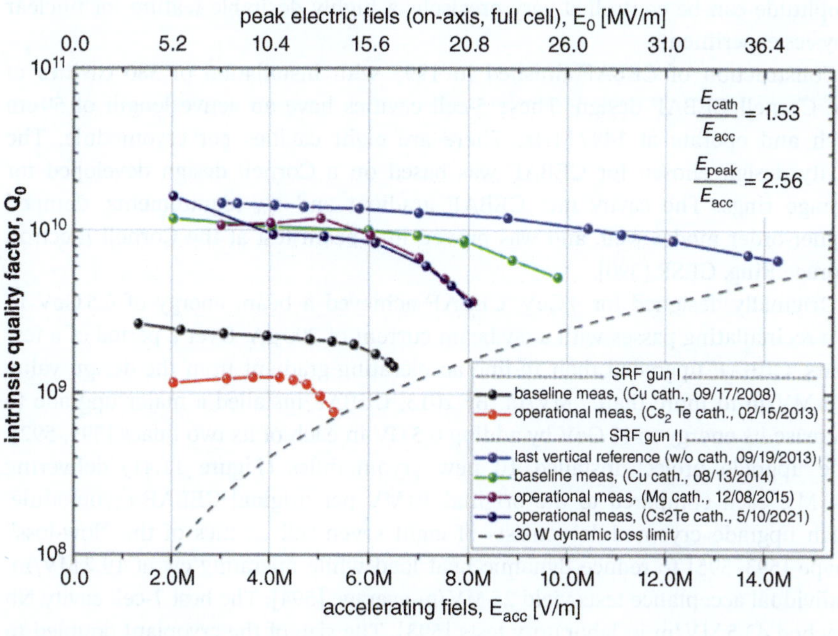

Figure 11.40 SRF Gun cavity performance with and without various cathodes present. Source: A. Arnold et al., Rossendorf [588]/JACoW/CC BY 3.0.

PKU achieved $E_{pk} = 47$ MV/m at Q_0 of 1×10^{10} in the bare cavity. During the beam test a quantum efficiency of 4% was first seen with a Cs2Te cathode but degraded to 2% by the end of the test. This is still an acceptable value.

At ELBE (electron accelerator for beams with high brilliance and low emittance), the first beam test was carried out with a copper photocathode with gradients up to 9 MV/m. For SRF gun 1, installation of a Cs_2Te photocathode was not successful as it resulted in very low-quantum efficiency and contamination of the gun cavity. SRF gun 2 was more successful. More than 3400 hours of user beam have been generated since 2019. A total of 20 cathodes (2 Cu,12 Mg, and 6 Cs_2Te) were used, but no serious cavity degradation was observed. Figure 11.40 shows the performance with and without cathodes for two versions of the gun [588].

11.7 Electrons for Nuclear Physics, Nuclear Astrophysics, Radio-Isotope Production

11.7.1 CEBAF at Jefferson Lab

At the frontier of nuclear science, the continuous electron beam accelerator facility (CEBAF) at Jefferson Lab has been the largest SRF installation in the USA [589]. CEBAF improves basic understanding of nuclear matter by elucidating the quark and gluon structure of protons and neutrons. SRF cavities offer special advantages: high average current, low peak current, continuous beam, and excellent beam quality with low energy spread and low transverse emittance. Because of the highly stable operation possible with a cw superconducting linac, the RF phase and amplitude can be controlled very precisely, a highly desirable feature for nuclear physics experiments.

Construction of CEBAF finished in 1993 with installation of 380 cavities of the Cornell/CEBAF design. These 5-cell cavities have an active length of 50 cm each and operate at 1497 MHz. There are eight cavities per cryomodule. The cavity design chosen for CEBAF was based on a Cornell design developed for storage rings. The cavity met CEBAF gradient and Q_0 requirements, damped higher-order modes well, and was proven in a beam test at the Cornell Electron Storage Ring, CESR [590].

Originally designed for 4 GeV, CEBAF achieved a beam energy of 6.5 GeV in five recirculating passes with a cw beam current of 200 μA. Over a period of a few years, CEBAF upgraded their in-line accelerating gradient from the design value of 5 MV/m to more than 7 MV/m. By 2015, CEBAF installed a major upgrade to increase its energy to 12 GeV by adding 0.5 GV in each of its two linacs [591, 592]. The upgrade project installed 10 new cryomodules (Figure 11.41) delivering 100 MV each compared to the original 30 MV per original CEBAF cryomodule. Each upgrade cryomodule consists of eight seven-cell cavities of the "low-loss" shape [593–595] to reduce dynamic heat load while operating cw at 19.2 MV/m. Individual acceptance tests yield 25 MV/m average [594]. The best 7-cell cavity Nb reached 43.5 MV/m in laboratory tests [593]. The size of the cryoplant doubled to provide 10 kW total at about 2 K.

Figure 11.41 JLab 12 GeV Upgrade linac installation. Source: Jefferson Lab News [596]/Photo Courtesy of Jefferson Lab/U.S. Department of Energy Office of Science.

11.7.2 ARIEL at TRIUMF

The Advance Rare Isotope Laboratory (ARIEL) is a MW class cw superconducting electron linac under installation at TRIUMF as a driver for radioactive beam and isotope production [597–599]. The rare isotope facility is unique in using photofission generated by a high-β electron linac instead of the more usual low-β ion linacs. In its final configuration, the e-linac is planned to consist of five 1.3 GHz nine-cell cavities housed in three cryomodules with one single-cavity injector cryomodule (EINJ) and two double-cavity accelerating cryomodules (Figure 11.42) to accelerate up to 10 mA of electrons to 50 MeV. The e-linac is being installed in stages. A demonstrator phase (2014) consisting of a 300 kV electron gun, EINJ, and a partially outfitted accelerating module with just one accelerating cavity was installed for initial technical and beam tests to 23 MeV. A Stage 2 upgrade has completed the acceleration module with two cavities to reach an operational goal of 3 mA of electrons to 30 MeV for first science from the ARIEL ISOL targets. After RF conditioning, both cavities achieved an acceleration gradient of 10 MV/m at Q_0 of 10^{10}, meeting the design goal. 31 MeV has been achieved to deliver 10 kW beam. The goal is for 100 kW by 2024.

11.7.3 ERL for LHeC at CERN

CERN is developing ERL technology for the LHeC [601] to turn the LHC into a precision Higgs factory (Section 12.1), or for further frontier physics discoveries. LHeC will use two one km long electron linear accelerators arranged in a racetrack configuration, tangential to the LHC tunnel. The electron energy will be 60 GeV

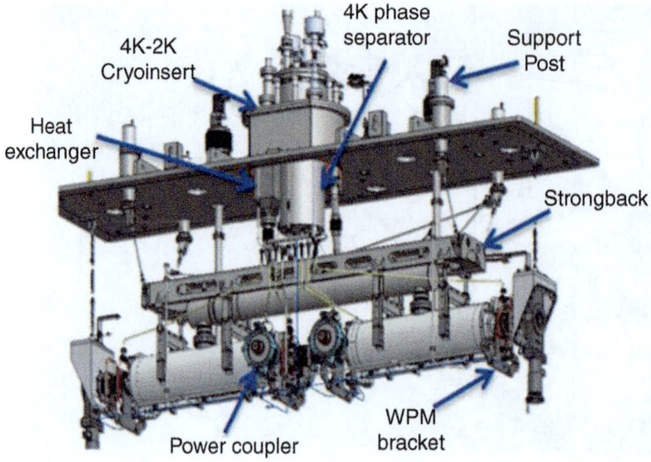

Figure 11.42 ARIEL two-cavity box cryomodule. Source: R. Laxdal et al., TRIUMF [600]/JACoW/CC BY 3.0.

in a 3-turn operation with beam current of 10–20 mA. A demonstrator facility (PERLE – Powerful ERL for Experiments) is proposed by a multi-lab collaboration [602]. PERLE will develop an SRF cavity cryomodule that comprises four 5-cell 801 MHz cavities with a 15 – 20 MV/m gradient operated in cw mode. The cavities will have a high current cell shape to efficiently propagate HOMs, and HOM dampers will be developed to handle 1 kW HOM power.

11.8 Crab Cavities for LHC High Luminosity

CERN (LHC) and collaborators are developing two types of 400 MHz crab cavities (Section 10.1) to be installed in LHC by 2025 to increase LHC luminosity [444, 454]. The large beam aspect ratio combined with the finite crossing angle reduces the overlap between colliding bunches, and thereby reduces the luminosity. The crab cavities rotate the bunches before and after the collision point to recover the optimum overlap of head-on collisions between colliding bunches. The expected peak luminosity gain by crabbing the beams is 65% for HL-LHC.

Very tight transverse space constraints in the HL-LHC triggered the development of novel and compact RF deflecting structures (Section 10.1). One cavity geometry (double quarter wave – DQW) is for vertical beam crossing and other (RF dipole – RFD) is for horizontal beam crossing. Two cavities each will be needed per beam per IP for the two detectors for a total of 16 cavities in 8 CMs. A full crab compensation in HL-LHC requires a kick voltage of up to 9 MV per beam per side of each collision point. 4 DQW and 2 RFD cavities are prepared and tested to reach kick values of 4.4–5.8 MV as compared to design for 3.4 MV (corresponding $E_{pk} < 40$ MV/m and $H_{pk} < 70$ mT). The dynamic heat load expected is 5 watt per cavity.

A first demonstration experiment of crabbing with hadron beams was successfully carried out with high-energy protons in the SPS [454]. The bunch rotation was measured to be the expected 17 mrad. It was possible to drive both cavities in counterphase to demonstrate that the cavities can be made transparent to the beam.

11.9 Ongoing and Near-Future Projects Summary

Summarizing the status of SRF applications (2010 – 2022) in the three categories of accelerators, the heavy-ion low-β accelerators have completed installation of about 50 CM with about 350 cavities, the majority for FRIB. The plans described in this section call for future installations of an additional 150 CM with about 600 cavities for projects such as RAON, HIAF, and others described above.

Among the high-intensity proton accelerators, the completed installations total about 30 CM with about 130 cavities, the majority for SNS. With ESS and PIP-II well on the way, along with ambitious plans at CiADS, PIP-III, and others, the total SRF installations forthcoming add up to an additional 260 CM with more than 1200 medium to high-β cavities.

Finally, the $\beta = 1$ electron accelerators installed in the decade after 2010 include 135 CM with more than 1080 cavities, mostly for E-XFEL and LCLS-II. For the many linac-based light sources underway, such as SHINE, or under discussion, such as PERLE for the Higgs factory, the impressive totals reach more than 380 CM with nearly 2000 cavities.

Thus, the grand total number of cavities for all three classes would approach 4000, which is five times the number of cavities installed for E-XFEL, the largest SRF facility in operation.

12

Future Prospects for Large-Scale SRF Applications

12.1 The International Linear Collider (ILC) for High-Energy Physics

The goal of high-energy physics is to advance our understanding of nature at the most fundamental level. Comprehending the basic features of energy, matter, and space–time remain the focus. Current knowledge is captured by the standard model (SM) of particle physics which became the remarkably successful paradigm through a vast range of measurements and theories. The SM is our best knowledge with strong predictive capabilities about the elementary building blocks of matter, i.e. quarks and leptons. Tested rigorously to a part in a billion, it is the best answer to the question – what are we made of?

But there is still major unfinished business in several realms. Many mysteries demand a deeper explanation and a more unified picture. The SM does not account for gravity. Astrophysical data clearly establishes the existence of a new form of matter, called dark matter, which holds galaxies together, and is present everywhere, but so far has remained a mystery. The mass of dark matter present is roughly five times the mass of familiar, ordinary matter. There must be an elementary particle, or particles, that account for dark matter just as all matter is made up of the SM particles.

The existence of dark energy has also been established by astrophysical data from the accelerating expansion of the universe. It seems to be driving our universe apart. It dominates (70%) of the mass-energy content of the universe. As a candidate explanation, the "vacuum-energy" of quantum fields in the SM is excessively too large to account for the dark energy.

There are many other lingering questions. How did matter in the universe survive over antimatter, when both types were created equally in the big bang? Why did one matter particle out of a billion particle/antiparticle annihilations survive, so that matter-based beings like us could exist? What caused this small asymmetry between matter and antimatter? Why are there so many types of particles, three quarks, three leptons, and three neutrinos, each, together with their corresponding antiparticles? Why do the particles have such a large range of masses? If neutrinos are massless matter particles in the SM, why do neutrinos show mass?

There are many theories proposed to confront the mysteries: supersymmetry, extra spatial dimensions, and others. Many predict the existence of new particles close to

Superconducting Radiofrequency Technology for Accelerators: State of the Art and Emerging Trends, First Edition. Hasan Padamsee.

the Higgs mass, which provides excellent discovery prospects for the new accelerators discussed today. New, light particles that couple only very weakly to ordinary matter including axions, axion-like particles, and dark photons, also arise in extensions of the SM. But no such particles beyond those in the SM have been detected directly. Clever ways to search for such exotic particles go on.

In 2012, the LHC discovered the Higgs boson. It was the last missing puzzle piece in the SM, long thought to exist along with the SM particles. The Higgs boson is the physical manifestation of the Higgs field present everywhere. The Higgs field is responsible for endowing particles with mass. A particle such as a heavy quark with a strong coupling to the Higgs field shows high mass as if it has to move through thick molasses. A particle with low mass, such as the electron, glides through more easily and shows weak coupling to the Higgs field. Measuring the couplings of the Higgs field to all the known particles becomes a powerful method to test some of the new theories beyond the SM.

The couplings of the Higgs boson to the decay particles detected at the LHC agree with those predicted in the SM, at the available level of 20% accuracy. This level is insufficient to distinguish between predictions of new theories beyond the SM. For a physics discovery beyond the SM, the deviations are expected to be of the order of 5%, so that a 1% precision is desirable. Needed is a massive, factory-like production of Higgs bosons in a clean environment, with as few stray events as possible, to successfully detect the decay of the Higgs boson into many of the lighter particles.

Such high levels of precision will be possible with the international linear collider (ILC). The ILC will also carry out searches for postulated new particles in regions that are very difficult for the LHC to access.

There is general agreement in the world high energy physics community that an electron–positron collider Higgs factory is one of the highest priorities [603], and that the ILC with center-of-mass energy of 250 GeV is the best option available to realize the Higgs factory in a timely fashion. To quote from the ICFA report [604], "the discovery of a Higgs boson in 2012 at the large hadron collider (LHC) at CERN is one of the most significant recent breakthroughs in science. Precision studies of the Higgs boson will further deepen our understanding of the most fundamental laws of matter, and its interactions, helping to elucidate presently known deviations from the standard model of high energy physics. The ILC operating at 250 GeV center-of-mass energy will provide excellent science from precision studies of the Higgs boson. Therefore, ICFA considers the ILC a key science project complementary to the LHC and its luminosity upgrade. ICFA emphasizes the extendibility of the ILC to higher energies and notes that there is "large discovery potential with important additional measurements accessible at energies beyond the initial 250 GeV."

The ILC will be an international project, anticipated to be led by a Japanese initiative. The ILC will be an excellent tool for both the understanding of the newly discovered Higgs, and the exploration of new frontiers. Continued R&D on SRF science and technology discussed extensively in this book shows strong potential for increased gradients, increased Q_0's, and for the emergence of new technologies to make the energy upgrades of ILC affordable and realizable in the future.

Figure 12.1 Conceptual layout of the ILC tunnel, cryomodules, and RF sources. Source: T. Behnke et al. [609]/Springer Nature/CC BY 4.0.

The ILC-250 will be based on 800 9-cell, 1.3 GHz cavities installed into 1000 cryomodules of the EXFEL type (Figure 12.1). The unprecedented scale of this SRF cavity application will require worldwide laboratory and industrial participation. Substantial SRF capabilities and infrastructure exist at laboratories all over the world through realization of the accelerator applications described in previous Chapter 11.

The main challenge is to reproducibly obtain or exceed the technical design report (TDR) [605] cryomodule specification of 31.5 MV/m in operation. This demands that the cavities demonstrate at qualification at least 10% higher gradient i.e. 35 MV/m with a high yield before they are installed into the CM. The TDR assumed a first pass (preparation) yield of 75% for 35 MV/m. EXFEL experience (Section 11.4.1) shows a substantial performance gain and yield increase on re-treatment of contaminated cavities by HPR. ILC TDR assumed a yield after second pass of 90% for 35 MV/m.

The expected cavity yield has been demonstrated by the EXFEL production. The cavity performance for the EXFEL cavities is close to the requirements for the ILC TDR. The yield at 35 MV/m maximum gradient for the ILC recipe treatment (i.e. final EP) cavities was 95%, after re-treatment by HPR of a small number of cavities below 20 MV/m. For the usable gradient, the yield was 82% for 33.4 MV/m. The 417 cavities from one vendor that followed the ILC EP recipe for final treatment succeeded in reaching a maximum gradient of 34.7 (±4.4) MV/m. More than 10% of cavities from this vendor exceeded 40 MV/m.

The average cryomodule gradient reached was 27.5 MV/m (±4.8). Clearly, cryomodule performance needs to improve further to reach ILC specs (31.5 MV/m). It is very encouraging that 18 out of 97 cryomodules reached the administrative limit of 30 MV/m, which is close to the ILC-TDR specification. 47% of all cavities tested in 97 modules reached the same 30 MV/m. At Saclay, where the CMs were assembled, there was a strong effort to improve cryomodule yield by careful scrutiny of all procedures. Resulting benefits were very encouraging. The cavity experience from DESY and the cryomodule experience from Saclay are very encouraging for ILC and will be very valuable to ILC realization.

Two labs have sought to demonstrate the higher performance ILC CM goal, but on a smaller scale. An ILC demonstration cryomodule prepared by Fermilab reached 31.5 MV/m [606]. A beam through the CM reached 255 MV which is equivalent to 31.9 MV/m for 8 active meters of cavity. Eight out of twelve cavities in a demonstration cryomodule at KEK test facility (STF) reached the ILC goal of 33 MV/m pulsed operation [607, 608]. Two out of eight cavities in the CM achieved 36 MV/m.

The demonstrated success from the EXFEL cavity and cryomodule production is very encouraging for reaching the ILC gradient goals. It is rare that a ≅ 10% prototype demonstration exists for a new high-energy physics (HEP) machine, and for its core technology. None of the other HEP machine proposals have reached a comparably mature level of technology.

Extensive SRF infrastructure exists worldwide for cavity fabrication, surface treatment, clean assembly, cold testing, and cryomodule assembly. Major facilities for electron accelerators are available at DESY, CERN, INFN, Saclay, Orsay, INFN, KEK, JLab, Cornell, Fermilab, MSU, and at several industries around the world. New infrastructure is available for upcoming projects such as for ESS in Europe, at PAPS, and at Wuxi in China. New industries in S. Korea, China, and Japan are rapidly growing familiar with SRF technology, and its related production needs. The projects described above show a wide utilization of SRF technology which will keep the facilities in high working order to be ready for ILC needs in the 2020's.

The status and prospects for SRF together with the availability of the complete ILC TDR and recent staging studies make ILC an attractive, technology-mature proposal. However, long delays in decision-making by Japan could result in a situation where SRF activity passes its levels of high peak activity and possibly starts to decline.

Other paths to the Higgs factory are the future circular collider, electron–positron (FCCee) [610] or the compact linear collider (CLIC) [611] in Europe, and the circular electron–positron collider (CepC) [612] in China. All these future accelerators are in the conceptual design report (CDR) stage. With a technical design report (TDR) completed some years ago, the ILC remains the most technologically ready and mature of all possible Higgs factories options for an expeditious start, a proposal that falls in the category of "shovel-ready" as compared to competing approaches.

12.2 Future Circular Collider FCCee

The future circular collider (FCC) envisions a staged approach. The first stage will be a circular electron–positron collider (FCCee), With a 100 km circumference tunnel, it will run in Z-factory, W-factory, H-factory, and top-factory modes. This will be followed in the subsequent decades by a 100 TeV proton–proton collider with advanced technology superconducting magnets.

A conceptual design of the FCC SRF system is proposed along with highlights of specific R&D topics to reach the design performance. Many challenges related to structure design, power input, and higher order modes must be addressed. The FCCee SRF system must satisfy significantly different requirements for the many stages of operation. A single-system design to meet all four running states is not efficient [613]. The highest energy will be 175 GeV (Top factory) with a few mA. The highest beam loaded case will be at 45.5 GeV (Z factory) with 1.45 A, where higher order mode power will be a major issue. The RF voltage requirement spans from 0.1 to 11 GV.

For the high beam current Z-factory, the cavity shape must be optimized with respect to higher order modes (HOM). This favors low frequency, low shunt resistance, and a low number of cells per cavity. Each of the two beams will be supported by a 400 MHz system of Nb/Cu cavities. The low frequency will also be compatible later for the FCChh, where the LHC as the injector also operates at this frequency. A large part of the hardware and infrastructure can be reused.

The 400 MHz single-cell system could be built with today's technology near 5 MV/m for Nb–Cu, but improvements to 10 MV/m will offer considerable savings. An R&D program on Nb thin-film coated Cu cavities (Section 9.1) operating at 4.5 K is underway to decrease the surface resistance at high RF fields by factors of two to three. R&D is focusing on Nb/Cu produced by high-power impulse magnetron sputtering (HIPIMS), which will improve the microstructure of the coating due to the larger energy available during film growth. Any progress on substrate manufacturing and preparation will also have an impact, as demonstrated by the seamless cavities produced for the HIE-ISOLDE project (Section 9.1.3), where the Q slope with increasing field was reduced compared to the welded counterparts [614]. With the highest acceleration voltage need, the top factory (175 GeV) and the booster will need bulk Nb 4-cell, 800 MHz cavities at 20 MV/m.

Table 12.1 [363] shows the present plan for the various stages with two alternatives each for the Z, W, and H factories, the choices depending on the level of development

Table 12.1 Plan (with alternatives) for SRF cavities and cryomodules for FCCee [363].

	Z	**Z (alt)**	**W**	**W (alt)**	**H**	**H (alt)**	**T**	**Booster**
Number cryomodules	72	36	56	28	102	102	159	159
Material	Nb/Cu	Nb/Cu	Nb/CU	Nb/Cu	Nb/Cu	bulk NB	bulk NB	bulk NB
Temperature (K)	4.5	4.5	4.5	4.5	4.5	2	2	2
Frequency (MHz)	400	400	400	400	400	800	800	800
Acc. gradient (MV/m)	5	10	5	10	10	20	20	20
Number of cells	1	1	4	4	4	4	4	4
Number of cavities/CM	4	4	4	4	4	4	4	4
Number of cavities	288	144	224	112	408	408	636	636

achieved. The maximum number of cavities for FCCee and the Booster will be 2192, if the lower gradient (fall back) options have to be used.

12.3 China Electron–Positron Collider, CEPC

In China, CEPC [612] has developed a similar concept for a 100 km circular electron–positron collider operating at 90–240 GeV center-of-mass energy of Z, W, and Higgs Bosons physics. The top factory is not included. After a decade or more running for lepton physics, CEPC will be replaced by a 100 TeV center-of-mass proton–proton collider with advanced technology superconducting magnets. Significant technology demonstration is underway.

The SRF system [615] is one of the most challenging accelerator systems due to the wide range of beam energy and current. CEPC will use a 2-cell, 650 MHz SRF cavity system with 672 cavities for the collider, as the maximum needed for the Higgs factory, and a 9-cell, 1.3 GHz SRF system with 256 cavities for the booster. The grand total SRF system will be 144 CM and 928 cavities. Each of the 11m long collider cryomodules contains six 650 MHz 2-cell cavities, and each of the 12 m long Booster cryomodules contains eight 1.3 GHz 9-cell cavities. Maximum operating gradient for 650 MHz will be 14 MV/m at Q of 1×10^{10}. For the booster, the maximum gradient will be 17 MV/m at a Q of 2×10^{10}. Table 12.2 shows the main ring SRF system parameters and Table 12.3 shows parameters for the booster.

CEPC SRF R&D and industrialization will be carried out in synergy with ongoing projects in China. A large SRF infrastructure facility (PAPS) is being built at the Huairou science city near Beijing and another facility at Wuxi for SHINE. The SRF prototype program (to 2022) is to build two small test cryomodules (650 MHz 2×2-cell, 1.3 GHz 2×9-cell) followed by two full-scale prototype cryomodules (650 MHz 6×2-cell, 1.3 GHz 8×9-cell) (Figure 12.2, Tables 12.2 and 12.3).

Table 12.2 CepC Main Ring SRF Parameters.

CepC stage	H	W	Z-HL
RF voltage (GV)	2.1	0.41	0.14
Beam current/beam (mA)	19.2	97.1	466
Cavity number in use/beam (650 MHz 2-cell)	336	96	48
Gradient (MV/m) (with margin for HV-H)	14	9.3	6.3
Input power/cavity (kW) (with margin for HL-H)	190	333	335
HOM power/cavity (kW)	0.4	0.3	1.8
Cryomodule number (6 cavities/module)	56	32	16
Qo @ 2 K at operating gradient (long term)	1E10	1E10	1E10
Total wall loss @ 4.5 Keq. (kW)	23	6	1

Table 12.3 CepC SRF parameters for the Booster.

CepC stage	H	W	Z-HL
Cavity number in use (1.3 GHz TESLA 9-cell)	160	64	32
Gradient (MV/m)	16.9	15.1	12.0
Q_L	2E+07	2E+07	2E+07
Input power per cavity (kW) (remained detuning 10 Hz)	5.8	4.0	2.5
HOM power per cavity (W)	0.4	0.2	1.6
Cryomodule number in use (8 cavities per module)	20	8	4
Qo @ 2 K at operating gradient (long term)	2E+10	2E+10	2E+10
Total wall loss @ 4.5 K eq. (kW) (assume CW)	8.4	2.7	0.9

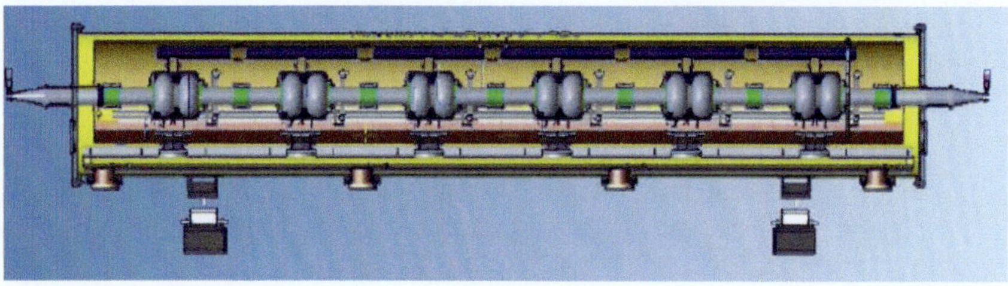

Figure 12.2 CEPC collider ring 2-cell cavity and CM. Source: J. Zhai et al. [615]/JACoW/CC BY 3.0.

Mid-T baking with its N-doping collateral effect is under exploration for 650 MHz cavities (Section 4.3) with encouraging results.

In summary, compared to the ILC which will need 8000 × 1.3 GHz cavities, FCCee will need a 2192 cavities total, and CepC will need 928 cavities, with the majority of cavities at lower frequencies (400, 650 or, 800 MHz) for both rings.

13

Quantum Computing with SRF Cavities

The aim of this Section is to summarize in highly simple terms the status and potential applications of quantum computing based on SRF cavities, with a view to exciting prospects for the future. The arrival of SRF-based quantum computing opens pathways to transformative advances.

13.1 Introduction to Quantum Computing

Quantum computing is a fusion of quantum physics and computer science, harnessing the dual phenomena of quantum *superposition* and quantum *entanglement* for massive computations. In analogy with "classical computing circuits" called "bits," the unit of quantum computing is a "qubit." A bit can be 1 or a 0 since it is just a switch with two possible configurations, "on" or "off." A classical bit is like a coin, which is either heads (0) or tails (1). A set of *n* coins can be described as a probabilistic mixture of *2n* states, but the set is always in only one of these states.

13.2 Qubits

A quantum bit or *qubit* can take on a quantum superposition of an infinite range of values between 1 and 0. The Bloch sphere (Figure 13.1) provides a convenient way to visualize the superposition state of a single qubit, which can be a rotating bloch vector inside the sphere, starting at the origin and pointing to any value on the sphere's surface. As a simpler (but cruder) analogy, a qubit behaves like a spinning coin. Each coin can be *simultaneously* in both states, heads and tails, with each state having a probability (of 50%) when landed. If two coins spin at the same time they would represent a superposition of four states. Three coins would represent eight possible states. 50 spinning coins would represent 2^{50} (10^{15}) states, which is more states than possible with the largest supercomputer today. Three hundred coins would represent more states than there are atoms in the universe [616].

Superconducting Radiofrequency Technology for Accelerators: State of the Art and Emerging Trends, First Edition. Hasan Padamsee.

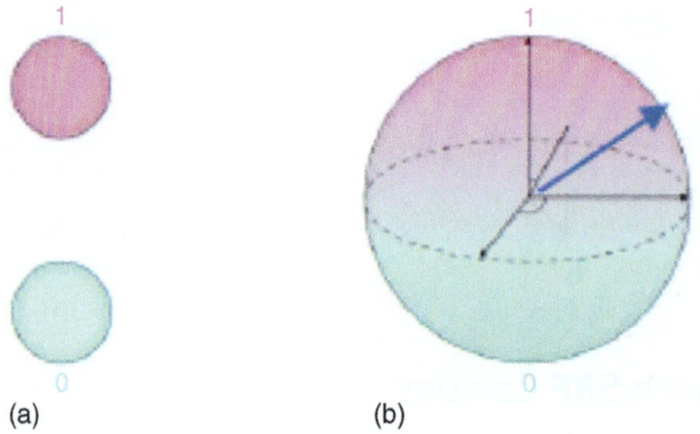

Figure 13.1 Comparison between (a) classical and (b) quantum bit.

13.3 Superposition and Coherence

Continuing the analogy to spinning coins, for successful quantum computing, it is necessary to keep the coins spinning in the superposition state for a long time to process the information by manipulating the qubit states. The longer the time, the higher the *coherence*, the longer the qubit states can be manipulated to carry out operations.

The quantum state of superposition is rather fragile due to interaction with the environment, and other factors. Quantum computation has to be completed before decoherence kicks in and scrambles the qubits, leading to errors. The *coherence time* of the qubit is a key performance measure. The slightest disturbance, such as vibration, or change in temperature, can cause qubits to tumble out of superposition before the job is done.

13.4 Entanglement

Entanglement is another essential requirement of quantum computing. Entanglement is an enigmatic characteristic of the quantum state of two or more interacting particles. When a pair of qubits become entangled, the two members exist in a single, correlated quantum state. Entangled qubits affect each other instantly when measured, no matter how far apart they are, even if (in principle) they are separated at opposite ends of a galaxy! Changing the state of one of the qubits instantaneously changes the state of the other qubit in a predictable way. Measuring one member of a pair of entangled qubits, directly gives the properties of its partner without having to test it. The entangled state contains information that the states are correlated.

Entanglement greatly enhances computational power and scalability. Entanglement makes it possible to change the state of multiple qubits simultaneously. Using entanglement, qubits work together, as multiple qubits with hidden correlations. With each additional qubit entangled, a quantum computer can explore double the number of possible solutions, an exponential increase not possible with classical machines. Entanglement becomes a computational multiplier. Like superposition, quantum entanglement is also fragile, subject to decoherence due to the coupling of the quantum system to the environment.

Quantum logic is performed via microwaves on a set of single-qubit and two-qubit gates. Single-qubit operations translate an arbitrary quantum state from one point on the Bloch sphere to another point by rotating the Bloch vector a certain angle about a particular axis. Two-qubit gates are generally "conditional" gates that take two qubits as inputs. Typically, the first qubit is the "control" qubit, and the second is the "target" qubit. Another type of transformation is a phase-shift operation, that changes the phase of qubits. Phase changes introduce a new dimension to quantum computation and eventually lead to phase cancellations via interference toward building the final result.

A central challenge in developing quantum computers that will eventually outperform classical computers is to increase the coherence time of qubits to increase the number of computations possible. *Quantum error correction* helps to counteract loss of coherence, but requires more qubits, and so introduces more complexity.

In classical computers, the zeros and ones so rarely switch by accident that little error correction is necessary. Error-correcting codes rely on copying the information many times to ensure correct representations. Errors can be corrected, for example, using the *majority rule*: A desired bit, whether 1 or 0, is first triplicated as 111 or 000. Later, even if one of the three bits gets corrupted, the other two "outvote" it and allow recovery of the original data. Binary data can also be easily checked for errors with a "checksum."

Quantum error correction has many challenges. Looking at a quantum state destroys the information contained within, making it impossible to copy a quantum state. No cloning prevents backup of quantum data. There can also be no checking similar to checksum because checking requires measurement, which destroys the quantum state of superposition.

But entanglement can be used for error correction. It is possible to measure *relationships* between entangled qubits without measuring the values stored by the qubits. The strategy as with classical error correction is to add additional qubits and gates. Instead of using a single qubit of data, use three entangled physical qubits with correlated states. It is possible to determine whether the first and second qubit maintain their correlation and whether the second and third qubit have the same relation, without determining what that state is, and without destroying the state. These two measurements suffice to determine whether one qubit is flipped, and which of the three it is, without providing any information about what the actual values are. If one of the qubits turns out to disagree with the other two, it can be reset to its proper value by performing an appropriate gate operation to flip the state, even without knowing its value. The method resembles the "majority rule" method for classical computing.

Integrating a large number of qubits into a single chip, requires more electronic controls, which means more interactions, which reduces the overall coherence time. While applying quantum gates, the interaction between components causes errors to quickly spread through the system. The challenge is to apply error correcting codes to a huge number of qubits in a single system and to maintain a low decoherence rate. These strategies impose a huge debt of computational overhead – all computing power goes to correcting errors, and not to running algorithms. Unless coherence times are long and error detection and correction rates get small, error correction is what a quantum computer will do most of the time. The best two-qubit quantum gates have an error rate of around 0.5%, meaning one error for every 200 operations, compared to the standard classical circuit, where there's about one error for every 10^{17} operations.

13.5 2D SRF Qubits

Any quantum system with two distinguishable basis states can be used for a qubit. There are many types of qubits. In the conceptually simplest case, a qubit uses two energy levels of an atom that interacts with light to activate the transition between the ground state and the excited state. Quantum electrodynamics (QED) governs the physics of the interaction. But the coupling between the atom and the interacting light is normally very weak. An optical cavity with mirrors is used to enhance the interaction. Photons pass by the atom multiple times while bouncing between the cavity mirrors. A single atom can induce large enough changes in the electromagnetic fields of the optical cavity light waves to allow measurements.

Electrical circuits have been devised that behave exactly like the optical cavity–atom system. In the 2D circuit version, a harmonic oscillator circuit consists of inductors and capacitors. The inductors are made from superconducting elements for high Q, leading eventually to higher coherence. But the basic 2D harmonic oscillator circuit cannot provide qubit states. All the different states form a ladder with *equal spacing* so that photons drive the system from one state to the next higher energy state, or all the states up the ladder. To operate as a qubit, two adjacent

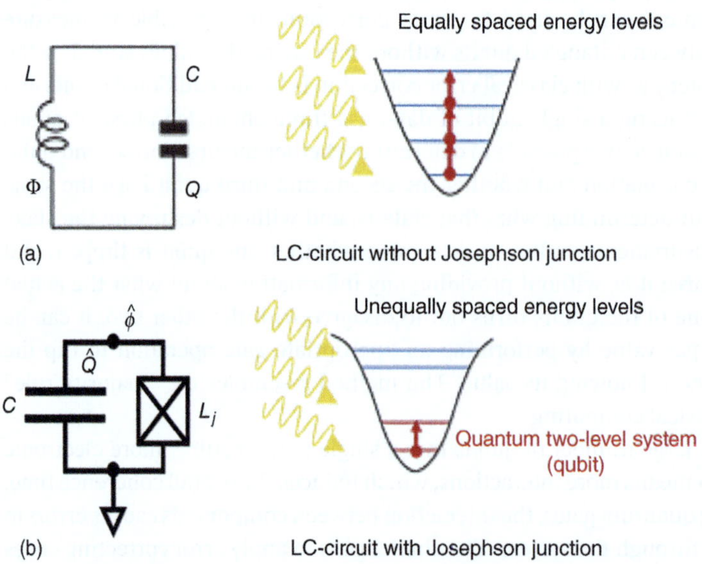

Figure 13.2 (a) An LC circuit with uniformly spaced energy levels. (b) When a Josephson junction replaces the inductor, it provides and nonlinear inductance to make the level spacings uneven [617] Courtesy of J. Lykken, Fermilab.

resonant energy levels must be made uneven. With uneven levels, RF control signals can be used to only make changes between two chosen quantum states, and avoid promoting the first quantum state to an arbitrary higher level. These two levels then serve as the two states of the qubit, $|0\rangle$ and $|1\rangle$ (Figure 13.2).

To arrive at two-level qubits, a key circuit element is needed. Replacing the circuit inductor with a *Josephson junction* converts the oscillator into the "transmon qubit" [199]. The Josephson junction behaves like a nonlinear inductor due to the Josephson effect. Along with a capacitive element that functions as a dipole antenna, the Josephson junction forms a nonlinear oscillator with a periodic potential. This circuit no longer contains evenly spaced levels. The nonlinear inductance changes the step between the levels, making the spacings unique. The energy differences decrease up the ladder. The first two levels can be uniquely addressed to form a qubit. The circuit behaves like an artificial atom embedded into a resonating structure, and couples to the modes of the resonator with the dipole antenna. GHz frequency microwaves transfer the circuit from a "ground" state to the" first excited state" ($|0\rangle$ to $|1\rangle$) without accidentally energizing other states to cause confusion. The first two states of the 2D superconducting circuit encode qubit information.

The superconducting circuit-based qubit with the transmon is called an "artificial atom" because it behaves like an atom with two quantum energy levels, and the circuit is called 2D circuit quantum electrodynamics (cQED).

When the drive frequency (e.g. 4–6 GHz) is resonant with the fundamental transition frequency of the transmon, the transmon undergoes a *Rabi oscillation* – sequential stimulated absorption and emission of photons. By using different pulse lengths, Rabi oscillations provide essential tools to prepare arbitrary quantum states of the transmon qubit. The duration of the pulse controls the angle of rotation of the qubit state around the bloch sphere representing the superposition state of the qubit. Different pulses form different quantum gates to manipulate the qubit states.

To describe a quantum computation in very simple terms, the computation starts by putting the qubits in a succession of states. Multiple qubits in superposition explore many computational paths at once. Then the states are modified with chosen quantum gate operations to assume other states or change phases; then modified again; and so on. Interference leads to cancellations between the paths, so only the correct answers survive. The final state represents the result of the computation after measurement. To make the measurement, the electronic system sends a microwave signal to the qubit and analyzes the signal it reflects back. The amplitude and phase of the reflected signal determine the final qubit state.

13.6 Josephson Junctions

When two superconducting islands are bridged by a weak barrier, such as a thin insulator, the Cooper pairs tunnel coherently through the barrier, and the wave functions of the two superconductors overlap (Figure 13.3). Despite the insulating barrier, the super-current flows without resistance. There is a phase drop across the

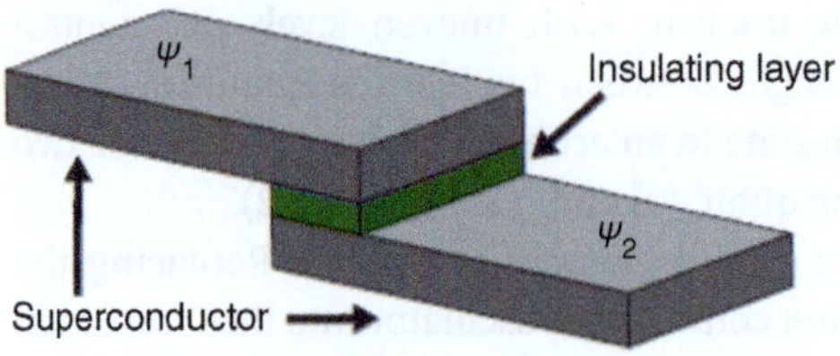

Figure 13.3 The Josephson effect is due to the overlapping wave functions of the two superconductors. Due to the weak coupling, there is a finite phase difference across the Josephson junction [192].

junction. The phase difference is related to the current flowing through the junction via

$$I(t) = I_c \sin(\varphi(t)) \tag{13.1}$$

where I_c is the critical tunnel current of the junction above which the junction becomes normal and resistive. From the derivative of the phase, we get

$$V(t) = \frac{\Phi_0}{2\pi}\frac{\partial\varphi(t)}{\partial t} \tag{13.2}$$

Differentiating Eq. (13.2) and substituting into Eq. (13.1) yields

$$V(t) = \frac{\Phi_0}{2\pi I_c \cos(\varphi(t))}\frac{\partial I(t)}{\partial t} \tag{13.3}$$

Thus the voltage across the junction is proportional to the change in phase and therefore dependent on the change in current. Compared to the voltage at a classical inductance $V = L\frac{\partial I(t)}{\partial t}$, the Josephson inductance is

$$L = \frac{\Phi_0}{2\pi I_c \cos\varphi(t)} \tag{13.4}$$

A Josephson junction (JJ) behaves as a nonlinear inductance, where the inductance depends on the current flowing. This nonlinear element enables an anharmonic quantum oscillator to be used as a qubit, making the Josephson junction a central building block for qubits.

Another important application of the Josephson Junction is the a superconducting quantum interference device (SQUID). When a superconducting loop is interrupted by two junctions, the flux enclosed in the loop can only take integer multiples of Φ_0. Screening currents in the loop start to circulate when an external magnetic flux is applied and hence tune the effective critical current for an additional bias current across the SQUID. This setup can be used as very sensitive sensor for magnetic fields. SQUID sensors are used for the sensing of miniscule magnetic fields created by the human body in biomedical applications, or for the detection of cosmic radiation.

13.7 Dilution Refrigerator for Milli-Kelvin Temperatures

To minimize decoherence from thermal interactions, a quantum computer based on superconducting circuits needs to operate at a hair above absolute zero, i.e. about 10–20 milli-Kelvin, and provide microwave signals via multiple connections for control and readout. The ultra-low temperatures also reduce the quasiparticle density in the superconductor, which increases qubit coherence. The low temperatures are reached by a helium-dilution refrigerator.

The dilution refrigerator operates with a mixture of 3 He and 4 He. The two isotopes behave differently below 2.17 K. The 4 He atoms are bosons that fall into a superfluid condensate with zero viscosity. 3 He atoms however are fermions that stay in the liquid state with normal viscosity. Below 2.17 K there is a separation between two phases, one with a high, and one with a low concentration of 3 He. Due to different mass densities, the two phases separate spatially.

A small amount of 3 He atoms can move into the 4 He superfluid liquid, almost as into a vacuum. The transition of 3 He atoms from the concentrated to the dilute phase is equivalent to an evaporation process, with cooling, where the transition energy is taken from the thermal energy of the environment.

The equilibrium concentration of about 6% of 3 He atoms in the dilute phase is quickly reached to end cooling. However, the 3 He concentration can be artificially decreased by connecting the dilute phase to a still chamber where a turbomolecular pump is connected, so as to push the system away from equilibrium and get continuous cooling. Due to the different partial pressure of the two phases, mainly 3 He evaporates in the still. The 3 He is then cleaned in a liquid nitrogen cold trap, compressed, and liquefied on its way back to the dilution unit, where it is inserted into the 3 He-rich phase.

A simplified schematic of the flow is shown in Figure 13.4. There are many heat exchanger segments along the way to provide effective cooling of the 3 He in stages.

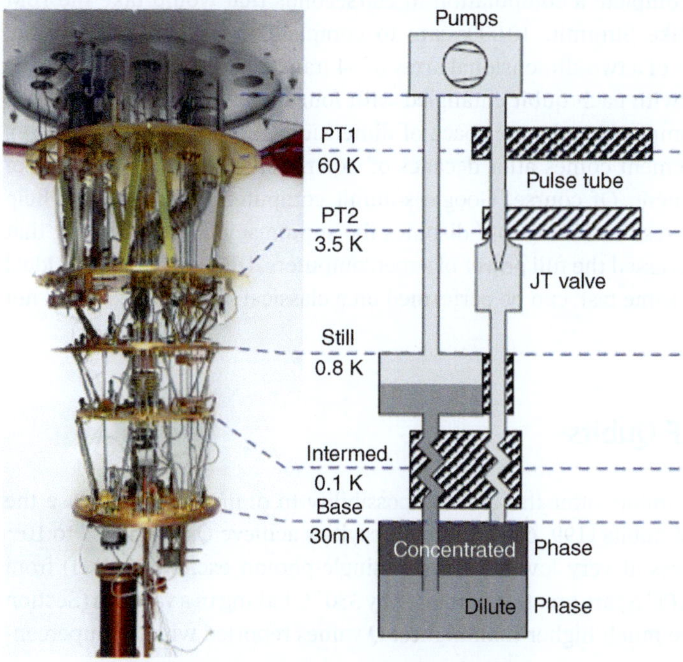

Figure 13.4 A dilution refrigerator showing the schematics of operation. Source: A. Schneider [618]/KIT Scientific Publishing/CC BY-SA 4.0.

13.8 Quantum Computing Examples

The 2D transmon-based superconducting qubit has been one of the most widely used versions for quantum computers. Superconducting circuit quantum electrodynamics (cQED) is rapidly progressing toward small and medium-scale quantum computation. Superconducting circuits consisting of lattices of Josephson junction qubits have successfully realized quantum information processors.

As one successful application, quantum algorithms for simulation of electronic structure of molecules are accepted as the most promising early application. Quantum simulation efforts can efficiently find the lowest-energy states of molecules such as H_2, LiH, or BeH_2. [616] Quantum calculations so far can handle only very small molecules with few electrons, which classical computers can also simulate accurately. But with 200 high-quality (low error) qubits, quantum chemistry could accomplish tasks beyond classical computing. Modeling the base state energy of penicillin would be an appropriate quantum challenge since the large molecule is composed of 41 atoms. A quantum computer calculation would be possible with as few as 286 high-quality qubits. In contrast, International Business Machines (IBM) has estimated that accurately modeling the base-state energy of the penicillin molecule would require a classical computer with more transistors than there are atoms in the observable universe. However the best quantum computers today are based on 50–100 (noisy) qubits [619] with coherence times less than one millisecond.

In 2019, Google announced a breakthrough, declaring that its 54-qubit Sycamore processor could complete a computation in 200 seconds that would take the IBM supercomputer, like Summit, 10 000 years to complete the same computation. Sycamore consists of a two-dimensional array of 54 transmon qubits (one qubit was not operational), with each qubit entangled with four nearest neighbors. Google's chip created a computational state space of dimension 253 (about 10^{16}). Google's landmark achievement comes after decades of experimentation, many billions of dollars in investment. Of course, Google summit computer participated to help verify Sycamore's results. (Rival IBM disputes the supremacy claim, asserting that Google had not accessed the full power of supercomputers. IBM argues that an ideal simulation of the same task can be performed on a classical system in 2.5 days, not 10 000 years.)

13.9 3D SRF Qubits

SRF cavity-based qubits offer the exciting possibility to dramatically increase the coherence time of qubits [199, 620]. Niobium cavities achieve Qs from 10^{10} to 10^{11} when the RF losses at very low fields (near single-photon excitation level) from two-level systems (TLS) are greatly attenuated by 350 °C baking in a vacuum (Section 4.7). These Q's are much higher than highest Q values reported with 2D supercon-

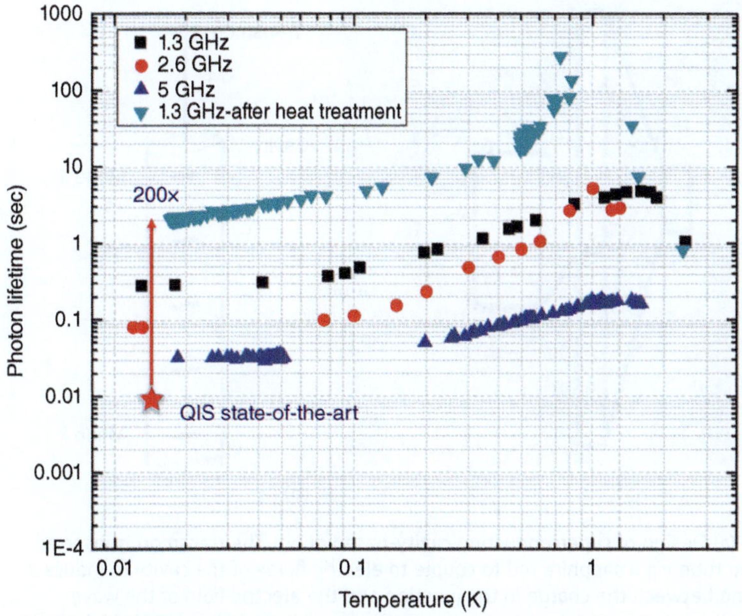

Figure 13.5 Photon lifetimes with Nb cavities after reducing TLS losses with 340 °C baking. Even prior to baking record lifetimes can be achieved. Source: A. Romanenko et al., Fermilab [191]/JACoW/CC BY 3.0.

ducting circuits with coherence times approaching one msec. Figure 13.5 shows photon lifetimes corresponding to the high Q's before and after getting rid of a significant part of the TLS losses by 340 °C heat-treatment.

In a 3D cavity, the increased volume occupied by the electromagnetic fields for similar surface currents allows higher Q's. Therefore, niobium cavities provide a promising approach for 3D circuit QED architecture for quantum computing, with the potential of a thousand-fold increase in coherence times. The enclosing superconducting cavity also serves as an effective shield for the embedded Josephson junction [621, 622].

Figure 13.6 shows the superconducting cavity-based qubit. The transmon is inserted from the beam tube via a sapphire rod to couple to electric fields of the single-cell Nb cavity to induce a strong coupling between the charge in the junction and the electric field of the cavity waves. One of the main challenges at this stage is that the transmon drags down the Q due to various sources of loss from the transmon device, the substrate and the capacitive pads (often made of niobium). Studies are underway to understand and compare the various loss mechanisms.

A first set of transmons from Rigetti [624] company has been successfully integrated with SRF cavities to reach combined system photon lifetimes of ~35 ms, which is substantially above coherence lifetimes of 2D circuit qubits [625]. By lowering TLS and other loss mechanisms in qubits, higher coherence times are anticipated in the near future.

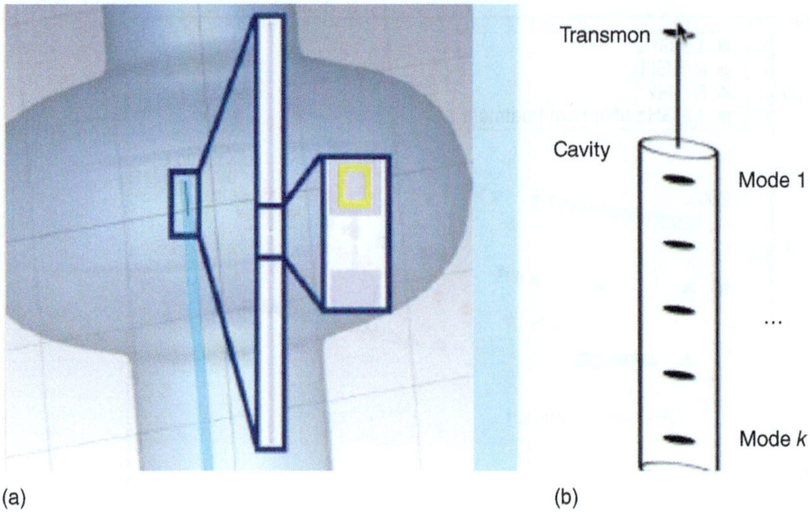

Figure 13.6 (a) Design of superconducting cavity-based qubit. The transmon is inserted from the beam tube via a sapphire rod to couple to electric fields of the cavity to induce a strong coupling between the charge in the junction and the electric field of the wave. Expanded detail on the right shows the transmon. Source: Adapted from [620]. (b) A single transmon controls and processes many qubits through the modes of the cavity. Source: [623]/Duckering et al. with permission of IEEE.

13.10 Cavity QED Quantum Processors and Memories

An outstanding challenge in cavity QED instead of 2D circuit QED is the realization of architectures with high qubit connectivity for the rapidly rising numbers of qubits. A single cavity provides the additional possibility to encode several qubits inside each of the cavity-transmon modules. This approach permits a substantial decrease in the required number of microwave channels for system and control manipulations.

Classical computation architectures typically address this challenge by using a central processor that randomly accesses a large memory. The two elements often comprise two distinct physical systems. A quantum analog of this architecture realizes a random access quantum information processor using cavity-QED as processor and quantum memories based on cavity modes [626].

But cavity-mode-based quantum memory qubits are incapable of logic operations on their own. A single logic-capable processor qubit, such as the transmon, supports many cavity-mode-based memory qubits. This architecture allows control of many highly coherent qubits with minimal electronic overhead to build larger systems.

With the best quality junctions of today, it is in principle possible to encode with a single transmon up to 14 qubits inside each of the single-cell cavity-transmon modules, substantially decreasing the required number of microwave channels for system control and manipulation for a large number of qubits [627]. Hence, the very long (seconds) coherence time in SRF cavity modes ultimately allows creation,

Figure 13.7 Conceptual representation of a 9-cell cavity with controlling transmon providing the possibility of a 100 qubit processor [625] Courtesy of A. Romanenko, Fermilab.

control, and manipulation up to 2^{14} (16 000) microwave photon energy levels, limited by residual nonlinearities and photon loss.

For a 9-cell cavity, this translates to 126 qubits (9×14) for a single mode with states in the 9-cell SRF cavity (Figure 13.7). By using two high-Q cavities, the corresponding number of cavity qubits could be $2 \times 126 = 252$. This could bring the capability of such a quantum computer to realize quantum advantage, i.e. solve problems not soluble with classical computers. However, introducing higher dimensional system with nine modes raises issues of how to realize complete control over the system, needing cavity geometries better optimized for quantum computing along with more complex gates and longer gate times.

The superconducting quantum materials and systems (SQMS) center at Fermilab [628] is developing 2D circuit-based and 3D cavity-based quantum processing units to be made available via the Fermilab computing portal, called HEPCloud, to a wide community of internal and external users for chemistry, materials, finance, medicine, and other societal applications.

Today's quantum processors should be regarded as a stepping-stone to more capable technologies; their primary near-term value is to provide a platform for the development of quantum algorithms and applications that will be useful over the long term. The quantum information community has taken gigantic strides in understanding the potential applications of a quantum computer and laid the foundational requirements for building one. With anticipated transformational advances over the next decade, quantum computing will become game-changing for every industry and will have a huge impact in the way we do business, invent new medicine and materials, safeguard our data, explore space, understand the universe, predict weather and climate change. Sometime in the very far future, perhaps even consciousness emulation will fall within the exponentially growing capabilities.

References

1 Onnes, H.K. (1991). Further experiments with liquid helium. C. On the change of electric resistance of pure metals at very low temperatures etc. IV. The resistance of pure mercury at helium temperatures. In: *Through Measurement to Knowledge* (ed. K. Gavroglu and Y. Goudaroulis), 261–263. Springer. https://doi.org/10.1007/978-94-009-2079-8_15.

2 Onnes, H.K. (1991). Further experiments with liquid helium. H. On the electrical resistance of pure metals etc. VII. The potential difference necessary for the electric current through mercury below 4°. 19 K. In: *Through Measurement to Knowledge* (ed. K. Gavroglu and Y. Goudaroulis), 273–314. Springer. https://doi.org/10.1007/978-94-009-2079-8_18.

3 Meissner, W. and Ochsenfeld, R. (1933). Ein neuer effekt bei eintritt der supraleitfähigkeit. *Naturwissenschaften* 21 (44): 787–788. http://dx.doi.org/10.1007/BF01504252.

4 Padamsee, H., Knobloch, J., and Hays, T. (1998). *RF Superconductivity for Accelerators*, 2nd edition, 2008 Wiley.

5 Padamsee, H. (2009). *RF Superconductivity: Science, Technology, and Applications*, 2e 2008. Wiley.

6 Kelly, M. (2012). Superconducting radio-frequency cavities for low-beta particle accelerators. *Rev. Accel. Sci. Technol.* 5: 185–203. https://doi.org/10.1142/9789814449953_0007.

7 De Silva, S.U. (2019). Superconducting cavities of interesting shapes (non-elliptical cavities). *Proceedings of the 19th International Conference on RF Superconductivity (SRF 2019)*, Dresden, Germany. http://accelconf.web.cern.ch/srf2019/talks/frtu1_talk.pdf.

8 Laxdal, R.E. (2021). Low/medium beta cavity design, tutorial. *Proceedings of the 20th International Conference on RF Superconductivity (SRF 2021)*, East Lansing, MI, USA. Slides FRTU1.

9 JACoW.org. https://www.jacow.org.

10 Knobloch, J. (1997). *Advanced thermometry studies of superconducting RF cavities*. PhD thesis. 1997. Cornell University. https://www.classe.cornell.edu/rsrc/Home/Research/SRF/JensKnobloch/Jens_Knobloch_PhD.pdf.

11 CST Studio Suite (2008). CST Microwave Studio. http://www.cst.com.

Superconducting Radiofrequency Technology for Accelerators: State of the Art and Emerging Trends, First Edition. Hasan Padamsee.

12 Bafia, D. (2020). *Exploring and understanding the limitations of Nb SRF cavity performance*. PhD thesis. Illinois Institute of Technology. https://www.proquest.com/openview/ff2b13e1aaacc72bf5389a759b0653ca/1?pqorigsite=gscholar&cbl=18750&diss=y.

13 Zhang, P. and Venturini Delsolaro, W. (2015). The multipacting study of niobium sputtered high-beta quarter-wave resonators for HIE-ISOLDE. *Proceedings of 17th International Conference on RF Superconductivity (SRF 2015)*, Whistler, BC, Canada, p. 760. https://accelconf.web.cern.ch/SRF2015/papers/tupb076.pdf.

14 Padamsee, H. (2020). History of gradient advances in SRF. https://doi.org/10.48550/ARXIV.2004.06720.

15 Padamsee, H., Reece, C., Noer, R. et al. (1987). Field emission studies in superconducting cavities. *Proceedings of the 1987 Particle Accelerator Conference (PAC 1987)*, Chicago, IL, USA, p. 1824. https://accelconf.web.cern.ch/p87/PDF/PAC1987_1824.PDF.

16 Knobloch, J., Muller, H., and Padamsee, H. (1994). Design of a high speed, high resolution thermometry system for 1.5 GHz superconducting radio frequency cavities. *Rev. Sci. Instrum.* 65 (11): 3521–3527. https://doi.org/10.1063/1.1144532.

17 Lilje, L., Reschke, K. Twarowski, D. et al. (1999). Electropolishing and in-situ baking of 1.3 GHz niobium cavities. *Proceedings of the 1999 Workshop on RF Superconductivity (SRF 1999)*, Santa Fe, New Mexico, p. 74.

18 Ciovati, G. et al. (2005). Temperature mapping system for single cell cavities. JLAB-TN-05-059. Internal Report.

19 Antipov, S., Efimenko, E., Romanenko, A., and Sergatskov, D.A. (2013). Time-resolved measurements of high field quench in SRF cavities. *Proceedings of the 16th International Conference on RF Superconductivity (SRF 2013)*, France, p. 743. https://accelconf.web.cern.ch/SRF2013/papers/tup112.pdf.

20 Padamsee, H. (2021). Superconducting RF (SRF) acceleration for frontier accelerators, TRIUMF seminar. *General Physics Seminar*, TRIUMF. https://www.triumf.ca/home/upcoming-events/about-seminars-lectures/upcoming-seminars-lectures/previous-seminars-lectures?expand=2295.

21 Tinkham, M. (2004). *Introduction to Superconductivity*. Courier Corporation.

22 Chao, A.W., Tigner, M., Mess, K.H., and Zimmerman, F. (2013). *Handbook of Accelerator Physics and Engineering*, 2e. World Scientific Publishing.

23 Singer, W., Brinkmann, A., Proch, D., and Singer, X. (2003). Quality requirements and control of high purity niobium for superconducting RF cavities. *Physica C* 386: 379–384. https://doi.org/10.1016/S0921-4534(02)02208-6.

24 Singer, W., Proch, D., and Brinkmann, A. (1998). Diagnostic of defects in high purity niobium. *Part. Accel.* 60: 83–102. https://cds.cern.ch/record/1120315/files/p83.pdf.

25 Aune, B., Gandelmann, R., Bloess, D. et al. (2000). Superconducting TESLA cavities. *Phys. Rev. Spec. Top. Accel. Beams* 3 (9): 092001. https://doi.org/10.1103/PhysRevSTAB.3.092001.

26 Lengeler, H. (1994). Modern Technologies in RF Superconductivity. https://inis.iaea.org/collection/NCLCollectionStore/_Public/26/004/26004572.pdf?r=1.

27 Kirchgessner, J.L. (1987). Forming and welding of niobium for superconducting cavities. *Proceedings of the 3rd Workshop on RF Superconductivity Argonne*

National Laboratory (SRF 1987), Illinois, USA, p. 533. https://accelconf.web.cern.ch/srf87/papers/srf87e05.pdf.

28 Iwashita, Y., Tajima, Y., and Hayano, H. (2008). Development of high resolution camera for observations of superconducting cavities. *Phys. Rev. Spec. Top. Accel. Beams* 11 (9): 093501. https://doi.org/10.1103/PhysRevSTAB.11.093501.

29 Saito, K., Kojima, Y., and Furuya, T. (1990). R and D of superconducting cavities at KEK. *Proceedings of the 4th Workshop on RF Superconductivity (SRF 1989)*, p. 85. https://accelconf.web.cern.ch/SRF89/papers/srf89g18.pdf.

30 Higuchi, T., Noguchi, S., Ono, M. et al. (1996). Investigation on barrel polishing for superconducting niobium cavities. *Proceedings of the 1995 Workshop on RF Superconductivity*, Gif-sur-Yvette, France, p. 723. accelconf.web.cern.ch/SRF95/papers/srf95f34.pdf.

31 Saito, K., Noguchi, S., Ono, M. et al. (1998). Superiority of electropolishing over chemical polishing on high gradients. *Part. Accel.* 60 (KEK-98-4): 193–217. https://cds.cern.ch/record/394033/files/p193.pdf.

32 Morgan, A.V., Romanenko, A., and Windsor, A. (2007). Surface studies of contaminants generated during electropolishing. *2007 IEEE Particle Accelerator Conference (PAC 2007)*, pp. 2346–2348. https://accelconf.web.cern.ch/p07/PAPERS/WEPMS010.PDF.

33 van der Horst, B., Matheisen, A., Petersen, B. et al. (2007). Update on cavity preparation for high gradient superconducting multi cell cavities at DESY. *Proceedings of SRF 2007*, Peking Univ., Beijing, China, p. 196. https://accelconf.web.cern.ch/srf2007/PAPERS/TUP30.pdf.

34 Mammosser, J., Bice, D., and Afanador, R. (2007). ILC cavity qualifications–Americas. *TESLA Technology Collaboration Meeting at Fermilab (TTC 2007)*. Batavia, IL, USA. https://indico.desy.de/event/200/contributions/37452/subcontributions/1191/attachments/24128/30737/TTC_JLab_S0_Efforts.pdf.

35 Popielarski, L. (2019). Cavity processing and cleanroom techniques, tutorial. SRF 2019. https://accelconf.web.cern.ch/srf2019/talks/frtu2_talk.pdf.

36 Cooper, C., Bullock, B., Joshi, S., and Palczewski, A. (2011). Centrifugal barrel polishing of cavities worldwide. *Proceedings of the 15th International Conference on RF Superconductivity (SRF 2011)*, Chicago, IL, USA, pp. 571–575. https://accelconf.web.cern.ch/SRF2011/papers/weioa02.pdf.

37 Conway, Z.A., Chojnacki, E.P., Hartill, D.L. et al. (2009). ILC testing program at Cornell University. *Proceedings of the 14th International Conference on RF Superconductivity (SRF 2009)*, Berlin, Germany, p. 176. https://accelconf.web.cern.ch/SRF2009/papers/tuoaau05.pdf.

38 Kneisel, P., Lewis, B., and Turlington, L. (1993). *Experience with High Pressure Ultrapure Water Rinsing of Niobium Cavities. Proceedings of the 6th Conference on RF Superconductivity (SRF 1993)*, Newport News, VA. USA, p. 628. https://accelconf.web.cern.ch/SRF93/papers/srf93i09.pdf.

39 Saito, K., Noguchi, S., Ono, M. et al. (1993). Study of ultra-clean surface for niobium SC cavities. P00024607, *Proceedings of the 6th Workshop on RF Superconductivity (SRF 1993)*, Newport News, VA, pp. 1151–1159. https://accelconf.web.cern.ch/SRF93/papers/srf93j03.pdf.

40 Reschke, D. (2005). Cleanliness techniques. *Proceedings of the 12th International Workshop on RF Superconductivity*, Cornell University, Ithaca, NY, USA, pp. 71–77. https://accelconf.web.cern.ch/SRF2005/papers/sup03.pdf.

41 Charrier, J.P., Coadou, B., and Visentin, B. (1998). Improvements of superconducting cavity performances at high accelerating gradients. *Proceedings of the 6th European Particle Accelerator Conference (EPAC 1998)*, Stockholm, Sweden. https://accelconf.web.cern.ch/e98/PAPERS/TUP07B.PDF.

42 Grassellino, A. (2019). Pushing bulk Nb limits (high Q, high gradient, reliable SRF accelerators) tutorial. *19th International Conference on RF Superconductivity (SRF 2019)*, Dresden, Germany. https://accelconf.web.cern.ch/srf2019/talks/frtu3_talk.pdf.

43 London, F. and London, H. (1935). The electromagnetic equations of the supraconductor. *Proc. R. Soc. London Ser. - Math. Phys. Sci.* 149 (866): 71–88. https://doi.org/10.1098/rspa.1935.0048.

44 London, F. (1961). *Superfluids: 2d Rev. Ed., with a New Epilogue, Theoretical Developments 1950–1960*, vol. 2. Dover Publications.

45 Gorter, C.J. and Casimir, H. (1934). On supraconductivity I. *Physica* 1 (1–6): 306–320. https://doi.org/10.1016/S0031-8914(34)90037-9.

46 Drude, P. (1900). Zur elektronentheorie der metalle; II. Teil. galvanomagnetische und thermomagnetische effecte. *Ann. Phys.* 308 (11): 369–402. https://doi.org/10.1002/andp.19003081102.

47 Drude, P. (1900). Zur elektronentheorie der metalle. *Ann. Phys.* 306 (3): 566–613. https://doi.org/10.1002/andp.19003060312.

48 Pippard, A.B. and Bragg, W.L. (1953). An experimental and theoretical study of the relation between magnetic field and current in a superconductor. *Proc. R. Soc. London Ser. - Math. Phys. Sci.* 216 (1127): 547–568. https://doi.org/10.1098/rspa.1953.0040.

49 Martinello, M. (2016). *The path to high Q-factors in superconducting accelerating cavities: flux expulsion and surface resistance optimization*. Thesis, Illinois Institute of Technology. https://www.osti.gov/biblio/1352032.

50 Chambers, R.G. (1950). Anomalous skin effect in metals. *Nature* 165 (4189): 239–240. https://doi.org/10.1038/165239b0.

51 Reuter, G.E.H. and Sondheimer, E.H. (1948). The theory of the anomalous skin effect in metals. *Proc. R. Soc. London Ser. - Math. Phys. Sci.* 195 (1042): 336–364. https://doi.org/10.1098/rspa.1948.0123.

52 Harlingen, D. (2019) Models and theories of superconductivity. *Physics 498 SQD – Lecture 3*, University of Illinois at Urbana-Champaign, IL, USA. https://courses.physics.illinois.edu/phys498SQD/fa2019/lecture-notes.html.

53 Bardeen, J., Cooper, L.N., and Schrieffer, J.R. (1957). Microscopic theory of superconductivity. *Phys. Rev.* 106 (1): 162. https://doi.org/10.1103/PhysRev.106.162.

54 Bardeen, J., Cooper, L.N., and Schrieffer, J.R. (1957). Theory of superconductivity. *Phys. Rev.* 108 (5): 1175. https://doi.org/10.1103/PhysRev.108.1175.

55 Cooper, L.N. (1956). Bound electron pairs in a degenerate Fermi gas. *Phys. Rev.* 104 (4): 1189. https://doi.org/10.1103/PhysRev.104.1189.

56 Corak, W.S., Goodman, B.B., Satterthwaite, C.B., and Wexler, A. (1954). Exponential temperature dependence of the electronic specific heat of superconducting vanadium. *Phys. Rev.* 96 (5): 1442. https://doi.org/10.1103/PhysRev.96.1442.2.

57 Fröhlich, H. (1950). Theory of the superconducting state. I. The ground state at the absolute zero of temperature. *Phys. Rev.* 79 (5): 845. https://doi.org/10.1103/PhysRev.79.845.

58 Sheahen, T.P. (1966). Rules for the energy gap and critical field of superconductors. *Phys. Rev.* 149 (1): 368. https://doi.org/10.1103/PhysRev.149.368.

59 Sonier, J. (1994). MuSR measurement of the temperature dependence of the magnetic penetration depth in $YBa_2Cu_3O_{6.95}$ single crystals. vol. MSc thesis. Vancouver, Canada: University of British Columbia. https://doi.org/10.14288/1.0099133.

60 Mattis, D.C. and Bardeen, J. (1958). Theory of the anomalous skin effect in normal and superconducting metals. *Phys. Rev.* 111 (2): 412. https://doi.org/10.1103/PhysRev.111.412.

61 Halbritter, J. (1970). Fortran-program for the computation of the surface impedance of superconductors. Kernforschungszentrum. Karlsruhe (West Germany): Karlsruher Institut für Technologie.

62 Liepe Research Group: Research. https://www.classe.cornell.edu/~liepe/webpage/researchsrimp.html.

63 Ciovati, G. (2004). Effect of low-temperature baking on the radio-frequency properties of niobium superconducting cavities for particle accelerators. *J. Appl. Phys.* 96 (3): 1591–1600. https://doi.org/10.1063/1.1767295.

64 Kneisel, P. (2006). R&D paths towards achieving ultimate capabilities. *ICFA Beam Dynamics Newsletter, No. 39*, pp. 14–23. https://icfa-usa.jlab.org/archive/newsletter/icfa_bd_nl_39.pdf.

65 Slater, J.C. (1946). Microwave electronics. *Rev. Mod. Phys.* 18 (4): 441. https://doi.org/10.1103/RevModPhys.18.441.

66 Dynes, R.C., Narayanamurti, V., and Garno, J.P. (1978). Direct measurement of quasiparticle-lifetime broadening in a strong-coupled superconductor. *Phys. Rev. Lett.* 41 (21): 1509. https://doi.org/10.1103/PhysRevLett.41.1509.

67 Ginzburg, V.L. and Landau, L.D. (1950). On the theory of superconductivity. *Zh. Eksp. Teor. Fiz.* 20: 1064.

68 Checchin, M. (2016). *Physics of limiting phenomena in superconducting microwave resonators: vortex dissipation, ultimate quench and quality factor degradation mechanisms*. PhD Thesis. Illinois Institute of Technology. https://lss.fnal.gov/archive/thesis/2000/fermilab-thesis-2016-34.pdf.

69 Shubnikov, L.V., Khotkevich, V.I., Shepelev, Y.D., and Ryabinin, Y.N. (1937). Magnetic properties of superconducting metals and alloys. *Zh. Eksp. Teor. Fiz.* 7: 221.

70 Abrikosov, A.A. (1957). On the magnetic properties of superconductors of the second group. *Sov. Phys. JETP* 5: 1174–1182.

71 Abrikosov, A.A. (1957). The magnetic properties of superconducting alloys. *J. Phys. Chem. Solids* 2 (3): 199–208. https://doi.org/10.1016/0022-3697(57)90083-5.

72 Caroli, C., De Gennes, P.G., and Matricon, J. (1964). Bound fermion states on a vortex line in a type II superconductor. *Phys. Lett.* 9 (4): 307–309. https://doi.org/10.1016/0031-9163(64)90375-0.

73 Huebener, R. (2001). *Magnetic Flux Structures in Superconductors: Extended Reprint of a Classic Text*, vol. 6. Springer Science & Business Media. ASIN : B000UFP99M. ISBN-13 978-3540679530.

74 Kleiner, W.H., Roth, L.M., and Autler, S.H. (1964). Bulk solution of Ginzburg-Landau equations for type ii superconductors: upper critical field region. *Phys. Rev.* 133: A1226. https://doi.org/10.1103/PhysRev.133.A1226.

75 Bishop, D.J., Gammel, P.L., and Huse, D.A. (1993). Resistance in high-temperature superconductors. *Sci. Am.* 268 (2): 48–55. https://ui.adsabs.harvard.edu/link_gateway/1993SciAm.268b.48B/doi:10.1038/scientificamerican0293-48.

76 Gammel, P.L., Bishop, D.J., Dolan, G.J. et al. (1987). Observation of hexagonally correlated flux quanta in $YBa_2Cu_3O_7$. *Phys. Rev. Lett.* 59 (22): 2592. https://doi.org/10.1103/PhysRevLett.59.2592.

77 Hess, H.F., Robinson, R.B., Dynes, R.C. et al. (1989). Scanning-tunneling-microscope observation of the Abrikosov flux lattice and the density of states near and inside a fluxoid. *Phys. Rev. Lett.* 62 (2): 214. https://doi.org/10.1103/PhysRevLett.62.214.

78 Hudson, W.R. (1970). *The Mixed State of Superconductors*, vol. 240. Scientific and Technical Information Division, National Aeronautics and Space Administration. https://ntrs.nasa.gov/citations/19710000941.

79 Goodstein, D.L. (2014). *States of Matter*. Courier Corporation.

80 Ginzburg, V.L. (1958). On the destruction and the onset of superconductivity in a magnetic field. *Sov. Phys. JETP* 34: 78–87. http://www.jetp.ras.ru/cgi-bin/dn/e_007_01_0078.pdf.

81 Garfunkel, M.P. and Serin, B. (1952). The formation of a boundary between normal-conducting and superconducting metal. *Phys. Rev.* 85 (5): 834. https://doi.org/10.1103/PhysRev.85.834.

82 Saint-James, D. and de Gennes, P. (1963). Onset of superconductivity in decreasing fields. *Phys. Lett.* 7 (5): 306–308.

83 Matricon, J. and Saint-James, D. (1967). Superheating fields in superconductors. *Phys. Lett. A* 24 (5): 241–242. https://doi.org/10.1016/0031-9163(63)90047-7.

84 Bean, C.P. and Livingston, J.D. (1964). Surface barrier in type-II superconductors. *Phys. Rev. Lett.* 12 (1): 14. https://doi.org/10.1103/PhysRevLett.12.14.

85 Bardeen, J. and Stephen, M.J. (1965). Theory of the motion of vortices in superconductors. *Phys. Rev.* 140 (4A): A1197. https://doi.org/10.1103/PhysRev.140.A1197.

86 Brewer, D.F. (1966). *Orsay Group on Superconductivity in Quantum Fluids*. North-Holland.

87 Tholfsen, P. and Meissner, H. (1968). Cylindrically symmetric solutions of the Ginzburg-Landau equations. *Phys. Rev.* 169 (2): 413. https://doi.org/10.1103/PhysRev.169.413.

88 Harden, J.L. and Arp, V. (1963). The lower critical field in the Ginzburg-Landau theory of superconductivity. *Cryog. Engl.* 3: 105–108. https://doi.org/10.1016/0011-2275(63)90029-8.

89 Dolgert, A.J., Di Bartolo, S.J., and Dorsey, A.T. (1996). Superheating fields of superconductors: asymptotic analysis and numerical results. *Phys. Rev. B* 53 (9): 5650. https://doi.org/10.1103/physrevb.53.5650.

90 Transtrum, M.K., Catelani, G., and Sethna, J.P. (2011). Superheating field of superconductors within Ginzburg-Landau theory. *Phys. Rev. B* 83 (9): 094505. https://doi.org/10.1103/PhysRevB.83.094505.

91 Finnemore, D.K., Stromberg, T.F., and Swenson, C.A. (1966). Superconducting properties of high-purity niobium. *Phys. Rev.* 149 (1): 231. https://doi.org/10.1103/PHYSREV.149.231.

92 Gor'kov, L.P. (1959). Microscopic derivation of the Ginzburg-Landau equations in the theory of superconductivity. *Sov. Phys. JETP* 9 (6): 1364–1367.

93 Orlando, T.P., McNiff, E.J. Jr., Foner, S., and Beasley, M.R. (1979). Critical fields, Pauli paramagnetic limiting, and material parameters of Nb_3Sn and V_3Si. *Phys. Rev. B* 19 (9): 4545. https://doi.org/10.1103/PhysRevB.19.4545.

94 Hein, M. (1999). *High-Temperature-Superconductor Thin Films at Microwave Frequencies*, vol. 155. Springer Science & Business Media.

95 Gonnella, D.A. (2016). *The Fundamental Science of Nitrogen-Doping of Niobium Superconducting Cavities*. Thesis. Cornell University.

96 Maxfield, B.W. and McLean, W.L. (1965). Superconducting penetration depth of niobium. *Phys. Rev.* 139 (5A): A1515. https://doi.org/10.1103/PhysRev.139.A1515.

97 Novotny, V. and Meincke, P.P.M. (1975). Single superconducting energy gap in pure niobium. *J. Low Temp. Phys.* 18 (1): 147–157. https://doi.org/10.1007/BF00116976.

98 Padamsee, Z. (2022). Derivation of surface resistance and reactance.

99 Grassellino, A., Romanenko, A., Sergatskov, D. et al. (2013). Nitrogen and argon doping of niobium for superconducting radio frequency cavities: a pathway to highly efficient accelerating structures. *Supercond. Sci. Technol.* 26 (10): 102001. https://doi.org/10.1088/0953-2048/26/10/102001.

100 Galayda, J.N. (2014). The LCLS-II project. *Proc. IPAC* 2014: 935. https://doi.org/10.18429/JACoW-IPAC2014-TUOCA01.

101 Raubenheimer, T. (2016). LCLS-II-HE FEL facility overview. *Workshop Scientific Opportunities Ultrafast Hard X-rays High Rep Rate*, p. Talk.

102 Bellandi, A., Branlard, J., Eschke, J. et al. (2021). Overview of CW R&D with a European XFEL cryomodule. *SRF 2021*, Virtual (MSU), p. MOPFAV001. https://epaper.kek.jp/srf2021/html/auth0053.htm.

103 Derwent, P., Holmes, S., and Lebedev, V. (2015). An 800-Mev Superconducting Linac to Support Megawatt Proton Operations at Fermilab. *ArXiv Prepr. ArXiv150201728*. https://arxiv.org/ftp/arxiv/papers/1502/1502.01728.pdf.

104 Klebaner, A., Boffo, C., Chandrasekaran, S. K., Passarelli, D. et al. (2021). Proton Improvement Plan-II: Overview of progress in the construction. *Proceedings* *of the 20th International Conference on RF Superconductivity (SRF 2021)*, p. MOOFAV05. https://doi.org/10.18429/JACoW-SRF2021-MOOFAV05.

105 Reschke, D., Decking, W., Walker, N., and Weise, H. (2017). The commissioning of the European-XFEL Linac and its performance. *18th International Conference on RF Superconductivity (SRF 2017)*, Lanzhou, China. https://accelconf.web.cern.ch/srf2017/talks/moxa02_talk.pdf.

106 Dhakal, P., Ciovati, G., and Myneni, G.R. (2012). A path to higher Q0 with large grain niobium cavities. *Proceedings of IPAC 2012*, New Orleans, USA, 2012. p. 3. https://accelconf.web.cern.ch/IPAC2012/papers/weppc091.pdf.

107 Dhakal, P., Ciovati, G., Myneni, G.R. et al. (2013). Effect of high temperature heat treatments on the quality factor of a large-grain superconducting radio-frequency niobium cavity. *Phys. Rev. Spec. Top. Accel. Beams* 16 (4): 042001. https://doi.org/10.1103/PhysRevSTAB.16.042001.

108 Dhakal, P. (2020). Nitrogen doping and infusion in SRF cavities: a review. *Phys. Open* 5: 100034. https://doi.org/10.1016/j.physo.2020.100034.

109 Ciovati, G., Dhakal, P., and Myneni, G.R. (2016). Superconducting radio-frequency cavities made from medium and low-purity niobium ingots. *Supercond. Sci. Technol.* 29 (6): 064002. https://doi.org/10.1088/0953-2048/29/6/064002.

110 Grassellino, A., Romenenko, D., Posen, S. et al. (2015). N doping: progress in development and understanding. *Proceedings of the 17th International Conference on RF Superconductivity (SRF 2015)*, Whistler, BC, Canada, pp. 48–54. https://accelconf.web.cern.ch/SRF2015/talks/moba06_talk.pdf.

111 Liepe, M., Bishop, P., Conklin, H. et al. (2015). LCLS-II SRF cavity processing protocol development and baseline cavity performance demonstration. *Proceedings of the 17th International Conference on RF Superconductivity (SRF 2015)*, MB. https://doi.org/10.18429/JACOW-SRF2015-MOPB033.

112 Liepe, M., Eichhorn, R.G., Furuta, F. et al. (2014). The joint high Q_0 R&D program for LCLS-II. *Proceedings of the 5th International Particle Accelerator Conference (IPAC 2014)*, Dresden, Germany, p. 2627. https://doi.org/10.18429/JACoW-IPAC2014-WEPRI062.

113 Trenikhina, Y., Grassellino, A., Melnychuk, O., and Romanenko, A. (2015). Characterization of nitrogen doping recipes for the Nb SRF cavities. *Proceedings of the 17th International Conference on RF Superconductivity (SRF 2015)*, BC Canada. MOPB055, pp. 223–227. https://accelconf.web.cern.ch/SRF2015/papers/mopb055.pdf.

114 Dangwal Pandey, A., Dalla Lana Semione, G., Prudnikava, A. et al. (2018). Surface characterization of nitrogen-doped Nb (100) large-grain superconducting RF cavity material. *J. Mater. Sci.* 53 (14): 10411–10422. https://dx.doi.org/10.1007/s10853-018-2310-8.

115 Yang, Z., Lu, X., Tan, W. et al. (2018). XPS studies of nitrogen doping niobium used for accelerator applications. *Appl. Surf. Sci.* 439: 1119–1126. https://doi.org/10.1016/j.apsusc.2017.12.214.

116 Brauer, G. (1960). Nitrides, carbonitrides and oxynitrides of niobium. *J. Less Common Met.* 2 (2–4): 131–137. https://doi.org/10.1016/0022-5088(60)90008-4.

117 Cansever, N., Danışman, M., and Kazmanlı, K. (2008). The effect of nitrogen pressure on cathodic arc deposited Nbn thin films. *Surf. Coat. Technol.* 202 (24): 5919–5923. https://doi.org/10.1016/j.surfcoat.2008.06.158.

118 Mayr, W., Lengauer, W., Buscaglia, V. et al. (1997). Nitridation of Ti/Nb alloys and solid-state properties of δ-(Ti,Nb)N. *J. Alloys Compd.* 262: 521–528. https://doi.org/10.1016/S0925-8388(97)00367-8.

119 Clenny, J.T. and Rosa, C.J. (1980). Nitridation kinetics of niobium in the temperature range of 873 to 1273 K. *Metall. Trans. A* 11 (9): 1575–1580. https://doi.org/10.1016/j.apsusc.2017.12.214.

120 Palczewski, A., Kelley, M., Reece, C., and Tuggle, J. (2017). Investigation of nitrogen absorption rate and nitride growth on SRF cavity grade RRR niobium as a function of furnace temperature. *28th Linear Accelerator Conference (LINAC 2016)*, East Lansing, MI, USA (25–30 September 2016), pp. 744–747. https://accelconf.web.cern.ch/linac2016/papers/thop02.pdf.

121 Gonnella, D., Eichhorn, R., Furuta, F. et al. (2016). Improved N-doping protocols for SRF cavities. *Proceedings of IPAC2016*, Busan Korea. WEPMR025. pp. 2323–2326. http://jacow.org/ipac2016/papers/wepmr025.pdf

122 Dhakal, P., Ciovati, G., Kneisel, P., and Myneni, G.R. (2014). Enhancement in quality factor of SRF niobium cavities by material diffusion. *IEEE Trans. Appl. Supercond.* 25 (3): 1–4. https://doi.org/10.1109/TASC.2014.2359640.

123 Tuggle, J., Pudasaini, U., Stevie, F.A. et al. (2018). Secondary ion mass spectrometry for superconducting radiofrequency cavity materials. *J. Vac. Sci. Technol., B* 36 (5): 052907. https://doi.org/10.1116/1.5041093.

124 Gonnella, D., Aderhold, S., Burrill, A. et al. (2019). Industrial cavity production: lessons learned to push the boundaries of nitrogen-doping. *19th International Conference on RF Superconductivity (SRF 2019)*, Dresden, Germany (30 June-05 July 2019), pp. 1199–1205. https://doi.org/10.18429/JACoW-SRF2019-FRCAA3.

125 Lechner, E. (2020). Developing an improved model of N diffusion in Nb. *TESLA Technology Collaboration Meeting (TTC 2020)*, Geneva, Switzerland. https://indico.desy.de/event/27572/contributions/94293/attachments/63535/77660/LechnerTTC2021VF.pdf.

126 Dhakal, P., Ciovati, G., Kneisel, P. et al. (2015). Nitrogen doping study in ingot niobium cavities. *6th International Particle Accelerator Conference (IPAC 2015)*, Richmond, VA, USA (3–8 May 2015), pp. 3506–3508. https://doi.org/10.18429/JACoW-IPAC2015-WEPWI009.

127 Goodman, B.B. and Kuhn, G. (1968). Influence des défauts étendus sur les propriétés supraconductrices du niobium. *J. Phys.* 29 (2–3): 240–252. https://doi.org/10.1051/jphys:01968002902-3024000.

128 Singer, W., Ermakov, A., and Singer, X. (2010). RRR-measurement techniques on high purity niobium. *TESLA Technology Collaboration Meeting (TTC 2010)*, Fermilab, Batavia, IL, USA. https://bib-pubdb1.desy.de/record/91801/files/TTC-Report%202010-02.pdf.

129 Ciovati, G., Dhakal, P., and Gurevich, A. (2014). Decrease of the surface resistance in superconducting niobium resonator cavities by the microwave field. *Appl. Phys. Lett.* 104 (9): 092601. https://doi.org/10.1063/1.4867339.

130 Gonnella, D. and Liepe, M. (2014). New insights into heat treatment of SRF cavities in a low-pressure nitrogen atmosphere. *Proceedings of the 5th International Particle Accelerator Conference (IPAC 2014)*, Dresden, DE, pp. 2634–2637. https://doi.org/doi:10.18429/JACoW-IPAC2014-WEPRI064.

131 Romanenko, A., Grassellino, A., Barkov, F. et al. (2014). Strong Meissner screening change in superconducting radio frequency cavities due to mild baking. *Appl. Phys. Lett.* 104 (7): 072601. https://doi.org/10.1063/1.4866013.

132 Morenzoni, E., Kottmann, F., Maden, D. et al. (1994). Generation of very slow polarized positive muons. *Phys. Rev. Lett.* 72 (17): 2793. https://doi.org/10.1103/PhysRevLett.72.2793.

133 Romanenko, A. and Grassellino, A. (2013). Dependence of the microwave surface resistance of superconducting niobium on the magnitude of the RF field. *Appl. Phys. Lett.* 102 (25): 252603. https://doi.org/10.1063/1.4812665.

134 Vallet, C., Bolore, M., Bonin, B. et al. (1992). Flux trapping in superconducting cavities. *Proceedings of the European Particle Accelerator Conference (EPAC 1992)*, p. 1295. https://accelconf.web.cern.ch/e92/PDF/EPAC1992_1295.PDF.

135 Romanenko, A., Barkov, F., Cooley, L.D., and Grassellino, A. (2013). Proximity breakdown of hydrides in superconducting niobium cavities. *Supercond. Sci. Technol.* 26 (3): 035003. https://doi.org/10.1088/0953-2048/26/3/035003.

136 Trenikhina, Y. (2014). TEM studies of niobium hydrides participants in superconducting niobium cavitiey cutouts. presented at the *Thin Film SRF Conference (TFSRF 2014)*, INFN, Legnaro, Italy. https://surfacetreatments.infn.it/tem-studies-of-niobium-hydrides-participants-in-superconducting-niobium-cavitiey-cutouts-yulia-trenikhina-tfsrf-2014/.

137 Cano, A., Martinello, M., Sung, Z. et al. (2022). Surface structural studies of SRF Nb cutouts at cryogenic conditions using synchrotron radiation-based characterization techniques. *TESLA Technology Collaboration (TTC 2022)*, January 2022, Aomori, Japan. https://indico.desy.de/event/32219/contributions/116549/.

138 Cano, A., Romanenko, A., Karapetrova, J. et al. (2021). Microstructural changes observed in the near-surface of SRF cavity cutouts using Gi-synchrotron XRD. *20th International Conference on RF Superconductivity (SRF 2021)*, Dresden, Germany, p. mopfdv010. https://doi.org/10.18429/JACoW-SRF2021-MOPFDV010.

139 Richter, D., Rush, J.J., and Rowe, J.M. (1983). Localized modes and hydrogen trapping in niobium with substitutional impurities. *Phys. Rev. B* 27 (10): 6227. https://doi.org/10.1103/PhysRevB.27.6227.

140 Pfeiffer, G. and Wipf, H. (1976). The trapping of hydrogen in niobium by nitrogen interstitials. *J. Phys. F: Met. Phys.* 6 (2): 167. https://doi.org/10.1088/0305-4608/6/2/013.

141 Garg, P., Balachandran, S., Adlakha, I. et al. (2017). Role of nitrogen on hydride nucleation in pure niobium by first principles calculations. *Proceedings of the*

18th International Conference on RF Superconductivity (SRF 2017), pp. 741–745. https://doi.org/10.18429/JACoW-SRF2017-THPB002.

142 Sitaraman, N., Arias, T., Liepe, M., and Maniscalco, J.T. (2019). Ab initio calculations on impurity doped niobium and niobium surfaces. https://doi.org/10.18429/JACoW-SRF2019-TUP045.

143 Garg, P., Balachandran, S., Adlakha, I. et al. (2018). Revealing the role of nitrogen on hydride nucleation and stability in pure niobium using first-principle calculations. *Supercond. Sci. Technol.* 31 (11): 115007. https://doi.org/10.1088/1361-6668/aae147.

144 Groll, N.R., Ciovati, G., Grassellino, A. et al. (2018). Insight Into Bulk Niobium Superconducting RF Cavities Performances by Tunneling Spectroscopy. *ArXiv Prepr. ArXiv180506359*. https://doi.org/10.48550/arXiv.1805.06359.

145 Romanenko, A. (2014). Breakthrough technology for very high quality factors in SRF cavities. *Proceedings of the 27th Linear Accelerator Conference (LINAC 2014)*, Dresden, Germany. https://accelconf.web.cern.ch/LINAC2014/talks/tuioc02_talk.pdf.

146 Martinello, M., Checchin, M., Romanendo, A. et al. (2018). Field-enhanced superconductivity in high-frequency niobium accelerating cavities. *Phys. Rev. Lett.* 121 (22): 224801. https://doi.org/10.1103/PhysRevLett.121.224801.

147 Furuta, F., Ge, M., Gruber, T. et al. (2018). RF test result of a BNL N-Doped 500 MHz B-Cell cavity at Cornell. *Proceedings of the 9th International Particle Accelerator Conference (IPAC 2018)*, pp. 2440–2442. https://doi.org/10.18429/JACoW-IPAC2018-WEPMF036.

148 Maniscalco, J.T. (2022). *Studies of the Field-Dependent Surface Resistance of Nitrogen-Doped Niobium for Superconducting Accelerators*. Cornell University Library. https://hdl.handle.net/1813/70473 (accessed 03 May 2022).

149 Maniscalco, J.T., Ge, M., Koufalis, P. N. et al. (2019). The field-dependent surface resistance of doped niobium: new experimental and theoretical results. *Proceedings of the 19th International Conference on RF Superconductivity (SRF 2019)*, Dresden, Germany. https://doi.org/10.18429/JACoW-SRF2019-TUFUA1.

150 Xiao, B. and Reece, C.E. (2014). A new first-principles calculation of field-dependent RF surface impedance of BCS superconductor and application to SRF cavities. *International Conference on RF Superconductivity (SRF 2013)*, Paris, FR. MOIOC02. pp. 63–65.

151 Xiao, B.P., Reece, C.E., and Kelley, M.J. (2013). Superconducting surface impedance under radiofrequency field. *Physica C* 490: 26–31. https://doi.org/10.1016/j.physc.2013.04.003.

152 Gurevich, A. (2017). Theory of RF superconductivity for resonant cavities. *Supercond. Sci. Technol.* 30 (3): 034004. https://iopscience.iop.org/article/10.1088/1361-6668/30/3/034004/ampdf.

153 Gurevich, A. (2014). Reduction of dissipative nonlinear conductivity of superconductors by static and microwave magnetic fields. *Phys. Rev. Lett.* 113 (8): 087001. https://doi.org/10.1103/PhysRevLett.113.087001.

154 Chang, J.-J. and Scalapino, D.J. (1977). Kinetic-equation approach to nonequilibrium superconductivity. *Phys. Rev. B* 15 (5): 2651–2670. https://doi.org/10.1103/PhysRevB.15.2651.

155 Goldie, D.J. and Withington, S. (2012). Non-equilibrium superconductivity in quantum-sensing superconducting resonators. *Supercond. Sci. Technol.* 26 (1): 015004. https://doi.org/10.1088/0953-2048/26/1/015004.

156 De Visser, P.J., Goldie, D.J., Diener, P. et al. (2014). Evidence of a nonequilibrium distribution of quasiparticles in the microwave response of a superconducting aluminum resonator. *Phys. Rev. Lett.* 112 (4): 047004. https://doi.org/10.1103/PhysRevLett.112.047004.

157 Martinello, M. (2022). Accelerator physics and engineering seminar, Recent advances and breakthrough for SRF cavities at FNAL. presented at the *Seminar*, MSU. https://frib.msu.edu/_files/pdfs/science_ase_seminars_martinello.pdf.

158 Eliashberg, G.M. (1970). Film superconductivity stimulated by a high-frequency field. *Sov. J. Exp. Theory Phys. Lett.* 11: 114.

159 Gray, K.E. (2012). *Nonequilibrium Superconductivity, Phonons, and Kapitza Boundaries*, vol. 65. Springer Science & Business Media.

160 Kommers, T. and Clarke, J. (1977). Measurement of microwave-enhanced energy gap in superconducting aluminum by tunneling. *Phys. Rev. Lett.* 38 (19): 1091. https://doi.org/10.1103/PhysRevLett.38.1091.

161 Sridhar, S. and Mercereau, J.E. (1986). Nonequilibrium dynamics of quasiparticles in superconductors. *Phys. Rev. B* 34 (1): 203. https://doi.org/10.1103/PhysRevB.34.203.

162 Weingarten, W. (2018). N-doped surfaces of superconducting niobium cavitiesas a disordered composite. IEEE Trans. Appl. Supercond. 28 (6): 1–4.

163 Eichhorn, R., Gonnella, D., Hoffstaetter, G., Liepe, M., and Weingarten, W. (2014). On superconducting niobium accelerating cavities fired underN2-gas exposure. ArXiv preprint:1407.3220.

164 De Gennes, P.G. and Hurault, J.P. (1965). Proximity effects under magnetic fields. II-Interpretation of "breakdown". *Phys. Lett.* 17 (3): 181–182. https://doi.org/10.1016/0031-9163(65)90469-5.

165 Cherkez, V., Cuevas, J.C., Brun, C. et al. (2014). Proximity effect between two superconductors spatially resolved by scanning tunneling spectroscopy. *Phys. Rev. X* 4 (1): 011033. https://doi.org/10.1103/PhysRevX.4.011033.

166 Martinello, M., Aderhold, S., Chandrasekaran, S. et al. (2017). Advancement in the Understanding of the Field and Frequency Dependent Microwave Surface Resistance of Niobium. *ArXiv Prepr. ArXiv170707582.* https://doi.org/10.48550/arXiv.1707.07582.

167 Gonnella, D. (2020). Statistical analysis of nitrogen-doped cavity production for LCLS-II Feb. *TESLA Technology Collaboration Meeting (TTC 2020)*, Virtual (DESY, Hamburg). https://indico.cern.ch/event/817780/contributions/3632122/attachments/1980945/3299773/LCLS2_Stats_Gonnella.pptx.

168 Martinello, M. (2021). LCLS-II-HE R&D overview, SRF 2021. *20th International Conference on RF Superconductivity (SRF 2021)*, Virtual (MSU). East Lansing, MI, USA, THOTEV08.

169 Valles, N. (2014). *Pushing the frontiers of superconducting radio frequency science: from the temperature dependence of the superheating field of niobium to higher-order mode damping in very high quality factor accelerating structures*. PhD thesis. Cornell University.

170 Crawford, C., Möller, W.D., Schmueser, P. et al. (1995). High gradients in linear collider superconducting accelerator cavities by high pulsed power to suppress field emission. *Part. Accel.* 49 (SRF-940714-05): 1–13. https://inspirehep.net/files/3737f8271a58752a49bef1e94af2ee72.

171 Romanenko, A., Grassellino, A., Crawford, A.C. et al. (2014). Ultra-high quality factors in superconducting niobium cavities in ambient magnetic fields up to 190 mG. *Appl. Phys. Lett.* 105 (23): 234103. https://doi.org/10.1063/1.4903808.

172 Gonnella, D., Aderhold, S., Bafia, D. et al. (2019). The LCLS-II HE high Q and gradient R&D program. *19th International Conference on RF Superconductivity (SRF 2019)*, Dresden, Germany (30 June-05 July 2019), pp. 154–158. https://doi.org/10.18429/JACoW-SRF2019-MOP045.

173 Palczewski, A. (2019). High Q0/high gradient at JLab: LCLS-2 HE 3N60 doping, furnace issues and FNAL 75C retests. *TESLA Technology Collaboration Meeting (TTC 2019)*, Vancouver, Canada. https://indico.desy.de/event/21337/contributions/42662/.

174 Spradlin, J., Palczewski, A., Reece, C., and Tian, H. (2019). Analysis of surface nitrides created during 'doping' heat treatments of niobium. *Proceedings of the 19th International Conference on RF Superconductivity (SRF 2019)*. https://doi.org/10.18429/JACOW-SRF2019-MOP030.

175 Gonnella, D. (2022). LCLS-II-HE high Q0& gradient R&D program, first CM test results, and CM plasma processing results. *TESLA Technology Collaboration Meeting (2022)*, Aomori, Japan. https://indico.desy.de/event/32219/contributions/116543/.

176 Chen, S., Chen, M., Feng, L. et al. (2019). Nitrogen doping studies of superconducting cavities at Peking University. *Proceedings of the 19th International Conference on RF Superconductivity (SRF 2019)*. https://doi.org/10.18429/JACOW-SRF2019-MOP039.

177 Chen, J., Hou, H.T., Huang, Y. et al. (2019). N-doping studies with single-cell cavities for the SHINE project. *Proceedings of the 19th International Conference on RF Superconductivity (SRF 2019)*. https://doi.org/10.18429/JACOW-SRF2019-MOP029.

178 Okada, T., Hori, Y., Kako, E. et al. (2019). Improvement of cavity performance by nitrogen doping at KEK. *Proceedings of 29th Linear Accelerator Conference (LINAC 2018)*, Beijing, China, pp. 480–483. https://doi.org/10.18429/JACoW-LINAC2018-TUPO065.

179 Posen, S., Romanenko, A., Grassellino, A. et al. (2020). Ultralow surface resistance via vacuum heat treatment of superconducting radio-frequency cavities. *Phys. Rev. Appl.* 13 (1): 014024. https://doi.org/10.1103/PhysRevApplied.13.014024.

180 He, F., Pan, W., Sha, P. et al. (2020). Medium-Temperature Furnace Bake of Superconducting Radio-Frequency Cavities at IHEP. *ArXiv Prepr. ArXiv201204817*. https://doi.org/10.48550/arXiv.2012.04817.

181 Ito, H., Araki, H., Takahashi, K., and Umemori, K. (2021). Influence of furnace baking on Q–E behavior of superconducting accelerating cavities. *Prog. Theor. Exp. Phys.* (7): 071G01. https://doi.org/10.1093/ptep/ptab056.

182 Martinello, M., Grassellino, A., Checchin, M. et al. (2016). Effect of interstitial impurities on the field dependent microwave surface resistance of niobium. *Appl. Phys. Lett.* 109 (6): 062601. https://doi.org/10.1063/1.4960801.

183 Safa, H. (2001). High field behavior of SCRF cavities. *Proceedings of the 10th Workshop on RF Superconductivity (SRF 2001)*, Tsukuba, Ibaraki, Japan, MA008, pp. 279–286. https://accelconf.web.cern.ch/srf01/papers/ma008.pdf.

184 Kneisel, P., Röth, R.W., and Kuerschner, H.G. (1996). Results from a nearly'defect-free'niobium cavity. *Proceedings of the 1995 workshop on RF Superconductivity (SRF 1995)*, Gif-sur-Yvette, France, https://accelconf.web.cern.ch/SRF95/papers/srf95c17.pdf.

185 Palmer, F.L., Kirby, R.E., King, F.K., and Garwin, E.L. (1990). Oxide overlayers and the superconducting RF properties of yttrium-processed high purity Nb. *Nucl. Instrum. Methods Phys. Res., Sect. A* 297 (3): 321–328. https://doi.org/10.1016/0168-9002(90)91314-2.

186 Eremeev, G.V. (2008). *Study of the High Field Q-Slope Using Thermometry.* Thesis. Cornell University.

187 Delheusy, M. (2008). X-ray investigation of Nb/O interfaces. *Appl. Phys. Lett.* 92 (10). https://doi.org/10.1063/1.2889474.

188 Padamsee, H.S. (2014). Superconducting radio-frequency cavities. *Annu. Rev. Nucl. Part. Sci.* 64: 175–196. https://doi.org/10.1146/annurev-nucl-102313-025612.

189 Visentin, B. (2006). Low, medium, high field Q-slopes change with surface treatments. ICFA Beam Dynamics Newsletter, 39, 94. https://icfa-usa.jlab.org/archive/newsletter/icfa_bd_nl_39.pdf.

190 Romanenko, A. and Schuster, D.I. (2017). Understanding quality factor degradation in superconducting niobium cavities at low microwave field amplitudes. *Phys. Rev. Lett.* 119 (26): 264801. https://doi.org/10.1103/PhysRevLett.119.264801.

191 Romanenko, A. et al. (2019). Superconducting cavities for quantum computers - talk. p. frchc1 talk. *NAPAC2019*, Lansing, Michigan. FRCHC1. http://accelconf.web.cern.ch/napac2019/talks/frchc1_talk.pdf.

192 Kruger, P. (2021). The Josephson Parametric Amplifier. In Technische Universität München Walther Meissner Institut Bayerische Akademie der Wissenschaften. https://www.wmi.badw.de/fileadmin/WMI/Lecturenotes/FOPRA/Manual_FOPRA__104_20211029.pdf.

193 Reschke, D.W. and Roth, R.W. (1993). *Proceedings of the 6th workshop on RF superconductivity (SRF 1993)*, Newport News, VA, USA. https://accelconf.web.cern.ch/SRF93/papers/srf93i54.pdf.

194 Antoine, C. (2012). Materials and surface aspects in the development of SRF niobium cavities. http://cds.cern.ch/record/1472363/files/EuCARD-BOO-2012-001.pdf.

195 Martinis, J.M., Cooper, K.B., McDermott, R. et al. (2005). Decoherence in Josephson qubits from dielectric loss. *Phys. Rev. Lett.* 95 (21): 210503. https://doi.org/10.1103/PhysRevLett.95.210503.

196 Müller, C., Cole, J.H., and Lisenfeld, J. (2019). Towards understanding two-level-systems in amorphous solids: insights from quantum circuits. *Rep. Prog. Phys.* 82 (12): 124501. https://doi.org/10.48550/arXiv.1705.01108.

197 Von Schickfus, M. and Hunklinger, S. (1977). Saturation of the dielectric absorption of vitreous silica at low temperatures. *Phys. Lett. A* 64 (1): 144–146. https://doi.org/10.1016/0375-9601(77)90558-8.

198 Romanenko, A., Pilipenko, R., Zorzetti, S. et al. (2020). Three-dimensional superconducting resonators at $T < 20$ mK with photon lifetimes up to $\tau = 2$ s. *Phys. Rev. Appl.* 13 (3): 034032. https://doi.org/10.48550/arXiv.1810.03703.

199 Paik, H., Schuster, D.I., Bishop, L.S. et al. (2011). Observation of high coherence in Josephson junction qubits measured in a three-dimensional circuit QED architecture. *Phys. Rev. Lett.* 107 (24): 240501. https://doi.org/10.1103/PhysRevLett.107.240501.

200 Reagor, M., Pfaff, W., Axine, C. et al. (2016). Quantum memory with millisecond coherence in circuit QED. *Phys. Rev. B* 94 (1): 014506. https://doi.org/10.1103/PhysRevB.94.014506.

201 Reagor, M., Paik, H., Catelani, G. et al. (2013). Reaching 10 ms single photon lifetimes for superconducting aluminum cavities. *Appl. Phys. Lett.* 102 (19): 192604. https://doi.org/10.1063/1.4807015.

202 Miyazaki, A. and Delsolaro, W.V. (2018). Determination of the BCS Material Parameters of the HIE-ISOLDE Superconducting Resonator. *ArXiv Prepr. ArXiv180604443*. https://doi.org/10.48550/arXiv.1806.04443.

203 Hall, D.L., Liarte, D.B., Liepe, M., and Sethna, J.P. (2017). Impact of trapped magnetic flux and thermal gradients on the performance of Nb_3Sn cavities. *Proceedings of IPAC 2017*, pp. 1127–1129, MOPVA118. https://accelconf.web.cern.ch/ipac2017/papers/mopva118.pdf.

204 Checchin, M., Martinello, M., Grassellino, A. et al. (2018). Frequency dependence of trapped flux sensitivity in SRF cavities. *Appl. Phys. Lett.* 112 (7): 072601. https://doi.org/10.1063/1.5016525.

205 Gurevich, A. and Ciovati, G. (2013). Effect of vortex hotspots on the radio-frequency surface resistance of superconductors. *Phys. Rev. B* 87 (5): 054502. https://doi.org/10.1103/PhysRevB.87.054502.

206 Gonnella, D., Kaufman, J., and Liepe, M. (2016). Impact of nitrogen doping of niobium superconducting cavities on the sensitivity of surface resistance to trapped magnetic flux. *J. Appl. Phys.* 119 (7): 073904. https://doi.org/10.1063/1.4941944.

207 Abrikosov, A.A. (1957). On the magnetic properties of superconductors of the second group. *Zh. Eksp. Teor. Fiz.* 32: 1442.

208 Huebener, R.P. and Seher, A. (1969). Nernst effect and flux flow in superconductors. I. Niobium. *Phys. Rev.* 181 (2): 701. https://doi.org/10.1103/PhysRev.181.701.

209 Ciovati, G. and Gurevich, A. (2008). Evidence of high-field radio-frequency hot spots due to trapped vortices in niobium cavities. *Phys. Rev. Spec. Top. Accel. Beams* 11 (12): 122001. https://doi.org/10.1103/PhysRevSTAB.11.122001.

210 Gittleman, J.I. and Rosenblum, B. (1966). Radio-frequency resistance in the mixed state for subcritical currents. *Phys. Rev. Lett.* 16 (17): 734. https://doi.org/10.1103/PhysRevLett.16.734.

211 Matsushita, T. (2007). *Flux Pinning in Superconductors*, vol. 164. Springer.

212 Allen, L.H. and Claassen, J.H. (1989). Technique for measuring the elementary pinning force in thin films. *Phys. Rev. B* 39 (4): 2054. https://doi.org/10.1103/PhysRevB.39.2054.

213 Park, G.S., Cunningham, C.E., Cabrera, B., and Huber, M.E. (1992). Vortex pinning force in a superconducting niobium strip. *Phys. Rev. Lett.* 68 (12): 1920. https://doi.org/10.1103/PhysRevLett.68.1920.

214 Checchin, M., Martinello, M., Grassellino, A. et al. (2017). Electron mean free path dependence of the vortex surface impedance. *Supercond. Sci. Technol.* 30 (3): 034003. https://doi.org/10.1088/1361-6668/aa5297.

215 Benvenuti, C., Calatroni, S., Campisis, I.E. et al. (1999). Study of the surface resistance of superconducting niobium films at 1.5 GHz. *Physica C* 316 (3–4): 153–188. https://doi.org/10.1016/S0921-4534(99)00207-5.

216 Miyazaki, A., Languevergene, D., Kramer, F. et al. (2022). Recent progress of flux trapping/expulsion studies in TEM cavities. presented at the *TESLA Technology Collaboration Meeting (TTC 2022)*. https://indico.desy.de/event/32219/contributions/116565/.

217 Weingarten, W. (2020). Fluxon-induced losses in niobium thin-film cavities revisited. *IEEE Trans. Appl. Supercond.* 30 (8): 3500411.

218 Pathirana, W. and Gurevich, A. (2020). Nonlinear dynamics and dissipation of a curvilinear vortex driven by a strong time-dependent Meissner current. *Phys. Rev. B* 101 (6): 064504. https://doi.org/10.1103/PhysRevB.101.064504.

219 Calatroni, S. and Vaglio, R. (2019). Simple model for the RF field amplitude dependence of the trapped flux sensitivity in superconducting RF cavities. *Phys. Rev. Accel. Beams* 22 (2): 022001. https://doi.org/10.1103/PhysRevAccelBeams.22.022001.

220 Liarte, D.B., Hall, D., Koufalis, P.N. et al. (2018). Vortex dynamics and losses due to pinning: dissipation from trapped magnetic flux in resonant superconducting radio-frequency cavities. *Phys. Rev. Appl.* 10 (5): 054057. https://doi.org/10.1103/PhysRevApplied.10.054057.

221 Checchin, M. (2019). New insights into RF field amplitude and frequency dependence of vortex surface resistance. Presented at the *19th International Conference on RF Superconductivity (SRF 2019)*, Dresden, Germany. https://accelconf.web.cern.ch/srf2019/talks/tufua4_talk.pdf.

222 Romanenko, A., Grassellino, A., Melnychuk, O., and Sergatskov, D.A. (2014). Dependence of the residual surface resistance of superconducting radio frequency cavities on the cooling dynamics around T_c. *J. Appl. Phys.* 115 (18): 184903. https://doi.org/10.1063/1.4875655.

223 Kugeler, O., Neumann, A., Voronenko, S. et al. (2009). Manipulating the intrinsic quality factor by thermal cycling and magnetic fields. *Proceedings of the 14th International Conference on RF Superconductivity (SRF 2009)*, Berlin, Germany. pp. 370–372. https://accelconf.web.cern.ch/SRF2009/papers/tuppo053.pdf.

224 Vogt, J.-M., Kugeler, O., and Knobloch, J. (2013). Impact of cool-down conditions at T_c on the superconducting RF cavity quality factor. *Phys. Rev. Spec. Top. Accel. Beams* 16 (10): 102002. https://doi.org/10.1103/PhysRevSTAB.16.102002.

225 Crawford, A.C. (2014). A Study of Thermocurrent Induced Magnetic Fields in ILC Cavities. *ArXiv Prepr. ArXiv14037996*. https://doi.org/10.48550/arXiv.1403.7996.

226 Eichhorn, R. et al. (2016). Thermocurrents and their role in high Q cavity performance. *Phys. Rev. Accel. Beams* 19 (1): 012001. https://doi.org/10.1103/PhysRevAccelBeams.19.012001.

227 Grassellino, A., Aderhold, S., Checchin, M. et al. (2015). Preservation of very high quality factors of 1.3 GHz nine cell cavities from bare vertical test to dressed horizontal test. *Proceedings of 17th International Conference on RF Superconductivity (SRF 2015)*, Whistler, BC, Canada MOPB028. https://accelconf.web.cern.ch/SRF2015/papers/mopb028.pdf.

228 Kubo, T. (2015). Model of flux trapping cooling down process. *Proceedings of 17th International Conference on RF Superconductivity (SRF 2015)*, Whistler, BC, Canada, p. IMOPB009. https://accelconf.web.cern.ch/SRF2015/papers/mopb009.pdf.

229 Kubo, T. (2016). Flux trapping in superconducting accelerating cavities during cooling down with a spatial temperature gradient. *Prog. Theor. Exp. Phys.* 5: 053G01. https://doi.org/10.1093/ptep/ptw049.

230 Martinello, M., Romanenko, A., Checchin, M. et al. (2015). Cooling Dynamics Through Transition Temperature of Niobium SRF Cavities Captured by Temperature Mapping. *ArXiv Prepr. ArXiv150404441*. https://doi.org/10.48550/arXiv.1504.04441.

231 Martinello, M., Checchin, M., Grassellino, A. et al. (2015). Magnetic flux studies in horizontally cooled elliptical superconducting cavities. *J. Appl. Phys.* 118 (4): 044505. https://doi.org/10.1063/1.4927519.

232 Longuevergne, D. (2019). How is flux expulsion affected by geometry: experi-mental evidence and model. *19th International Conference on RF Superconductivity (SRF 2019)*, Dresden, Germany. p. TUP054. https://accelconf.web.cern.ch/srf2019/posters/tup054_poster.pdf.

233 Knobloch, J. and Padamsee, H. (1997). Flux trapping in niobium cavities during breakdown events. *Proceedings of the 8th Workshop on RF Superconductivity*, Abano Terme, Italy, p. 337. https://accelconf.web.cern.ch/SRF97/papers/srf97b06.pdf.

234 Checchin, M., Martinello, M., Romanenko, A. et al. (2016). Quench-induced degradation of the quality factor in superconducting resonators. *Phys. Rev. Appl.* 5 (4): 044019. https://doi.org/10.1103/PhysRevApplied.5.044019.

235 Terechkine, I., Khabiboulline, T., and Sergatskov, D. (2013). Performance degradation of a superconducting cavity quenching in magnetic field. *Proceedings of the 16th International Conference on Rf Superconductivity (SRF 2013)*, p. 1013. https://accelconf.web.cern.ch/SRF2013/papers/thp047.pdf.

236 Khabiboulline, T., Ozelis, J., Sergatskov, D., and Terechkine, I. (2011). Superconducting cavity quenching in the presence of magnetic field. FNAL TD note TD-11-020.

237 Khabiboulline, T., Sergatskov, D., and Terechkine, I. (2012). SSR1 cavity quenching in the presence of magnetic field. *FNAL TD Note T-12-007*.

238 Khabiboulline, T., Sergatskov, D.A., and Terechkine, I.M. (2013). 650 MHz elliptical cavity performance degradation induced by magnetic field of a test coil. FNAL Note TD-13-002.

239 Posen, S. and Team, F.S. (2017). Magnetic flux expulsion studies on niobium. *Proceedings of the 18th International Conference on RF Superconductivity (SRF 2017)*, Lanzhou, China. https://accelconf.web.cern.ch/srf2017/talks/tuxba02_talk.pdf.

240 Martinello, M. (2019). Microscopic investigation of flux trapping sites in bulk Nb. Presented at the *19th International Conference on RF Superconductivity (SRF 2019)*, Dresden, Germany. https://accelconf.web.cern.ch/srf2019/talks/tufub4_talk.pdf.

241 Posen, S., Checchin, M., Crawford, A.C. et al. (2016). Efficient expulsion of magnetic flux in superconducting radiofrequency cavities for high Q_0 applications. *J. Appl. Phys.* 119 (21): 213903. https://doi.org/10.1063/1.4953087.

242 Posen, S., Chandrasekaran, S.K., Crawford, A.C. et al. (2018). The effect of mechanical cold work on the magnetic flux expulsion of niobium. *ArXiv Prepr. ArXiv180407207*. https://doi.org/10.48550/arXiv.1804.07207.

243 Aull, S., Kugeler, O., and Knobloch, J. (2012). Trapped magnetic flux in superconducting niobium samples. *Phys. Rev. Spec. Top. Accel. Beams* 15 (6): 062001. https://doi.org/10.1103/PhysRevSTAB.15.062001.

244 Dhavale, A.S., Dhakal, P., Polyanskii, A.A., and Ciovati, G. (2012). Flux pinning characteristics in cylindrical niobium samples used for superconducting radio frequency cavity fabrication. *Supercond. Sci. Technol.* 25 (6): 065014. https://doi.org/10.1088/0953-2048/25/6/065014.

245 Singer, W., Aderhold, S., Ermakov, C. et al. (2013). Development of large grain cavities. *Phys. Rev. Spec. Top. Accel. Beams* 16 (1): 012003. https://doi.org/10.1103/PhysRevSTAB.16.012003.

246 Posen, S., Checchin, M., Crawford, A.C. et al. (2015). Flux expulsion variation in SRF cavities. *Proceedings of the 17th International Conference on RF Superconductivity (SRF 2015)*, Whistler, BC, Canada, MOPB104, p. 404. https://accelconf.web.cern.ch/SRF2015/papers/mopb104.pdf.

247 Kneisel, P. (2007). Progress on large grain and single grain niobium – ingots and sheet and review of progress on large grain and single grain niobium cavities. *Proceedings of SRF 2007*, Beijing, China, Peking University, p. TH102, pp. 728–733. https://accelconf.web.cern.ch/srf2007/PAPERS/TH102.pdf.

248 Kneisel, P., Ciovati, G., Singer, W. et al. (2008). Performance of single crystal niobium cavities. *Proceedings of EPAC08*, Genoa, Italy.

249 Ciovati, G., Myneni, G., Stevie, F. et al. (2010). High field Q slope and the baking effect: review of recent experimental results and new data on Nb heat treatments. *Phys. Rev. Spec. Top. Accel. Beams* 13 (2): 022002. https://doi.org/10.1103/PhysRevSTAB.13.022002.

250 Lilje, L. (2005). Overview on high field Q-slope. *Proceedings of the Workshop on Pushing the Limits of RF Superconductivity*, Argonne National Laboratory, March 2005, ANL 05/10, p. 43. Argonne National Laboratory. https://www3.aps.anl.gov/News/Conferences/2004/RFSC-Limits/presentations/Day2/Lilje.pdf.

251 Romanenko, A. (2007). Review of high field Q-slope, surface measurements. *Proceedings of SRF 2007*, Beijing, China, Peking University. https://accelconf.web.cern.ch/srf2007/PAPERS/TU103.pdf; https://accelconf.web.cern.ch/srf2007/TALKS/TU103_TALK.pdf.

252 Saito, K. et al. (2021). FRIB production cavities data analysis and a model. presented at the *TESLA Technology Collaboration Meeting 2021*, Virtual (DESY, Hamburg).

253 Romanenko, A., Grassellino, A., Barkov, F., and Ozelis, J.P. (2013). Effect of mild baking on superconducting niobium cavities investigated by sequential nanoremoval. *Phys. Rev. Spec. Top. Accel. Beams* 16 (1): 012001. https://doi.org/10.1103/PhysRevSTAB.16.012001.

254 Ciovati, G., Kneisel, P., and Gurevich, A. (2007). Measurement of the high-field Q drop in a high-purity large-grain niobium cavity for different oxidation processes. *Phys. Rev. Spec. Top. Accel. Beams* 10 (6): 062002. https://doi.org/10.1103/PhysRevSTAB.10.062002.

255 Vines, J., Xie, Y., and Padamsee, H. (2007). Systematic trends for the medium field Q-slope. *Proceedings of the 13th International Workshop on RF Superconductivity (SRF 2007)*, Beijing, China. TUP27, p. 178. https://accelconf.web.cern.ch/srf2007/PAPERS/TUP27.pdf.

256 Eremeev, G. and Padamsee, H. (2006). Change in high field Q-slope by anodizing of the baked cavity. *Physica C* 441: 134–136. https://doi.org/10.1016/j.physc.2006.03.046.

257 Grassellino, A., Beard, C., Kolb. P. et al. (2013). Muon spin rotation studies of niobium for superconducting RF applications. *Phys. Rev. Spec. Top. Accel. Beams* 16 (6): 062002. https://doi.org/10.1103/PhysRevSTAB.16.062002.

258 Junginger, T. and Laxdal, R.E. (2019). Review of muon spin rotation studies of SRF materials. *19th International Conference on RF Superconductivity (SRF 2019)*, Dresden, Germany, pp. 360–363. https://doi.org/10.18429/JACoW-SRF2019-TUFUA7.

259 Junginger, T. (2017). Dirty layers, Bi-layers and Multi-layers: insights-from muon spin rotation experiments. *18th International Conference on RF*

Superconductivity (SRF 2017), Lanzhou, China, wexa05. https://accelconf.web.cern.ch/srf2017/talks/wexa05_talk.pdf.

260 Kowalski, K., Bernasik, A., Singer, W. et al. (2003). In situ'XPS investigation of the baking effect on the surface oxide structure formed on niobium sheets used for superconducting RF cavity production. *Proceedings of the 11th Workshop on RF Superconductivity (SRF 2003)*, Lübeck/Travemünde, Germany. https://accelconf.web.cern.ch/SRF2003/papers/thp09.pdf.

261 Grassellino, A., Romanenko, A., Trenikhina, Y. et al. (2017). Unprecedented quality factors at accelerating gradients up to 45 MVm $^{-1}$ in niobium superconducting resonators via low temperature nitrogen infusion. *Supercond. Sci. Technol.* 30 (9): 094004. https://doi.org/10.1088/1361-6668/aa7afe.

262 Romanenko, A., Bafia, D., Grassellino, A. et al. (2019). First direct imaging and profiling TOF-SIMS studies on cutouts from cavities prepared by state-of-the-art treatments. *19th International Conference on RF Superconductivity (SRF 2019)*, Dresden, Germany, pp. 866–870. https://accelconf.web.cern.ch/srf2019/papers/thp014.pdf.

263 Kneisel, P. (1999). Preliminary experience with 'in-situ' baking of niobium cavities. *Proceedings of the 1999 Workshop on RF Superconductivity*, Santa Fe, New Mexico, USA, p. TuP044. https://accelconf.web.cern.ch/SRF99/papers/tup044.pdf.

264 Ciovati, G. (2006). Improved oxygen diffusion model to explain the effect of low-temperature baking on high field losses in niobium superconducting cavities. *Appl. Phys. Lett.* 89 (2): 022507. https://doi.org/10.1063/1.2220059.

265 Romanenko, A., Grassellino, A., and Bafia, D. (2020). Oxygen role in the high field Q-slope/mild baking - revealed by TOF-SIMS + cavity surface engineering studies. Presented at the *TESLA Technology Collaboration Meeting (TTC 2020)*, Geneva, Switzerland. https://indico.cern.ch/event/817780/contributions/3713977/attachments/1981257/3299473/Romanenko_TTC2020.pdf.

266 Benvenuti, C., Calatroni, S., and Ruzinov, V. (2001). Diffusion of oxygen in niobium during bake-out. *Proceedings of the 10th Workshop on RF Superconductivity (SRF 2001)*, Tsukuba, Japan, p. 441. https://accelconf.web.cern.ch/srf01/papers/pr025.pdf.

267 Bafia, D. (2022). The role of oxygen in enabling high gradients and high Q0. presented at the *TESLA Technology Collaboration Meeting (TTC 2022)*, Aomori, Japan. https://indico.desy.de/event/32219/contributions/116548/.

268 Trenikhina, Y., Romanenko, A., Kwon, J. et al. (2015). Nanostructural features degrading the performance of superconducting radio frequency niobium cavities revealed by transmission electron microscopy and electron energy loss spectroscopy. *J. Appl. Phys.* 117 (15): 154507. https://doi.org/10.1063/1.4918272.

269 Manchester, F.D. (2000). Phase diagrams of binary hydrogen alloys. *ASM Int. MemberCustomer Serv. Cent. Mater.*, Park OH 44073-0002 USA 2000 322.

270 Manchester, F.D. and Pitre, J.M. (2000). H-Nb(hydrogen-niobium). I. The incoherent phase diagram. *Phase Diagr. Bin. Hydrog. Alloys ASM Int. MemberCustomer Serv. Cent. Mater.*, Park OH 44073-0002 USA 2000, pp. 115–137.

271 Ford, D.C., Cooley, L.D., and Seidman, D.N. (2013). Suppression of hydride precipitates in niobium superconducting radio-frequency cavities. *Supercond. Sci. Technol.* 26 (10): 105003. https://doi.org/10.1088/0953-2048/26/10/105003.

272 Pundt, A. and Kirchheim, R. (2006). Hydrogen in metals: microstructural aspects. *Annu. Rev. Mater. Res.* 36: 555–608. https://doi.org/10.1146/annurev.matsci.36.090804.094451.

273 Antoine, C.Z., Aune, B., Bonin, B. et al. (1991). The role of atomic hydrogen in *Q*-degradation of niobium superconducting RF cavities: analytical point of view. *Proceedings of the 5th Workshop on RF Superconductivity (SRF 1991) DESY*, Hamburg, Germany, pp. 616–634. https://accelconf.web.cern.ch/SRF91/papers/srf91f02.pdf.

274 Tajima, T., Edwards, R., Krawczyk, F. et al. (2003). *Q* disease: insights from 350 MHz spoke cavities tests and ERD analyses of hydrogen in Niobium. *Proceedings of the 11th Workshop on RF Superconductivity (SRF 2003)*, Lübeck/Travemünder, Germany. https://accelconf.web.cern.ch/SRF2003/papers/thp19.pdf.

275 Romanenko, A. and Goncharova, L.V. (2011). Elastic recoil detection studies of near-surface hydrogen in cavity-grade niobium. *Supercond. Sci. Technol.* 24 (10): 105017. https://doi.org/10.1088/0953-2048/24/10/105017.

276 Aune, B., Bonin, B., Cavedon, J.M. et al. (1990). Degradation of niobium superconducting RF cavities during cooling times. *Proceedings of the 1990 Linac Conference*, pp. 253–255. https://accelconf.web.cern.ch/l90/papers/mo472.pdf.

277 Bonin, B. and Röth, R.W. (1991). Q degradation of niobium cavities due to hydrogen contamination. *Proceedings of the 5th Workshop on RF Superconductivity (SRF 1991)*, Volume 1, pp. 210–244. https://accelconf.web.cern.ch/SRF91/papers/srf91d01.pdf.

278 Halbritter, J., Kneisel, P., and Saito, K. (1993). Additional losses in high purity niobium cavities related to slow cooldown and hydrogen segregation. *Proceedings of the 6th workshop on RF Superconductivity (SRF 1993)*, Newport News, VA: Thomas Jefferson National Accelerator Facility (TJNAF). https://accelconf.web.cern.ch/SRF93/papers/srf93i08.pdf.

279 Knobloch, J. (2003). The 'Q disease' in superconducting niobium RF cavities. *AIP Conf. Proceed.* 671 (1): 133–150. https://doi.org/10.1063/1.1597364.

280 Enderlein, G., Körber, W., Matheisen, A. et al. (1991). Investigations on hydrogen contamination of superconducting cavities. *Proceedings of the 1991 IEEE Particle Accelerator Conference (PAC 1991)*, pp. 2432–2434. https://accelconf.web.cern.ch/p91/PDF/PAC1991_2432.PDF.

281 Byrns, R., Dwersteg, B., Enderlein, G. et al. (1990). Status of the superconducting cavity, program for HERA. *Proceedings of 2nd European Accelerator Conference (EPAC 1990)*, Nice, France. https://accelconf.web.cern.ch/e90/PDF/EPAC1990_1073.PDF.

282 Roth, R., Heinrichs, H., Kurakin, V.G. et al. (1990). Suppression of field emission in superconducting S-band accelerating structures. *2nd European Accelerator Conference (EPAC 1990)*, Nice, France, p. 1097. https://accelconf.web.cern.ch/e90/PDF/EPAC1990_1097.PDF.

283 Barkov, F., Romanenko, A., Trenikhina, Y., and Grassellino, A. (2013). Precipitation of hydrides in high purity niobium after different treatments. *J. Appl. Phys.* 114 (16): 164904. https://doi.org/10.1063/1.4826901.

284 Smith, J.F. (1983). The H-Nb (Hydrogen-Niobium) and D-Nb (Deuterium-Niobium) systems. *Bull. Alloy Phase Diagr.* 4 (1): 39–46.

285 Nørskov, J.K. and Besenbacher, F. (1987). Theory of hydrogen interaction with metals. *J. Common Met.* 130: 475–490. https://doi.org/10.1016/0022-5088(87)90145-7.

286 Barkov, F., Romanenko, A., and Grassellino, A. (1991). Fast table top niobium hydride investigation using direct imaging in a cryo-stage. *Proceedings of the 16th International Conference on RF Superconductivity (SRF 2013)*, Paris, France, pp. 447–450. https://accelconf.web.cern.ch/SRF2013/papers/tup014.pdf.

287 Barkov, F., Romanenko, A., and Grassellino, A. (2012). Direct observation of hydrides formation in cavity-grade niobium. *Phys. Rev. Spec. Top. Accel. Beams* 15 (12): 122001. https://doi.org/10.1103/PhysRevSTAB.15.122001.

288 Sung, Z.H., Romanenko, A., Martinello, M., and Grassellino, A. (2019). Direct correlation of state of the art cavity performance with surface Nb nano-hydrides cutouts observed via cryogenic-AFM. *Proceedings of the 19th International Conference on RF Superconductivity (SRF 2019)*, Dresden, Germany. https://epaper.kek.jp/srf2019/talks/tufub1_talk.pdf.

289 Bafia, D., Grassellino, A., Melnychuk, O. et al. (2019). Gradients of 50 MV/m in TESLA shaped cavities via modified low temperature bake. *19th International Conference on RF Superconductivity (SRF 2019)*, Dresden, Germany (30 June-05 July 2019), pp. 586–591. https://accelconf.web.cern.ch/srf2019/papers/tup061.pdf.

290 Rauch, G.C., Rose, R.M., and Wulff, J. (1965). Observations on microstructure and superconductivity in the Nb-H system. *J. Common Met.* 8 (2): 99–113. https://doi.org/10.1016/0022-5088(65)90101-3.

291 Romanenko, A., Edwardson, C.J., Coleman, P.G., and Simpson, P.J. (2013). The effect of vacancies on the microwave surface resistance of niobium revealed by positron annihilation spectroscopy. *Appl. Phys. Lett.* 102 (23): 232601. https://doi.org/10.1063/1.4811090.

292 Visentin, B., Barthe, M.F., Moineau, V., and Desgardin, P. (2010). Involvement of hydrogen-vacancy complexes in the baking effect of niobium cavities. *Phys. Rev. Spec. Top. Accel. Beams* 13 (5): 052002. https://doi.org/10.1103/PhysRevSTAB.13.052002.

293 Alekseeva, O.K., Bykov, V.N., Levdik, V.A. et al. (1979). Annealing of defects in irradiated niobium. *Phys. Scr.* 20 (5–6): 683. https://doi.org/10.1088/0031-8949/20/5-6/026.

294 Hautojarvi, P., Huomo, H., Puska, M., and Vehanen, A. (1985). Vacancy recovery and vacancy-hydrogen interaction in niobium and tantalum studied by positrons. *Phys. Rev. B* 32 (7): 4326. https://doi.org/10.1103/PhysRevB.32.4326.

295 Romanenko, A., Edwardson, C., and Coleman, P. (2012). Inner workings of the 120°C baking effect studied by positron annihilation. *Talk Thin Film SRF Workshop at JLab (TFSRF 2012)*. https://indico.jlab.org/event/22/

contributions/2221/attachments/1878/2369/Inner_workings_of_the_120C_baking_effect_studied_by_positron_annihilation.pdf.

296 Proslier, T., Zasadzinski, J.F., Cooley, L. et al. (2008). Tunneling study of cavity grade Nb: possible magnetic scattering at the surface. *Appl. Phys. Lett.* 92 (21): 212505. https://doi.org/10.1063/1.2913764.

297 Grassellino, A., Romenenko, A., Bice, D. et al. (2018). Accelerating fields up to 49 MV/m in TESLA-shape superconducting RF niobium cavities via 75C vacuum bake. *ArXiv Prepr. ArXiv180609824.* https://doi.org/10.48550/arXiv.1806.09824.

298 Wenskat, M., Bafia, D., Grassellino, A. et al. (2019). A cross-lab qualification of modified 120°C baked cavities. *Proceedings of the 19th International Conference on RF Superconductivity (SRF 2019)*, Dresden, Germany. pp. 90–94.

299 Martinello, M. (2020). High-Q/high-g TTC task force report. presented at the *TESLA Technology Collaboration Meeting (TTC 2020)*, Geneva, Switzerland: CERN.

300 Wenskat, M., Bate, C., Butterling, M. et al. (2019). Vacancy-hydrogen dynamics in samples during low temperature baking. *19th International Conference on RF Superconductivity (SRF 2019)*, Dresden, Germany (30 June-05 July 2019), pp. 83–86. https://accelconf.web.cern.ch/srf2019/papers/mop024.pdf.

301 Romanenko, A. (2019), Fermilab. Private communication.

302 Martinello, M. (2017). High-Q at high gradient at FNAL. presented at the *LCWS 2017*, Strasbourg, France. https://agenda.linearcollider.org/event/7645/contributions/40038/.

303 Dhakal, P. (2019). Recent development on nitrogen infusion work towards high Q and high gradient. *19th International Conference on RF Superconductivity (SRF 2019)*, Dresden, Germany (30 June-05 July 2019), pp. 355–359. https://accelconf.web.cern.ch/srf2019/papers/tufua5.pdf.

304 Umemori, K., Dohmae, T., Egi, M. et al. (2019). Study on nitrogen infusion using KEK new furnace. *Proceedings of the 19th International Conference on RF Superconductivy (SRF 2019)*, Dresden, Germany. https://accelconf.web.cern.ch/srf2019/papers/mop027.pdf.

305 Wenskat, M., Bate, C., Jeromin, A. et al. (2019). Cavity cut-out studies of a 1.3 GHz single-cell cavity after a failed nitrogen infusion process. *19th International Conference on RF Superconductivity (SRF 2019)*, Dresden, Germany. https://doi.org/10.18429/JACoW-SRF2019-MOP025.

306 Umemori, K. (2018). First success of N-infusion at KEK/J-PARC 2018/Jan/11. presented at the *TESLA Technology Collaboration Meeting (TTC 2018)*, Milan, Italy: INFN. https://indico.desy.de/event/20010/contributions/37516/attachments/24201/30818/N-infusion_KEK_v1.pdf.

307 Umemori, K., Dohmae, T. Egi, M. et al. (2018). Study on nitrogen infusion for 1.3 GHz SRF cavities using J-PARC furnace. *Proceedings of the 29th Linear Accelerator Conference (LINAC 2018)*. https://doi.org/10.18429/JACOW-LINAC2018-TUPO071.

308 Checchin, M., Martinello, M. J., Melnychuk, O.S. et al. (2018). New insight on nitrogen infusion revealed by successive nanometric material removal.

9th International Particle Accelerator Conference, Vancouver, BC, Canada pp. 2665–2667. https://doi.org/10.18429/JACoW-IPAC2018-WEPMK016.

309 Checchin, M. and Grassellino, A. (2020). High-field Q-slope mitigation due to impurity profile in superconducting radio-frequency cavities. *Appl. Phys. Lett.* 117 (3): 032601. https://doi.org/10.1063/5.0013698.

310 Grassellino, A. (2019). Progress in high Q and high gradient R&D presented at the *TESLA Technology Collaboration Meeting (TTC 2019)*, Vancouver, Canada. https://indico.desy.de/event/21337/contributions/42673/attachments/27317/34419/Anna-TTCVancouver.pdf.

311 Dhakal, P. (2019). JLab work on low temperature near-surface diffusion aka 'infusion. presented at the *TESLA Technology Collaboration Meeting (TTC 2019)*, Vancouver, Canada. https://indico.desy.de/event/21337/contributions/42666/.

312 Conway, Z.A., Hartill, D.L., Padamsee, H., and Smith, E.N. (2009). Defect location in superconducting cavities cooled with He-II using oscillating superleak transducers. *Proceedings of the 23rd Particle Accelerator Conference (PAC 2009)*, Vancouver, British Columbia, Canada. https://accelconf.web.cern.ch/PAC2009/papers/tu5pfp044.pdf.

313 Checchin, M., Grassellino, A., Martinello, M. et al. (2016). Ultimate gradient limitation in niobium superconducting accelerating cavities. *7th International Particle Accelerator Conference (IPAC 2016)*, Busan, Korea (8–13 May 2016), pp. 2254–2257. http://dx.doi.org/10.18429/JACoW-IPAC2016-WEPMR002.

314 Catelani, G. and Sethna, J.P. (2008). Temperature dependence of the superheating field for superconductors in the high-κ London limit. *Phys. Rev. B* 78 (22): 224509. https://doi.org/10.1103/PhysRevB.78.224509.

315 Gurevich, A. (2015). Maximum screening fields of superconducting multilayer structures. *AIP Adv.* 5 (1): 017112. https://doi.org/10.1063/1.4905711.

316 Kubo, T., Iwashita, Y., and Saeki, T. (2014). Radio-frequency electromagnetic field and vortex penetration in multilayered superconductors. *Appl. Phys. Lett.* 104 (3): 032603. https://doi.org/10.1063/1.4862892.

317 Kubo, T. (2016). Multilayer coating for higher accelerating fields in superconducting radio-frequency cavities: a review of theoretical aspects. *Supercond. Sci. Technol.* 30 (2): 023001. https://doi.org/10.1088/1361-6668/30/2/023001.

318 Posen, S., Transtrum, M.K., Catelani, G. et al. (2015). Shielding superconductors with thin films as applied to RF cavities for particle accelerators. *Phys. Rev. Appl.* 4 (4): 044019. https://doi.org/10.1103/PhysRevApplied.4.044019.

319 Kubo, T. (2014). Superconducting nano-layer coating without insulator. *ArXiv Prepr. ArXiv14101248*. https://doi.org/10.48550/arXiv.1410.1248.

320 Kubo, T. (2014) Superconducting nano-layer coating without insulator. *Proceedings of the 27th International Linear Accelerator Conference (LINAC 2014)*, Geneva, pp. THPP074, 1026. https://accelconf.web.cern.ch/linac2014/papers/thpp074.pdf.

321 Checchin, M., Martinello, M., Grassellino, A. et al. (2016). Enhancement of the accelerating gradient in superconducting microwave resonators. *Proceedings of the 28th Linear Accelerator Conference (LINAC 2016)*, East Lansing, MI, USA. https://doi.org/10.18429/JACoW-LINAC2016-TUPLR024.

322 Piel, H. (1980). Diagnostic methods of superconducting cavities and identification of phenomena. *Proceedings of the 1st Workshop on RF Superconductivity (SRF 1980)*, Karlsruhe, Germany, p. 85. https://accelconf.web.cern.ch/srf80/papers/srf80-5.pdf.

323 Shepard, K., Scheibelhut, C., Markovich, P. et al. (1979). Development and production of superconducting resonators for the Argonne heavy ion LINAC. *IEEE Trans. Magn.* 15 (1): 666–669. https://accelconf.web.cern.ch/p81/PDF/PAC1981_3248.PDF.

324 Conway, Z.A., Chojnacki, E.P., Hartill, D.L. et al. (2009). Multi-cell reentrant cavity development and testing at Cornell. presented at the *Particle Accelerator Conference (PAC 2009)*, Vancouver, Canada RU5PFP047 p. 930. https://accelconf.web.cern.ch/PAC2009/papers/tu5pfp047.pdf.

325 Sherlock, R.A. and Edwards, D.O. (1970). Oscillating superleak second sound transducers. *Rev. Sci. Instrum.* 41 (11): 1603–1609. https://doi.org/10.1063/1.1684354.

326 Williams, R., Beaver, S.E.A., Fraser, J.C. et al. (1969). The velocity of second sound near $T\lambda$. *Phys. Lett. A* 29 (5): 279–280. https://doi.org/10.1016/0375-9601(69)90383-1.

327 Crawford, A.C. (2017). Extreme diffusion limited electropolishing of niobium radiofrequency cavities. *Nucl. Instrum. Methods Phys. Res., Sect. A* 849: 5–10. https://doi.org/10.1016/j.nima.2017.01.006.

328 Checchin, M. (2020). LCLS-II-HE, Talk, Workshop on Linear Colliders 2020, SLAC. https://agenda.linearcollider.org/event/8622/contributions/46626/attachments/35985/55925/AWLC_2020.pdf (accessed 19 January 2023).

329 Tian, H., Corcoran, S.G., Reece, C.E., and Kelley, M.J. (2008). The mechanism of electropolishing of niobium in hydrofluoric–sulfuric acid electrolyte. *J. Electrochem. Soc.* 155 (9): D563. https://doi.org/10.1149/1.2945913.

330 Diepers, H., Schmidt, O., Martens, H., and Sun, F.S. (1971). A new method of electropolishing niobium. *Phys. Lett. A* 37 (2): 139–140. https://doi.org/10.1016/0375-9601(71)90095-8.

331 Saito, K. (2003). Development of electropolishing technology for superconducting cavities. *Proceedings of the 2003 Particle Accelerator Conference*, Volume 1, pp. 462–466. https://accelconf.web.cern.ch/p03/PAPERS/ROAA002.PDF.

332 Furuta, F. (2021). Extra-cold electro-polishing process at Fermilab. *Proceedings of the 20th International Conference on RF Superconductivity (SRF 2021)*, East Lansing, MI, USA. Virtual (MSU), p. MOPTEV012. https://accelconf.web.cern.ch/srf2021/posters/moptev012_poster.pdf.

333 Zhao, X., Geng, R., Funahashi, Y. et al. (2009). Surface characterization of niobium samples electro-polished together with real cavities. *Proceedings of the 14th International Conference on SRF Superconductivity (SRF 2009)*, Berlin, Germany. TUPPO086. pp. 441–445. https://accelconf.web.cern.ch/SRF2009/papers/tuppo086.pdf.

334 Palczewski, A. (2021). A novel approach to producing high gradient doped cavities in non-ideal furnaces. *Proceedings of the 20th International*

Conference on RF Superconductivity (SRF 2021)*, Virtual (MSU), p. FROFDV02. https://indico.frib.msu.edu/event/38/attachments/160/1253/Ari_P.pdf.

335 Cooley, L.D., Burk, D., Cooper, C. et al. (2010). Impact of forming, welding, and electropolishing on pitting and the surface finish of SRF cavity niobium. *IEEE Trans. Appl. Supercond.* 21 (3): 2609–2614. https://doi.org/10.1109/TASC.2010.2083629.

336 Geng, R. (2007). Latest results of ILC high-gradient R&D 9-cell cavities at JLAB. *13th International Conference on RF Superconductivity (SRF 2007)*, Beijing, China. https://accelconf.web.cern.ch/srf2007/PAPERS/WEP28.pdf.

337 Watanabe, K., Hayano, H., Kako, E. et al. (2010). Repair techniques of superconducting cavity for improvement cavity performance at KEK-STF. *Proceedings of the 2010 International Particle Accelerator Conference (IPAC 2010)*, Kyoto. https://accelconf.web.cern.ch/IPAC10/papers/wepec033.pdf.

338 Aderhold, S. (2015). Study of field-limiting defects in superconducting RF cavities for electron-accelerators. Deutsches Elektronen-Synchrotron (DESY). Thesis. University of Hamburg, DE. https://bib-pubdb1.desy.de/record/207742/files/desy-thesis-15-004.pdf.

339 Popielarski, L. (2017). FRIB cryomodule production. *Americas Workshop on Linear Colliders (AWLC 2017)*, Menlo Park, CA: SLAC. https://agenda.linearcollider.org/event/7507/contributions/39208/attachments/31801/47959/Cryomodule_Production_for_FRIB_AMLC_2017_NO_VIDEO.pdf.

340 Malloch, I., Metzgar, E., Popielarski, L., and Stanley, S. (2016). Design and implementation of an automated high-pressure water rinse system for FRIB SRF cavity processing. *Proceedings of the 28th Linear Accelerator Converence (LINAC 2016)*, East Lansing U.S.A. https://doi.org/10.18429/JACoW-LINAC2016-TUPRC024.

341 Guo, H., Chu, Q., He, Y. et al. (2019). The design of an automated high-pressure rinsing system for SRF cavity and the outlook for future automated cleanroom on strings assembly. *Proceedings of the 19th International Conference on RF Superconductivity (SRF 2019)*. https://doi.org/10.18429/JACOW-SRF2019-FRCAB3.

342 Doleans, M. (2015). Plasma processing to improve SRF accelerating gradient. *Proceedings of the 17th International Conference on RF Superconductivity (SRF 2015)*, Whistler, BC, Canada. https://accelconf.web.cern.ch/SRF2015/talks/thba01_talk.pdf.

343 Doleans, M., Tyagi, P.V., Afanador, R. et al. (2016). In-situ plasma processing to increase the accelerating gradients of superconducting radio-frequency cavities. *Nucl. Instrum. Methods Phys. Res., Sect. A* 812: 50–59. https://doi.org/10.1016/j.nima.2015.12.043.

344 Massahi, S. Investigation of Photolithography Process on SPOs for the ATHENA Mission, DOI: 10.1117/12.2186810.

345 Schwettman, H.A., Turneaure, J.P., and Waites, R.F. (1974). Evidence for surface-state-enhanced field emission in RF superconducting cavities. *J. Appl. Phys.* 45 (2): 914–922. https://doi.org/10.1063/1.1663338.

346 Knobloch, J. and Padamsee, H. (1998). Explosive field emitter processing in superconducting RF cavities. *Part. Accel.* 61: 169–204. https://inspirehep.net/files/d74436e2233b67feb2b7cebdb719876c.

347 Bonin, B. (1996). Field emission in RF cavities. CEA Centre d'Etudes de Saclay. https://www-project.slac.stanford.edu/lc/wkshp/RFBreakdwn/references/bonin.pdf.

348 Giaccone, B. (2021). Overview on recent progress of plasma processing. Presented at the *20th International Conference on RF Superconductivity (SRF 2021)*, East Lansing, MI, USA. https://accelconf.web.cern.ch/srf2021/talks/weotev01_talk.pdf.

349 Giaccone, B., Berrutti, P., Doleans, M. et al. (2019). Plasma processing to reduce field emission in LCLS-II 1.3 GHz SRF cavities. presented at the *International Conference on RF Superconductivity (SRF 2019)*, Dresden, Germany. https://doi.org/10.18429/JACoW-SRF2019-FRCAB7.

350 Galayda, J. (2015). LCLS-II Final Design Report. *Technical Report*. LCLSII-1.1-DR-0251-R0 (SLAC 2015).

351 Powers, T., Brock, N., and Ganey, T. (2022). In Situ Plasma Processing of Superconducting Cavities at Jefferson Lab. arXiv:2204.02903. https://doi.org/10.48550/ARXIV.2204.02903.

352 Wang, Z.I., Wang, F.F., Huang, G.R. et al. (2019). The status of CiADS superconducting LINAC. *Proceedings of the 10th International Particle Accelerator Conference (IPAC 2019)*. https://doi.org/10.18429/JACOW-IPAC2019-MOPTS059.

353 Huang, S., Chu, Q., He, Y. et al. (2019). The effect of helium processing and plasma cleaning for low beta HWR cavity. *Proceedings of the 19th International Conference on RF Superconductivity (SRF 2019)*. https://doi.org/10.18429/JACOW-SRF2019-FRCAB6.

354 Benvenuti, C., Calatroni, S., Hakovirta, M. et al. (2001). CERN studies on niobium-coated 1.5 GHz copper cavities. *Proceedings of the 10th Workshop on RF Superconductivity (SRF 2001)*, Tsukuba, Japan. https://accelconf.web.cern.ch/srf01/papers/ma002.pdf.

355 Calatroni, S. (2019). Nb Films. presented at the *CERN Presentation* (21 May 2019). https://slideplayer.com/slide/16978892/ (accessed 1 August 2022).

356 Benvenuti, C., Calatroni, S., Darriulat, P. et al. (2001). Study of the residual surface resistance of niobium films at 1.5 GHz. *Physica C* 351 (4): 421–428. https://doi.org/10.1016/S0921-4534(00)01645-2.

357 Calatroni, S., Neupert, H., Barbero-Soto, E. et al. (2003). Progress of Nb/Cu technology with 1.5 GHz cavities. *11th Workshop on RF Superconductivity (SRF 2003)*, Luebeck/Travemuende, Germany. https://accelconf.web.cern.ch/SRF2003/talks/weo09_talk.pdf.

358 Junginger, T. (2015). Field dependent surface resistance of niobium on copper cavities. *Phys. Rev. Spec. Top. Accel. Beams* 18 (7): 072001. https://doi.org/10.1103/PhysRevSTAB.18.072001.

359 Weingarten, W. (1995). Progress in thin film techniques. *Part. Accel.* 53 (CERN-SL-95-122-RF-8): 199–215. https://accelconf.web.cern.ch/SRF95/papers/SRF95D02.pdf?n=SRF95/papers/SRF95D02.pdf.

360 Darriulat, P., Durand, C., Janot, P. et al. (1996). Dependence of the surface resistance of niobium coated copper cavities on the coating temperature. *Proceedings of the 7th Workshop on RF Superconductivity (SRF 1995)*, Gif-sur-Yvette, France. https://accelconf.web.cern.ch/SRF95/papers/srf95c20.pdf.

361 Weingarten, W. (2011). Field-dependent surface resistance for superconducting niobium accelerating cavities. *Phys. Rev. Spec. Top.-Accel. Beams* 14 (10): 101002. https://doi.org/10.1103/PhysRevSTAB.14.101002.

362 Valente-Feliciano, A.-M. (2019). Nb/Cu by energetic condensation and SIS multilayers. presented at the *TESLA Technology Collaboration Meeting (TTC 2019)*, Vancouver, Canada. https://indico.desy.de/event/21337/contributions/42596/.

363 Aull, S., Brunner, O., Butterworth, A., and Schwerg, N. (2017). Set of RF parameters for the FCC-ee machines. *Proceedings of the 18th International Conference on RF Superconductivity (SRF 2017)*, Lanzhou, China. Paper MOYA 04. https://accelconf.web.cern.ch/srf2017/talks/moya04_talk.pdf.

364 Benvenuti, C. (1991). Superconducting coatings for accelerating RF cavities: past, present, future. *Part. Accel.* 40 (CERN-MT-DI-91-6): 43–57. *Proceedings of the 5th Workshop on RF Superconductivity, DESY*, Hamburg, Germany. https://accelconf.web.cern.ch/SRF91/papers/srf91c04.pdf.

365 Chiaveri, E. (1999). The CERN Nb/Cu programme for the LHC and reduced-b superconducting cavities. *Proceedings of the 9th Workshop on RF Superconductivity (SRF 1999)*, Santa Fe, NM. https://accelconf.web.cern.ch/SRF99/papers/wea003.pdf.

366 Sublet, A., Aviles, I., Bártóva, B. et al. (2014). Nb coated HIE-ISOLDE QWR superconducting accelerating cavities: from process development to series production. *5th International Particle Accelerator Conference (IPAC 2014)*, Dresden, Germany (15–20 June 2014), pp. 2571–2573. https://doi.org/10.18429/JACoW-IPAC2014-WEPRI042.

367 Sublet, A., Zhang, P., Aviles, I. et al. (2013). Preliminary results of Nb thin film coating HIE-ISOLDE SRF cavities obtained by magnetron sputtering. *Proceedings of the 2013 International Conference on RF Superconductivity (SRF 2013)*, Paris, France, p. TUP076. http://ipnwww.in2p3.fr/srf2013/papers/tup076.pdf.

368 Sublet, A., Aviles, I., Jecklin, N. et al. (2013). Thin film coating optimization for HIE-ISOLDE SRF cavities: coating parameters study and film characterization. *Proceedings of the 16th International Conference on RF Superconductivity (SRF 2013)*, Paris, France, p. TUP077. https://accelconf.web.cern.ch/SRF2013/papers/tup077.pdf.

369 Jecklin, N., Sublet, A., Mondino, I. et al. (2013). Niobium coatings for the Hie-Isolde QWR superconducting accelerating cavities. *Proceedings of the 16th International Conference on RF Superconductivity (SRF 2013)*, Paris, France. TuP073. https://accelconf.web.cern.ch/SRF2013/papers/tup073.pdf.

370 Delsolaro, W.V., Calatroni, S., Delaup, B. et al. (2013). Nb sputtered quarter wave resonators for the HIE-ISOLDE. *Proceedings of the 16th International Conference on RF Superconductivity (SRF 2013)*, Paris France. https://accelconf.web.cern.ch/SRF2013/talks/weioa03_talk.pdf.

371 Delsolaro, W.V., Arnaudon, L., Artoos, K. et al. (2015). Status of the SRF systems at HIE-ISOLDE. *Proceedings of the 17th International Conference on RF Superconductivity (SRF 2015)*, pp. 481–487. https://accelconf.web.cern.ch/SRF2015/talks/tuba01_talk.pdf.

372 Delsolaro, W.V., Montesinos, E., Therasse, M. et al. (2017). Lessons learned from the HIE-Isolde cavity production and cryomodule commissioning. *Proceedings of the 18th International Conference on RF Superconductivity (SRF 2017)*, Lanzhou, China, p. Talk TUXAA02. https://doi.org/10.18429/JACoW-SRF2017-TUXAA02.

373 Lopez, S.T., Fraser, M., Garlasché, M. et al. (2017). A seamless quarter-wave resonator for HIE-ISOLDE. *Proceedings of the 18th International Conference on RF Superconductivity (SRF 2017)*, p. 686. https://doi.org/10.18429/JACOW-SRF2017-WEYA03.

374 Delsolaro, W.V. and Miyazaki, A. (2018). Seamless quarter wave resonators for HIE ISOLDE. *Proceedings of the 29th Linear Accelerator Conference (LINAC 2018)*, Beijing, China, pp. 292–296. https://doi.org/10.18429/JACoW-LINAC2018-TU1A05.

375 Miyazaki, A. and Delsolaro, W.V. (2019). Two different origins of the Q-slope problem in superconducting niobium film cavities for a heavy ion accelerator at CERN. *Phys. Rev. Accel. Beams* 22 (7): 073101. https://journals.aps.org/prab/pdf/10.1103/PhysRevAccelBeams.22.073101.

376 Junginger, T., Salman, Z., Proslier, T. et al. (2015). Low energy muon spin rotation and point contact tunneling applied to niobium films for SRF cavities. *Proceedings of 17th International Conference on RF Superconductivity (SRF 2015)*, Whistler, BC, Canada, p. TUPB042. https://accelconf.web.cern.ch/SRF2015/papers/tupb042.pdf.

377 Wu, G., Valente, A.-M., Phillips, H.L. et al. (2005). Studies of niobium thin film produced by energetic vacuum deposition. *Thin Solid Films* 489 (1–2): 56–62. https://doi.org/10.1016/j.tsf.2005.04.099.

378 Valente-Feliciano, A.-M., Aull, S., Spradlin, J. et al. (2015). Material quality & SRF performance of Nb films grown on Cu via ECR plasma energetic condensation. *Proceedings of the 17th International Conference on RF Superconductivity (SRF 2015)*, Whistler, BC, Canada, TUPB029. https://accelconf.web.cern.ch/SRF2015/papers/tupb029.pdf.

379 Purandare, Y.P., Ehiasarian, A.P., and Hovsepian, P.E. (2008). Deposition of nanoscale multilayer CrN/NbN physical vapor deposition coatings by high power impulse magnetron sputtering. *J. Vac. Sci. Technol. Vac. Surf. Films* 26 (2): 288–296. https://doi.org/10.1116/1.2839855.

380 Rosaz, G.G. et al. (2015). Biased HiPIMS technology for superconducting RF accelerating cavities coating. *6th International Conference on Fundamentals and Industrial Applications of HIPIMS 2015*, Braunschweig, Germany. http://cds.cern.ch/record/2155295.

381 Avino, F., Sublet, A., and Taborelli, M. (2019). Evidence of ion energy distribution shift in HiPIMS plasmas with positive pulse. *Plasma Sources Sci. Technol.* 28 (1): 01LT03. https://doi.org/10.1088/1361-6595/aaf5c9.

382 Burton, M.C., Palczewski, A.D., Phillips, H.L. et al. (2019). RF results of Nb coated SRF accelerator cavities via HiPIMS. *Proceedings of the 19th International Conference on RF Superconductivity (SRF 2019)*, Dresden, Germany. https://doi.org/10.18429/JACoW-SRF2019-TUP079.

383 Aull, S., Venturini Delsolaro, W., Junginger, R. et al. (2015). On the understanding of Q-Slope of niobium thin films. *Proceedings of 17th International Conference on RF Superconductivity (SRF 2015)*, Whistler, BC, Canada, pp. 494, TUBA03. https://accelconf.web.cern.ch/SRF2015/papers/tuba03.pdf.

384 Magnus, F., Ingason, A.S., Sveinsson, O.B. et al. (2011). Morphology of TiN thin films grown on SiO_2 by reactive high power impulse magnetron sputtering. *Thin Solid Films* 520 (5): 1621–1624. https://doi.org/10.1016/j.tsf.2011.07.041.

385 Terenziani, G., Junginger, T., Santillana, I.A. et al. (2013). Nb coating developments with HIPIMS for SRF applications. *Proceedings of the 16th International Conference on RF Superconductivity (SRF 2013)*, Paris, France, SRF 2013, p. TUP078. https://accelconf.web.cern.ch/SRF2013/papers/tup078.pdf.

386 Valente-Feliciano, A.-M. (2021). SRF thin film activities @ JLAB. presented at the *9th International Conference on Thin Films and New Ideas for Pushing the Limits of RF Superconductivity (TFSRF 2021)*, Virtual. (JLab). https://indico.jlab.org/event/405/contributions/8145/attachments/6606/8963/TF%40JLab-AMVF-TFSRF21.pdf.

387 Cryogenic Society Photo (2015). HIPIMS JLAB research advances niobium film deposition. Cold Facts, 31 October 2015. www.cryogenicsociety.or.

388 Andersson, J. and Anders, A. (2009). Self-sputtering far above the runaway threshold: an extraordinary metal-ion generator. *Phys. Rev. Lett.* 102 (4): 045003. https://doi.org/10.1103/PhysRevLett.102.045003.

389 Anders, A. (2011). Deposition of niobium and other superconducting materials with high power impulse magnetron sputtering: concept and first results. *Proceedings of the 15th International Conference on RF Superconductivity (SRF 2011)*, Chicago, IL, USA, p. TUIOA06 302. https://accelconf.web.cern.ch/SRF2011/papers/tuioa06.pdf.

390 Valente-Feliciano, A.-M. (2016). Superconducting RF materials other than bulk niobium: a review. *Supercond. Sci. Technol.* 29 (11): 113002. https://doi.org/10.1088/0953-2048/29/11/113002.

391 Antoine, C. (2019). Materials for superconducting accelerators: beyond bulk Nb tutorial. presented at the *19th International Conference on RF Superconductivity (SRF 2019)*, Dresden, Germany. https://accelconf.web.cern.ch/srf2019/talks/thfua3_talk.pdf.

392 Posen, S., Valles, N., and Liepe, M. (2015). Radio frequency magnetic field limits of Nb and Nb_3Sn. *Phys. Rev. Lett.* 115 (4): 047001. https://doi.org/10.1103/PhysRevLett.115.047001.

393 Godeke, A. (2006). A review of the properties of Nb_3Sn and their variation with A15 composition, morphology and strain state. *Supercond. Sci. Technol.* 19 (8): R68. https://doi.org/10.1088/0953-2048/19/8/R02.

394 Becker, C., Posen, S., Groll, N. et al. (2015). Analysis of Nb_3Sn surface layers for superconducting radio frequency cavity applications. *Appl. Phys. Lett.* 106 (8): 082602. https://doi.org/10.1063/1.4913617.

395 Posen, S. (2015). *Understanding and overcoming limitation mechanisms in Nb_3Sn superconducting RF cavities.* PhD thesis. Ithaca, NY: Cornell University. https://www.classe.cornell.edu/rsrc/Home/Research/SRF/SrfDissertations/Posen_thesis.pdf.

396 Posen, S. and Hall, D.L. (2017). Nb3Sn superconducting radiofrequency cavities: fabrication, results, properties, and prospects. *Supercond. Sci. Technol.* 30 (3): 033004. https://doi.org/10.1088/1361-6668/30/3/033004.

397 Posen, S. Frequency dependence studies of Nb_3Sn cavities. *TESLA Technology Collaboration (TTC 2019)*, Vancouver, Canada, Th WG3 Talk7. https://indico.desy.de/event/21337/contributions/42600/.

398 Posen, S., Lee, J., Seidman, D.N. et al. (2021). Advances in Nb_3Sn superconducting radiofrequency cavities towards first practical accelerator applications. *Supercond. Sci. Technol.* 34 (2): 025007. https://doi.org/10.1088/1361-6668/abc7f7.

399 Keckert, S., Junginger, T., Buck, T. et al. (2019). Critical fields of Nb_3Sn prepared for superconducting cavities. *Supercond. Sci. Technol.* 32 (7): 075004. https://doi.org/10.1088/1361-6668/ab119e.

400 Trenikhina, Y., Posen, S., Romanenko, A. et al. (2017). Performance-defining properties of Nb_3Sn coating in SRF cavities. *Supercond. Sci. Technol.* 31 (1): 015004. https://doi.org/10.48550/arXiv.1707.05856.

401 Posen, S., Lee, J., Melnychuk, O. et al. (2019). Nb_3Sn at Fermilab: exploring performance. *19th International Conference on RF Superconductivity (SRF 2019)*, Dresden, Germany (30 June-05 July 2019), pp. 818–822. https://doi.org/10.18429/JACoW-SRF2019-THFUB1.

402 Eremeev, G., Clemens, W., Macha, K. et al. (2020). Nb_3Sn multicell cavity coating system at Jefferson Lab. *Rev. Sci. Instrum.* 91 (7): 073911. https://doi.org/10.1063/1.5144490.

403 Pudasaini, U., Eremeev, G.V., Angle, J.W. et al. (2019). Growth of Nb_3Sn coating in tin vapor-diffusion process. *J. Vac. Sci. Technol. Vac. Surf. Films* 37 (5): 051509. https://doi.org/10.1116/1.5113597.

404 Posen, S. (2021). Efforts towards first applications of Nb_3Sn SRF cavities. presented at the *20th International Conference on RF Superconductivity (SRF 2021)*, Virtual (MSU). https://accelconf.web.cern.ch/srf2021/talks/frofdv05_talk.pdf.

405 Floris, A., Sanna, A., Lüders, M. et al. (2007). Superconducting properties of MgB_2 from first principles. *Physica C* 456 (1–2): 45–53. https://doi.org/10.1016/j.physc.2007.01.026.

406 Choi, H.J., Roundy, D., Sun, H. et al. (2002). The origin of the anomalous superconducting properties of MgB_2. *Nature* 418 (6899): 758–760. https://doi.org/10.1038/nature00898.

407 Xi, X.X., Pogrebnyakov, A.V., Xu, S.Y. et al. (2007). MgB_2 thin films by hybrid physical–chemical vapor deposition. *Physica C* 456 (1–2): 22–37. https://doi.org/10.1016/j.physc.2007.01.029.

408 Xiao, B.P., Zhao, X., Spradin, J. et al. (2012). Surface impedance measurements of single crystal MgB_2 films for radiofrequency superconductivity applications. *Supercond. Sci. Technol.* 25 (9): 095006. https://doi.org/10.1088/0953-2048/25/9/095006.

409 Welander, P.B., Franzi, M., and Tantawi, S. (2015). Cryogenic RF characterization of superconducting materials at SLAC with hemispherical cavities. *Proceedings of (SRF 2015)*, Whistler, BC, Canada, pp. 735–738. https://accelconf.web.cern.ch/SRF2015/papers/tupb065.pdf.

410 Welander, P. (2014). TFSRF 2014 (Talk) Cryogenic RF Characterizations of Superconducting Materials at SLAC with Cu and Nb Coated Cavities, Thin Films and New Ideas on RF Superconductivity, TFSRF 2014, at INFN.

411 Withanage, W.K., Xi, X.X., Nassiri, A. et al. (2017). Growth of magnesium diboride films on 2 inch diameter copper discs by hybrid physical–chemical vapor deposition. *Supercond. Sci. Technol.* 30 (4): 045001. https://doi.org/10.1088/1361-6668/aa5999.

412 Guard, R.W., Savage, J.W., and Swarthout, D.G. (1967). Constitution of a portion of the niobium (columbium)–nitrogen system. *Trans. Met. Soc. ATIME* 239: 643.

413 Toth, L. (2014). *Transition Metal Carbides and Nitrides*. Elsevier.

414 Gemme, G., Fabbricatore, P., Musenich, R. et al. (1995). RF surface resistance measurements of binary and ternary niobium compounds. *J. Appl. Phys.* 77 (1): 257–264. https://doi.org/10.1063/1.359386.

415 Bosland, P., Guemas, F., Juillard, M. et al. (1992). NbTiN superconducting thin films prepared by magnetron sputtering. *3rd European Particle Accelerator Conference (EPAC 1992)*, Berlin, Germany. https://accelconf.web.cern.ch/e92/PDF/EPAC1992_1316.PDF.

416 Cantacuzène, S., Bosland, P., Martignac, J. et al. (1994). Elaboration and characterization of Nb and NbTiN superconducting thin films for RF applications. *European Particle Accelerator Conference (EPAC 1994)*, p. 2076.

417 Marino, M. (1997). Study of NbTiN coatings by reactive magnetron sputtering for the production of 1.5 GHz superconducting accelerating cavities. *Proceedings of the 1997 Workshop on RF Superconductivity (SRF 1997)*, Padova, Italy, SRF97D29. pp. 1074–1080.

418 Valente-Feliciano, A.-M. (2015). Growth and characterization of multi-layer NbTiN films. *Proceedings of International Conference on RF Superconductivity (SRF 2015)*, Whistler, BC, Canada, p. TUBA08 516. https://accelconf.web.cern.ch/SRF2015/talks/tuba08_talk.pdf.

419 Antoine, C.Z., Berry, S., Bouat, G. et al. (2010). Characterization of superconducting nanometric multilayer samples for superconducting RF applications: first evidence of magnetic screening effect. *Phys. Rev. Spec. Top. Accel. Beams* 13 (12): 121001. https://doi.org/10.1103/PhysRevSTAB.13.121001.

420 Antoine, C.Z., Berry, S., Aurino, M. et al. (2011). Characterization of field penetration in superconducting multilayers samples. *IEEE Trans. Appl. Supercond.* 21 (3): 2601–2604. https://doi.org/ 10.1109/TASC.2010.2100347.

421 Antoine, C.Z., Villegier, J.-C., and Martinet, G. (2013). Study of nanometric superconducting multilayers for RF field screening applications. *Appl. Phys. Lett.* 102 (10): 102603. https://doi.org/10.1063/1.4794938.

422 Ito, H., Hayano, H., Kubo, T., and Saeki, T. (2020). Lower critical field measurement system based on third-harmonic method for superconducting RF materials. *Nucl. Instrum. Methods Phys. Res., Sect. A* 955: 163284. https://doi.org/10.1016/j.nima.2019.163284.

423 Díaz-Palacio, I.G., Blick, R., Hillert, W. et al. (2021). ALD-Based NbTiN studies for SIS R&D. *12th International Particle Accelerator Conference (IPAC 2021)*, Campinas, SP, Brazil. https://doi.org/10.18429/JACoW-IPAC2021-THPAB320.

424 Sezgin, A.Ö., Vogel, M., Lakki Reddy Venkata, B.R. et al. (2021). NbN thin film-based multilayer(S-(I)-S) structures for SRF cavities NbN thin film-based multilayer (S-(I)-S) structures for SRF cavities. *9th International Conference on Thin Films (TFSRF 2021)*, Virtual (JLab). https://indico.jlab.org/event/405/contributions/8109/attachments/6624/8994/9th_TFSRF21_Presentation_AOS_20210317-6.pdf.

425 Sundahl, C., Makita, J., Welander, P. et al. (2021). Development and characterization of Nb_3Sn/Al_2O_3 superconducting multilayers for high-performance radio-frequency applications 9th International Workshop on Thin Films. *9th International Conference on Thin Films (TFSRF 2021)*, Virtual (JLab). https://indico.jlab.org/event/405/contributions/8099/attachments/6617/8985/Development and characterization of Nb3SnAl2O3 super conducting multilayers for high-performance radio-frequency applications - TFSRF 2021.pdf.

426 Valizadeh, R. et al. (2021). 9th International Workshop on Thin Film JLAB, 2021 synthesis of alternative superconducting film to Nb for SRF cavity. *9th International Conference on Thin Films (TFSRF 2021)*, Virtual. (Jlab). https://indico.jlab.org/event/405/contributions/8114/attachments/6611/8981/TF SRF workshop JLAB 17th 03 2021.pdf.

427 Sun, Z., Oseroff, T., Porter, R.D. et al. (2021). Study of alternative materials for next generation SRF cavities at Cornell University. presented at the *9th International Conference on Thin Films (TFSRF 2021)*, Virtual (Jlab).

428 Junginger, T., Wasserman, W., and Laxdal, R.E. (2017). Superheating in coated niobium. *Supercond. Sci. Technol.* 30 (12): 125012. https://doi.org/10.1088/1361-6668/aa8e3a.

429 Junginger, T. et al. (2018). Field of first magnetic flux entry and pinning strength of superconductors for RF application measured with muon spin rotation. *Phys. Rev. Accel. Beams* 21 (3): 032002. https://doi.org/10.1103/PhysRevAccelBeams.21.032002.

430 Palmer, R. (1988). Simulation studies for the LHC long-range beam-beam compensators, SLAC-PUB-4707. http://www.slac.stanford.edu/cgi-wrap/getdoc/slac-pub-4707.pdf (accessed 1 August 2022).

431 Furuya, T. (2008). Development of superconducting RF technology. *Rev. Accel. Sci. Technol.* 1 (01): 211–235. https://doi.org/10.1142/S1793626808000149.

432 Hosoyama, K., Hara, K., Homna, A. et al. (2007). Construction and commissioning of KEK-b superconducting crab cavities. *13th International Workshop on RF Superconductivity (SRF 2007)*, Beijing, China. pp. 63–69. https://accelconf.web.cern.ch/srf2007/HTML/AUTHOR.

433 Akai, K., Ebihara, K., Furuya, T. et al. (2007). Commissioning and beam operation of KEKB crab-RF system. *13th Workshop on RF Superconductivity (SRF 2007)*, Beijing, SRF 2007, pp. 632–636. https://accelconf.web.cern.ch/srf2007/PAPERS/WEP57.pdf.

434 Abe, T., Akai, K., Akemoto, M. et al. (2007). Compensation of the crossing angle with crab cavities at KEKB. *IEEE Particle Accelerator Conference (PAC 2007)*, pp. 27–31. https://accelconf.web.cern.ch/p07/PAPERS/MOZAKI01.PDF.

435 De Silva, S. (2019). Overview of SRF deflecting and crabbing cavities. *19th International Conference on RF Superconductivity (SRF 2019)*, Dresden, Germany (30 June-05 July 2019), pp. 1192–1198. https://accelconf.web.cern.ch/srf2019/papers/frcaa1.pdf.

436 Panofsky, W.K.H. and Wenzel, W.A. (1956). Some considerations concerning the transverse deflection of charged particles in radio-frequency fields. *Rev. Sci. Instrum.* 27 (11): 967–967. https://doi.org/10.1063/1.1715427.

437 De Silva, S. (2019). Superconducting cavities of inteteresting shapes (non-elliptical cavities), tutorial. *19th International Conference on RF Superconductivity (SRF 2019)*, Dresden, Germany. https://accelconf.web.cern.ch/srf2019/talks/frtu1_talk.pdf.

438 Hall, B. (2012). *Designing the four rod crab cavity for the high-luminosity LHC upgrade*. PhD thesis. Cockcroft Institute Accelerator Science and Technology. Thesis UK. https://cds.cern.ch/record/1742051/files/EuCARD-BOO-2014-007.pdf.

439 Miyazaki, A. (2019). Radiation shield design for the vertical cryostat GERSEMI. Uppsala University. http://uu.diva-portal.org (accessed 1 August 2022).

440 Verdú-Andrés, S., Calaga, R., Cano-Pleite, E. et al. (2019). Design of LHC crab cavities based on DQW cryomodule test experience. *Proceedings of the 19th International Conference on RF Superconductivity (SRF 2019)*, p. 4 pages, 4.987 MB. https://doi.org/10.18429/JACOW-SRF2019-THP035.

441 Padamsee, H., Barnes, P., Chen, C. et al. (1991). Crab cavity development for the Cornell B-factory, CESR-B. *1991 IEEE Particle Accelerator Conference (PAC 1991)*, pp. 2423–2425. https://accelconf.web.cern.ch/p91/PDF/PAC1991_2423.PDF.

442 Akai, K., Kirchgessner, J., Moffat, D. et al. (1993). Development of crab cavity for CESR-B. *Proceedings of International Conference on Particle Accelerators (PAC 1993)*, pp. 769–771. https://accelconf.web.cern.ch/p93/PDF/PAC1993_0769.PDF.

443 Wu, Q. (2015). Crab cavities: past, present, and future of a challenging device. *6th International Particle Accelerator Conference (IPAC 2015)*, Richmond, VA. https://accelconf.web.cern.ch/IPAC2015/papers/thxb2.pdf.

444 Calaga, R. (2017). Crab cavities for the high-luminosity LHC. *Proceedings of the 18th International Conference on RF Superconductivity (SRF 2017)*, 2018. https://doi.org/10.18429/JACOW-SRF2017-THXA03.

445 Arduini, G., Banfi, D., Barranco, J. et al. (2013). How to maximize the HL-LHC performance (HL-LHC)? presented at the *Review of LHC and Injector Upgrade Plans Workshop*, Archamps. http://dx.doi.org/10.5170/CERN–2014–006.81.

446 De Silva, S.U. and Delayen, J.R. (2013). Design evolution and properties of superconducting parallel-bar rf-dipole deflecting and crabbing cavities. *Phys. Rev. Spec. Top. Accel. Beams* 16 (1): 012004. https://doi.org/10.1103/PhysRevSTAB.16.012004.

447 Xiao, B., Alberty, L., Belomestnykh, S. et al. (2015). Design, prototyping, and testing of a compact superconducting double quarter wave crab cavity. *Phys. Rev. Spec. Top. Accel. Beams* 18 (4): 041004. https://doi.org/10.1103/PhysRevSTAB.18.041004.

448 Calaga, R., Wu, Q., Ben-Zvi, I., and Belomestnykh, S. (2012). A quarter wave design for crab crossing in the LHC. *International Particle Accelerator Conference (IPAC 2012)*, New Orleans, LA, USA, WEPPC027, pp. 2260–2262. https://accelconf.web.cern.ch/IPAC2012/papers/weppc027.pdf.

449 Gorelov, D., Grimm, T., De Silva, S.U., and Delayen, J.R. (2012). Engineering of a superconducting 400 MHz crabbing cavity for the LHC hilumi upgrade. *IPAC 2012-International Particle Accelerator Conference*, pp. 2411–2413. https://accelconf.web.cern.ch/IPAC2012/papers/weppc085.pdf.

450 Hall, B., Dexter, A.C., Gorelov, D. et al. (2012). Analysis of the four rod crab cavity for HL-LHC. *International Particle Accelerator Conference (IPAC 2012)*, New Orleans, LA, USA, pp. 2275–2277. https://accelconf.web.cern.ch/IPAC2012/papers/weppc032.pdf.

451 Carra, F. (2015). Crab cavity and cryomodule development for HL-LHC. *17th International Conference on RF Superconductivity*, Whistler, BC, Canada, p. FRBA02. https://accelconf.web.cern.ch/SRF2015/talks/frba02_talk.pdf.

452 De Silva, S.U. and Delayen, J.R. (2013). Superconducting RF-Dipole deflecting and crabbing cavities. *16th International Conference on RF Superconductivity (SRF 2013)*, Paris, France, pp. 1176–1182. https://accelconf.web.cern.ch/SRF2013/papers/frioa04.pdf.

453 Verdú-Andrés, S., Ben-Zvi, I., Wu, Q., and Xiao, Q. (2017). 338 MHz Crab cavity design for the eRHIC Hadron beam. Upton, NY (United States): Brookhaven National Lab. (BNL) (SRF 2017), pp. 382–384. https://accelconf.web.cern.ch/srf2017/papers/tupb001.pdf.

454 Calaga, R., Alekaou, A., Antoniou, F. et al. (2021). First demonstration of the use of crab cavities on hadron beams. *Phys. Rev. Accel. Beams* 24 (6): 062001. https://doi.org/10.1103/PhysRevAccelBeams.24.062001.

455 De Silva, S. (2019). Overview of SRF deflecting and crabbing cavities – talk. *19th International Conference on RF Superconductivity (SRF 2019)*, p. 7. https://doi.org/10.18429/JACOW-SRF2019-FRCAA1.

456 Zholents, A., Heimann, P., Zolotorev, M., and Byrd, J. (1999). Generation of subpicosecond X-ray pulses using RF orbit deflection. *Nucl. Instrum. Methods Phys. Res., Sect. A* 425 (1–2): 385–389. https://doi.org/10.1016/S0168-9002(98)01372-2.

457 Rimmer, R., Daly, E.F., Hicks, W.R. et al. (2005). Concepts for the JLab ampere-class CW cryomodule. *Proceedings of the 2005 Particle Accelerator Conference (PAC 2005)*, pp. 3588–3590. https://accelconf.web.cern.ch/p05/PAPERS/RPPE063.PDF.

458 Fuerst, J.D., Horan, D., Kaluzny, J. et al. (2012). Tests of SRF deflecting cavities at 2 K. *International Particle Accelerator Conference (IPAC 2012)*, New Orleans, LA. pp. 2300–2302. https://accelconf.web.cern.ch/IPAC2012/papers/weppc041.pdf.

459 Nassiri, A., Arnold, N., Berenc, T. et al. (2012). Status of the short-pulse X-ray (SPX) project at the advanced photon source (APS). *Proceedings of IPAC2012*, New Orleans, Louisiana, USA. pp. 2292–2294. https://accelconf.web.cern.ch/IPAC2012/papers/weppc038.pdf.

460 Zhang, P., Hao, X., Huang, T. et al. (2017). A 166.6 MHz superconducting RF system for the HEPS storage ring. *J. Phys. Conf. Ser.* 874 (1): 012091. https://doi.org/10.1088/1742-6596/874/1/012091.

461 Zhang, P. (2021). Development and vertical tests of a 166.6 MHz PoP SC $\beta = 1$ quarter wave cavity for HEPS. *International Conference on RF Superconductivity (SRF 2021)*, Virtual (MSU). https://indico.frib.msu.edu/event/38/attachments/160/1296/SRF2021-PeiZhang.pdf. published in: https://doi.org/10.1063/1.5119093.

462 He, F.S. (2017). The R&D on TEM-type SRF cavities for high-current applications at IHEP. *18th International Conference on RF Superconductivity (SRF 2017)*, Lanzhou, China, pp. WEYA04-Talk. https://accelconf.web.cern.ch/srf2017/talks/weya04_talk.pdf.

463 Avrakhov, P., Kanareykin, A., and Solyak, N. (2005). Traveling wave accelerating structure for a superconducting accelerator. *2005 Particle Accelerator Conference (PAC 2005)*, Knoxville, TN, USA, pp. 4296–4298. https://accelconf.web.cern.ch/p05/PAPERS/WPAT094.PDF.

464 Avrakhov, P., Kanareykin, A., Kazakov, S. et al. (2008). Progress towards development of an L-band SC traveling wave accelerating structure with feedback. *European Particle Accelerator Conference (EPCA 2008)*, eEPAC08, Genoa, Italy, p. 871. https://accelconf.web.cern.ch/e08/papers/mopp132.pdf.

465 Avrakhov, P., Kanareykin, A., Kazakov, S. et al. (2012). Three-cell traveling wave superconducting test structure. *Particle Accelerator Conference, ArXiv Prepr. ArXiv12031599*, & NAPAC 2011, pp. 940–942. https://accelconf.web.cern.ch/PAC2011/papers/tup066.pdf & https://doi.org/10.48550/arXiv.1203.1599.

466 Shemelin, V., Padamsee, H., and Yakovlev, V. (2022). Optimization of a traveling wave superconducting RF cavity for upgrading the International Linear Collider. *Phys. Rev. Accel. Beams* 25 (2): 021001. https://doi.org/10.1103/PhysRevAccelBeams.25.021001.

467 Belomestnykh, S. (2012). Superconducting radio-frequency systems for high-ß particle accelerators. In: *Reviews of Accelerator Science and Technology: Applications of Superconducting Technology to Accelerators* (eds. Choa, A and Chou, W.) vol. 5, 147–184. World Scientific. https://doi.org/10.1142/S179362681230006X.

468 Kostin, R., Avrakhov, P., Kanareykin, A. et al. (2015). A high gradient test of a single-cell superconducting radio frequency cavity with a feedback waveguide. *Supercond. Sci. Technol.* 28 (9): 095007. https://doi.org/10.1088/0953-2048/28/9/095007.

469 Tan, T. (2020). Progress of SRF technology in China. Presented at the *TESLA Technology Collaboration Meeting (TTC 2020)*, Geneva, Switzerland: CERN. https://indico.cern.ch/event/817780/contributions/3632127/.

470 Wang, D. (2021). New SRF infrastructures in China. Presented at the *TESLA Technology Collaboration Meeting (TTC 2021)*, Hamburg (Virtual): DESY. https://indico.desy.de/event/27572/contributions/94503/.

471 Conway, Z.A., Kelly, M.P., and Ostroumov, P.N. (2015). Advanced low-beta cavity development for proton and ion accelerators. *Nucl. Instrum. Methods Phys. Res., Sect. B* 350: 94–98. https://doi.org/10.1016/j.nimb.2015.01.012.

472 Fuerst, J.D. (2009). The ATLAS energy upgrade cryomodule. *14th International Conference on RF Superconductivity (SRF 2009)*, Berlin, Germany, MOOCAU04. pp. 52–56. https://accelconf.web.cern.ch/SRF2009/papers/moocau04.pdf.

473 Kelly, M.P. et al. (2012). Cold testing of superconducting quarter-wave cavities for ATLAS. *26th Linear Accelerator Conference (LINAC 2012)*, Tel-Aviv, Israel p. 348. https://accelconf.web.cern.ch/LINAC2012/papers/mopb073.pdf.

474 Kim, S.-h. (2014). Superconducting RF cavities development at Argonne National Laboratory. presented at the *AKPA - IBS Symposium on Special Topics in Physics* (9–10 May 2014), University of Chicago. https://doi.org/10.13140/RG.2.2.28627.17445.

475 Bisoffi, G., Bassato, G., Canella, S. et al. (2007). ALPI QWR and S-RFQ operating experience. *13th International Workshop on RF Superconductivity (SRF 2007)*, Beijing, China. MO4. https://accelconf.web.cern.ch/srf2007/PAPERS/MO404.pdf.

476 Laxdal, R.E. et al. (2007). Recent progress in the superconducting RF Program at TRIUMF/ISAC. *13th International Conference on RF Superconductivity (SRF 2007)*, Beijing, China. p. MO402. https://accelconf.web.cern.ch/srf2007/PAPERS/MO402.pdf.

477 Ferdinand, R., Bernaudin, P.E., Di Giacomo, M. et al. (2013). Status and challenges of Spiral2 SRF linac. *16th International Conference on RF Superconductivity (SRF 2013)*, Paris, France, pp. 11–17. https://accelconf.web.cern.ch/SRF2013/papers/moioa02.pdf.

478 Ghribi, A. (2021). In situ dynamic heat load measurements of the SPIRAL2 LINAC superconducting cavities. *TESLA Technology Collaboration Meeting (TTC 2021)*, DESY, Hamburg (Virtual), January 2021, p. Talk. https://indico.desy.de/event/27572/contributions/94171/.

479 Delsolaro, W.V., Artoos, K., Brunner, O. et al. (2017). HIE ISOLDE cavity production & cryomodule commissioning, lesson learned. *18th International Conference of RF Superconductivity (SRF 2017)*, Lanzhou, China.

480 Delsolaro, W.V. et al. (2018). HIE-ISOLDE cavity production & cryomodule commissioning: lessons learned. *18th International Conference on RF Superconductivity (SRF 2017)*, p. 7 pages, 1.967 MB. https://doi.org/10.18429/JACOW-SRF2017-TUXAA02.

481 Sakamoto, N., Kamgaito, O., Okuno, H. et al. (2017). Construction and performance tests of prototype quarter-wave resonator and its cryomodule at RIKEN.

18th International Conference on RF Superconductivity (SRF 2017), Lanzhou, China, p. 681. https://accelconf.web.cern.ch/srf2017/papers/weya02.pdf.

482 Sakamoto, N. (2021). Commissioning and operation of RIKEN heavy ion superconducting linac. *TESLA Technology Collaboration Meeting (TTC 2021)*, Hamburg: DESY, p. Talk. https://indico.desy.de/event/27572/contributions/94175/.

483 Sakamoto, N. (2019). Design, assembly, transportation of the cryomodule for SRILAC, heavi-ion, CW. *TESLA Technology Collaboration Meeting 2019*, Vancouver, Canada, p. Talk. https://indico.desy.de/event/21337/contributions/42647/.

484 Tsymbaliuk, A., Muraron, E., Borlolato, D. et al. (2021). Current Status of the ALPI Linac Upgrade for the SPES faciliities at INFN LNL Virtual. *20th International Conference on RF Superconductivity (SRF 2021)*. https://doi.org/doi:10.18429/JACoW-SRF2021-SUPCAV005.

485 Yamada, K. (2021). Successful beam commissioning of heavy-ion superconducting LINAC at RIKEN. *20th International Conference on RF Superconductivity (SRF 2021)*, Virtual (MSU). https://accelconf.web.cern.ch/srf2021/papers/moofav01.pdf.

486 Xu, T., Al-Mahmound, Y., Asciutto, J. et al. (2021). Completion of FRIB superconducting LINAC and phased beam commissioning. *Proceedings of the International Conference on RF Superconductivity (SRF 2021)*, East Lansing, MI, USA, pp. 197–202. https://doi.org/doi:10.18429/JACoW-SRF2021-MOOFAV10.

487 Xu, T., Ao, H., Bird, B. et al. (2016). FRIB cryomodule design and production. *Proceedings of the LINAC 2016*, East Lansing, MI, USA, pp. 673–678. https://accelconf.web.cern.ch/linac2016/papers/we2a02.pdf.

488 Xu, T., Saito, K., Ao, H. et al. (2017). Progress of FRIB SRF production. *Proceedings of the International Conference of RF Superconductivity (SRF 2017)*, Lanzhou, China, p. 345. https://accelconf.web.cern.ch/srf2017/papers/tuxaa03.pdf.

489 Wei, J., Ao, H., Beher, S. et al. (2019). The FRIB SC-LINAC-installation and phased commissioning. *19th International Conference on RF Superconductivity (SRF 2019)*, Dresden, Germany (30 June-05 July 2019), pp. 12–20. https://accelconf.web.cern.ch/srf2019/papers/mofaa3.pdf.

490 Yamazaki, Y., Ao, H., Bultman, N. et al. (2016). Beam physics and technical challenges of FRIB driver LINAC. *7th International Particle Accelerator Conference (IPAC 2016)*, Busan, Korea (8–13 May 2016), pp. 2039–2044. https://accelconf.web.cern.ch/ipac2016/papers/weya01.pdf.

491 Hartung, W., Chang, W., Kim, S.-H. et al. (2019). Large-scale dewar testing of FRIB production cavities: results. *Proceedings - North American Particle Accelerator Conference (NAPAC 2019)*. https://doi.org/10.18429/JACOW-NAPAC2019-MOPLO17.

492 Popielarski, J. (2021). Lessons learned from FRIB linac SRF commissioning: field calibration, mitigation of field emission, resonance control. presented at the *TESLA Technology Collaboration Meeting 2021*, Virtual (DESY, Hamburg). https://indico.desy.de/event/27572/contributions/94177/.

493 Ostroumov, P.N., Contreras, C., Plastun, A.S. et al. (2018). Elliptical superconducting RF cavities for FRIB energy upgrade. *Nucl. Instrum. Methods Phys. Res., Sect. A* 888: 53–63. https://doi.org/10.1016/j.nima.2018.01.001.

494 Martinello, M. (2021). Recent advances and breakthrough for SRF cavities at FNAL, MSU. https://frib.msu.edu/_files/pdfs/science_ase_seminars_martinello.pdf.

495 McGee, K. (2021). New results on N-doping applied to 5-cell elliptical b = 0.65 644 MHz cavities. presented at the *TESLA Technology Collaboration Meeting 2021*, Virtual (DESY, Hamburg). https://indico.desy.de/event/27572/contributions/94301/.

496 Jeong, S. (2016). Progress of the RAON heavy ion accelerator project in Korea. *7th International Particle Accelerator Conference (IPAC 2016)*, Busan, Korea. pp. 4261–4265 https://accelconf.web.cern.ch/ipac2016/papers/fryaa01.pdf.

497 Jeon, D.-O. (2017). Progress of the RAON (RISP). *International Conference on RF Superconductivity (SRF 2017)*, Lanzhou, China, p. MOYAO3 Talk. https://accelconf.web.cern.ch/srf2017/talks/moya03_talk.pdf.

498 Jung, H.C., Jang, H., Kim, Y., and Kwon, Y.K. (2021). SRF status of the RAON heavy ion accelerator project. *SRF 19*, Dresden, Germany, p. MOFAB3. https://accelconf.web.cern.ch/srf2019/talks/mofab3_talk.pdf.

499 Kwon, M., Chung, Y.S., Kwon, Y.K. et al. (2021). RAON, Korean heavy ion accelerator facility. *J. Part. Accel. Soc. Jpn.* 17 (4): 293–301. https://www.pasj.jp/kaishi/cgi-bin/kasokuki.cgi?articles/17/p293-301.pdf.

500 Sohn, Y.U. (2021). Status of the RAON superconducting linear accelerator, SRF 2021. presented at the *20th International Conference on RF Superconductivity (SRF 2021)*, Virtual (MSU). https://doi.org/10.18429/JACoW-SRF2021-MOOFAV02.

501 Yao, Z., Keir, J., Kishi, D. et al. (2018). Fabrication and test of $\beta = 0.3$ 325MHz balloon single spoke resonator. *9th International Particle Accelerator Conference (IPAC 2018)*, Vancouver, Canada, p. THPAL123. https://doi.org/10.18429/JACoW-IPAC2018-THPAL123

502 Kondo, Y. et al. (2021). Development of QWRS for future upgrade of JAEA tandem superconducting booster. *20th International Conference on RF Superconductivity (SRF 2021)*, Virtual (MSU), p. MOPCAV015. https://accelconf.web.cern.ch/srf2021/papers/mopcav015.pdf.

503 Yee-Rendón, B., Kondo, Y., Maekawa, F. et al. (2019). Electromagnetic design of the low beta cavities for the JAEA ADS. *J. Phys. Conf. Ser.* 1350 (1): 012197. https://doi.org/10.50868/pasj.17.4_293.

504 Kabumoto, H., Takeuchi, S., Matsuda, M. et al. (2010). Superconducting twin quarter wave resonator for acceleration of low velocity heavy ions. *Nucl. Instrum. Methods Phys. Res., Sect. A* 612 (2): 221–224. https://doi.org/10.1016/j.nima.2009.09.130.

505 Conrad, T. et al. (2019). Cavity designs for the CH3 to CH11 of the superconducting heavy ion accelerator HELIAC. *19th International Conference on RF Superconductivity (SRF 2019)*, Dresden, Germany. https://doi.org/10.18429/JACOW-SRF2019-TUP005.

506 Conrad, T., Basten, M., Busch, M. et al. (2021). Cavity designs for the CH3 to CH11 of the superconducting heavy ion accelerator HELIAC. *The 20th International Conference on RF Superconductivity (SRF 2021)*, Virtual (MSU), SUPCAV006. https://accelconf.web.cern.ch/srf2021/papers/supcav006.pdf.

507 Busch, M., Podlech, H., Amberg, M. et al. (2010). Superconducting CH-cavity development. *International Particle Accelerator Conference (IPAC 2010)*, pp. 753–755. https://accelconf.web.cern.ch/IPAC10/papers/mopd032.pdf.

508 Madec, C., Bazin, N., Cubizolles, R. et al. (2017). CEA cryomodules design for SARAF phase 2. *18th International Conference on RF Superconductivity (SRF 2017)*, Lanzhou, Cihina, pp. 70–73. https://doi.org/10.18429/JACoW-SRF2017-MOPB011.

509 Ferrand, G. (2021). SARAF low-beta and high-beta superconducting cavities. *20th International Conference on RF Superconductivity (SRF 2021)*, Virtual (MSU). https://doi.org/10.18429/JACoW-SRF2021-WEOCAV06.

510 Xu, M. and He, Y. (2021). CiADS and HIAF superconducting cavity development status and the transition to production stage. *20th International Conference on RF Superconductivity (SRF 2021)*, Virtual (MSU), p. MOPCAV008. https://doi.org/10.18429/JACoW-SRF2021-WEOCAV06.

511 Bazin, N., Devanz, G., Jenhani, H. et al. (2017). Status of the IFMIF LIPAc SRF LINAC. *Proceedings of the 18th International Conference on RF Superconductivity (SRF 2017)*, Lanzhou, China. https://doi.org/10.18429/JACoW-SRF2017-MOPB012.

512 Bazin, N. and Chel, S. (2021). Cryomodule development for materials irradiation facilities: from IFMIF/EVEDA to IFMIF-DONES. *20th International Conference on RF Superconductivity (SRF 2021)*, Virtual (MSU), p. WEPFAV001. doi.org/10.18429/JACoW-SRF2021-WEPFAV001

513 Kim, S.H., Afanador, R., Blokland, W. et al. (2013). The status of the superconducting LINAC and SRF activities at the SNS. *16th International Conference on RF Superconductivity (SRF 2013)*, Paris, France, pp. 83–88. https://accelconf.web.cern.ch/SRF2013/papers/mop007.pdf.

514 Kim, S.-H., Doleans, M., Galambos, J. et al. (2015). Superconducting LINAC upgrade plan for the second target station project at SNS. *17th International Conference on RF Superconductivity (SRF 2015)*, Whistler, BC, Canada, pp. 268–271. https://accelconf.web.cern.ch/SRF2015/papers/mopb069.pdf.

515 Mammosser, J. and Kim, S.H. (2019). Overview of 12 year operation of the SCL at SNS. *TESLA Technology Collaboration Meeting (TTC 2019)*, Vancouver, Canada, p. WG1Day1Talk2. https://indico.desy.de/event/21337/contributions/42591/attachments/27262/34358/WG1talk2day1.pptx.

516 Mammosser, J., Kim, S.-H., Howell, M. et al. (2021). Results from the proton power upgrade project cavity quality assurance plan. *20th International Conference on RF Superconductivity (SRF 2021)*, p. THPCAV008. https://doi.org/10.18429/JACoW-SRF2021-THPCAV008.

517 Dhakal, P. (2021). Status of SNS proton power upgrade SRF cavities production qualification. *20th International Conference on RF Superconductivity (SRF 2021)*, Virtual (MSU), p. MOPCAV005. https://doi.org/10.18429/JACoW-SRF2021-MOPCAV005.

518 Doleans, M., Afanador, R., Ball, J. et al. (2013). Plasma processing R&D for the SNS superconducting linac RF cavities. *Proceedings of the 16th International Conference on RF Superconductivity (SRF 2013)*, Paris, France, TUP057, pp. 551–557. https://accelconf.web.cern.ch/srf2013/papers/tup057.pdf.

519 Doleans, M. (2016). Ignition and monitoring technique for plasma processing of multicell superconducting radio-frequency cavities. *J. Appl. Phys.* 120 (24): 243301. https://doi.org/10.1063/1.4972838.

520 Schlander, F., Darve, C., Elias, N. et al. (2017). The superconducting accelerator for the ESS project. *18th International Conference on RF Superconductivity (SRF 2017)*, Lanzhou, China. https://doi.org/10.18429/JACoW-SRF2017-MOYA01.

521 Pierini, P. (2019). European spallation source SRF systems, overview and status. *19th International Conference on RF Superconductivity (SRF 2019)*, Dresden, Germany, p. mofaa4_talk. http://accelconf.web.cern.ch/srf2019/talks/mofaa4_talk.pdf.

522 Maiano, C. (2021). (ESS), status of the ESS accelerator modules and cold LINAC. *International Conference on RF Superconductivity (SRF 2021)*, Virtual (MSU). https://indico.frib.msu.edu/event/38/attachments/160/1248/ESS_SRF2021_CeciliaMaiano_V6.pdf.

523 Bertucci, M., Bosotti, A., D'Ambros, A. et al. (2019). Surface treatments for the series production of ESS medium beta cavities. *19th International Conference on RF Superconductivity (SRF 2019)*, Dresden, Germany. https://doi.org/10.18429/JACoW-SRF2019-MOP056.

524 Sertore, D., Bertucci, M., Bignami, A. et al. (2019). ESS medium beta activity at INFN LASA. *19th International Conference on RF Superconductivity (SRF 2019)*. https://doi.org/10.18429/JACOW-SRF2019-MOP058.

525 Cenni, E., Baudrier, M., Carbonnier, P. et al. (2019). ESS prototype cavities developed at CES Saclay. *19th International Conference on RF Superconductivity (SRF 2019)*, Dresden, Germany (30 June-05 July 2019), pp. 1143–1146. https://doi.org/10.18429/JACoW-SRF2019-THP096.

526 May, A.J. (2021). Commissioning of the UKRI STFC Daresbury vertical test facility for jacketed SRF cavities. *20th International Conference on RF Superconductivity (SRF 2021)*, East Lansing, MI, USA, p. MOPFAV002. https://doi.org/doi:10.18429/JACoW-SRF2021-MOPFAV002.

527 Miyazaki, A., Fransson, K., Gajewski, K. et al. (2019). Conditioning experience of the ESS spoke cryomodule prototype. *19th International Conference on RF Superconductivity (SRF 2019)*, Dresden, DE. https://doi.org/10.18429/JACOW-SRF2019-THP058.

528 Li, H. (2020). RF Performance and experience of the 1st series spoke cryomodule for ESS. *TESLA Technology Collaboration Meeting 2020*, CERN, Geneva, 202AD. https://indico.desy.de/event/27572/contributions/94168/attachments/63519/77635/20210119_TTC_HL.pdf.

529 Li, H., Miyazaki, A., Zhovner, M. et al. (2021). Progress and preliminary statistics for the ESS series spoke cryomodule test. *20th International Conference on RF Superconductivity (SRF 2021)*, Lansing, MI. p. TUPTEV012. https://doi.org/10.18429/JACoW-SRF2021-TUPTEV012.

530 Darve, C. (2016). The SRF LINAC components for the European spallation source: first test results-talk. *Linear Accelerator Conference (LINAC 2016)*, East Lansing, MI, USA. WE1A03. pp. 651–655. https://accelconf.web.cern.ch/linac2016/papers/we1a03.pdf.

531 He, Y. and Pan, W.M. (2015). Progress on Chinese ADS project. *SRF2015*, Whistler, BC, Canada, p. Talk MOAA03. https://accelconf.web.cern.ch/SRF2015/talks/moaa03_talk.pdf.

532 He, Y. (2017). Successful beam commissioning of Chinese ADS injector-II. *18th International Conference on RF Superconductivity (SRF 2017)*, Lanzhou, China, MOXA01. https://accelconf.web.cern.ch/srf2017/talks/moxa01_talk.pdf.

533 Fu, S. and Wang, S. (2019). Operation status and upgrade of CSNS. *10th International Particle Accelerator Conference (IPAC 2019)*, Melbourne, Australia. https://doi.org/10.18429/JACOW-IPAC2019-MOZPLM1.

534 Singh, P. (2017). India ADS programme. *EuCARD2 Workshop on Status of Accelerator Driven Systems Research and Technology Developments*, CERN, pp. 7–9. https://indico.cern.ch/event/564485/contributions/2379331/attachments/1403459/2149741/psingh_ADSIndia.pdf.

535 Sahni, V.C. and Singh, P. (2010). Indian accelerator program for ADS applications. In: *Applications of High Intensity Proton Accelerators* (ed. R. Raja and M. Mishra), 221–227. World Scientific. https://doi.org/10.1142/9789814317290_0028.

536 Wu, G. (2019). SRF cryomodules for PIP-II at Fermilab. *19th International Conference on RF Superconductivity (SRF 2019)*, Dresden, DE. https://accelconf.web.cern.ch/srf2019/talks/wetea2_talk.pdf.

537 Belomestnykh, S. et al. (2022). An 8 GeV linac as the Booster replacement in the Fermilab power upgrade: a Snowmass 2021 white paper. *ArXiv Prepr. ArXiv220305052*. https://doi.org/10.48550/arXiv.2203.05052.

538 Martinello, M., Bice, D.J., Boffo, C. (2021). 1Q-factor optimization for high-beta 650 MHz cavities for PIP-II. *J. Appl. Phys.* 130: 174501. https://doi.org/10.1063/5.0068531.

539 Neil, G.R. and Merminga, L. (2002). Technical approaches for high-average-power free-electron lasers. *Rev. Mod. Phys.* 74 (3): 685. https://doi.org/10.1103/RevModPhys.74.685.

540 Kondratenko, A.M. and Saldin, E.L. (1980). Generating of coherent radiation by a relativistic electron beam in an ondulator. *Part. Accel.* 10: 207–216. https://s3.cern.ch/inspire-prod-files-8/872da099a0e9c171a4dca19256f7ca0e.

541 Omet, M., Ayvazyan, V., Branlard, J. et al. (2019). Operation of the european XFEL towards the maximum energy. *19th International Conference on RF Superconductivity (SRF 2019)*, Dresden, Germany (30 June-05 July 2019), pp. 9–11. https://doi.org/10.18429/JACoW-SRF2019-MOFAA2.

542 Decking, W. and Weise, H. (2017). Commissioning of the European XFEL accelerator. *8th International Particle Accelerator Conference (IPAC 2017)*, no. 8, pp. 1–6. https://doi.org/10.18429/JACoW-IPAC2017-MOXAA1.

543 Berry, S. and Napoly, O. (2017). Assembly of XFEL cryomodules: lessons and results. *28th Linear Accelerator Conference (LINAC 2016)*. https://doi.org/10.18429/JACOW-LINAC2016-WE1A02.

544 Weise, H. (2014). How to produce 100 superconducting modules for the european XFEL in collaboration and with industry. *5th International Particle Accelerator Conference (IPAC 2014)*, p. 6 pages, 1.036 MB. https://doi.org/10.18429/JACOW-IPAC2014-WEIB03.

545 Kasprzak, K., Wiencek, M., Zwozniak, A. et al. (2017). Test results of the European XFEL serial-production accelerator modules. *18th International Conference on RF Superconductivity (SRF 2017)*, Lanzhou, China. pp. 312–316. https://doi.org/10.18429/JACoW-SRF2017-MOPB106.

546 Bellandi, A., Branlard, J., Eschke, J. et al. (2021). Overview of CW R&D with a European XFEL cryomodule. *20th International Conference on RF Superconductivity (SRF 2021)*, p. Poster ID: MOPFAV001. https://indico.frib.msu.edu/event/38/attachments/158/1015/MOPFAV001.pdf.

547 Kostin, D. (2017). European-XFEL summary: cavities/modules performance. presented at the *International Workshop on Future Linear Colliders (LCWS 2017)*, Strasbourg, Germany. https://agenda.linearcollider.org/event/7645/contributions/40026/attachments/32135/48698/LCWS2017_E-XFEL_DKostin.pdf.

548 Walker, N.J. and Kostin, D. (2017). The European XFEL–experience and lessons learned. *International Workshop on Future Linear Colliders (LCWS 2017)*, Strasbourg, France. https://agenda.linearcollider.org/event/7777/contributions/40232/attachments/32445/49307/LCWS17-NJW.pdf.

549 Singer, W., Brinkmann, A., Brinkmann, R. et al. (2016). Production of superconducting 1.3-GHz cavities for the European X-ray free electron laser. *Phys. Rev. Accel. Beams* 19 (9): 092001. https://doi.org/10.1103/PhysRevAccelBeams.19.092001.

550 Reschke, D., Gubarev. V., Schaffran, J. et al. (2017). Performance in the vertical test of the 832 nine-cell 1.3 GHz cavities for the European X-ray free electron laser. *Phys. Rev. Accel. Beams* 20 (4): 042004. https://doi.org/10.1103/PhysRevAccelBeams.20.042004.

551 Burrill, A. (2017). Status of the LCLS-II project. *Proceedings of the 18th International Conference on RF Superconductivity (SRF 2017)*, Lanzhou, CN. https://accelconf.web.cern.ch/srf2017/talks/moxa05_talk.pdf.

552 Raubenheimer, T.O. (2018). The LCLS-II-HE, a high energy upgrade of the LCLS-II. *60th ICFA Advanced Beam Dynamics Workshop on Future Light Sources*, pp. 6–11. https://doi.org/10.18429/JACoW-FLS2018-MOP1WA02.

553 Burrill, A. (2021). LCLSII status and progress. *20th International Conference on RF Superconductivity (SRF 2021)*. https://indico.frib.msu.edu/event/38/attachments/160/1255/Burrill_SRF_Talk_sent.pdf.

554 Ross, M. (2019). LCLS-II: Scope, status, issues and plans. *19th International Conference on RF Superconductivity (SRF 2019)*, Dresden, Germany. https://doi.org/10.18429/JACoW-SRF2019-MOFAA1.

555 Peterson, T. (2018). LCLS-II cryomodule transportation: failures, successes, and lessons learned. https://indico.fnal.gov/event/18092/contributions/45684/attachments/28389/35103/Peterson-LCLS-II-CryomoduleTransportation-7Sep2018.pptx.

556 Posen, S., Cravatta, A., Checchin, M. et al. (2021). LCLS-II-HE verification cryomodule high gradient performance and quench behavior. arXiv:2110.14580.

557 Huque, N., Daly, E., Hartsell, B. et al. (2019). Improvements to LCLS-II cryomodule transportation. *19th International Conference on RF Superconductivity (SRF 2019)*, Dresden, Germany (30 June-05 July 2019), pp. 684–689. https://doi.org/10.18429/JACoW-SRF2019-TUP094.

558 Zhu, Z.Y., Zhao, Z. T., Wang, D. et al. (2017). SCLF: an 8-GeV CW SCRF linac-based X-ray FEL facility in Shanghai. *Proceedings of the FEL2017 St. Fe NM USA*, pp. 20–25. https://doi.org/10.18429/JACoW-FEL2017-MOP055 or http://jacow.org/fel2017/papers/mop055.pdf.

559 Hou, H.T., Chen, J., He, F. et al. (2019). Prototypes fabrication of 1.3 GHz superconducting RF components for SHINE. *19th International Conference on RF Superconductivity (SRF 2019)*, Dresden, Germany (30 June-05 July 2019), pp. 164–167. https://doi.org/10.18429/JACoW-SRF2019-MOP049.

560 Hou, H.T. and Chen, J. (2019). SRF cavity R&D at SHINE. presented at the *TESLA Technology Collaboration Meeting (TTC 2019)*, Vancouver, Canada. https://indico.desy.de/event/21337/contributions/42618/attachments/27215/34295/TTC2019-SRF_Cavity_RD_at_SHINE-0204.pptx.

561 Hou, H.T. et al. (2019). SRF status of the SHINE project at Shanghai. *19th International Conference on RF Superconductivity (SRF 2019)*, Dresden, Germany, p. paper MOFAB2. https://accelconf.web.cern.ch/srf2019/talks/mofab2_talk.pdf.

562 Chen, J. (2022). High Q/Gradient R&D based on the completely new Wuxi platform. presented at the *TESLA Technology Collaboration Meeting (TTTC 2022)*, Aomori, Japan. https://indico.desy.de/event/32219/contributions/116544/.

563 Wang, D. (2022). Challenge of multi-laboratories and industries collaboration: SHINE experience. presented at the *TESLA Technology Collaboration Meeting (TTC 2022)*, Aomori, Japan. https://indico.desy.de/event/32219/contributions/116531/.

564 Nietubyć, R., Banaszkiewicz, T., Bartnik, A. et al. (2021). Status of the POLFEL project. *20th International Conference on RF Superconductivity (SRF 2021)*, MSU. p. THPFAV003. https://indico.frib.msu.edu/event/38/attachments/158/1069/THPFAV003.pdf.

565 Zhang, P., Zhang, X., Li, Z. et al. (2019). Development and vertical tests of a 166.6 MHz proof-of-principle superconducting quarter-wave beta= 1 cavity. Rev. Sci. Instrum. 90: 084705. https://doi.org/10.1063/1.5119093.

566 Belomestnykh, S. (2007). Superconducting RF in storage-ring-based light sources. presented at the *13th International Workshop on RF Superconductivity (SRF 2007)*, Beijing, China. https://accelconf.web.cern.ch/srf2007/PAPERS/MO302.pdf.

567 Nakanishi, K., Hirosawa, K., Kobayashi, T., and Nishiwaki, M. (2018). SRF system for KEKB and SuperKEKB. *Proceedings of the 62th ICFA ABDW High Luminosity Circular e+e Colliders*, Volume eeFACT2018, p. 8 pages, 4.228 MB. https://doi.org/10.18429/JACOW-EEFACT2018-WEYAA03.

568 Wang, C. (2018). Status of SRF operation at NSRRC. presented at the *CWRF Workshop*, Hsinchu, Taiwan. https://indico.cern.ch/event/688528/contributions/3035716/attachments/1677540/2693795/NSRRC16x9-201806-CWRF2018-09.ppt.

569 Pedrozzi, M., Gloor, W., Anghel, A. et al. (2003). First operational results of the 3rd harmonic superconducting cavities in SLS and ELETTRA. *Proceedings of the 2003 Particle Accelerator Conference*, Volume 2, pp. 878–880. https://accelconf.web.cern.ch/p03/PAPERS/MPPB024.PDF.

570 Diop, M., Baete, J.P., Cuoq, R. et al. (2018). Status of the SOLEIL superconducting RF system. *18th International Conference on RF Superconductivity (SRF 2017)* Langzhou, China. https://doi.org/10.18429/JACOW-SRF2017-MOPB041.

571 Hou, H.T., Liu, J.F., and Zhao, Z.T. (2013). Conceptual design of a superfluid superconducting third harmonic RF system for the SSRF storage ring. *4th International Particle Accelerator Conference (IPAC 2013)*. Shanghai, China. WEPWO036. pp. 2381-2383 https://accelconf.web.cern.ch/IPAC2013/papers/wepwo036.pdf.

572 Kelly, M.P., Bacikowski, A., Carwardine, J. et al. (2015). Superconducting harmonic cavity for the advanced photon source upgrade. *6th International Particle Accelerator Conference (IPAC 2015)*, Richmond, VA USA, WEPTY008. https://doi.org/10.18429/JACoW-IPAC2015-WEPTY008.

573 Vélez, A., Glock, H.-W., Knobloch, J., and Neumann, A. (2015). HOM damping optimization design studies for BESSY VSR cavities. *6th International Particle Accelerator Conference (IPAC 2015)*, Richmond, VA, USA (3–8 May 2015), pp. 2774–2776. https://doi.org/18429/JACoW-IPAC2015-WEPMA013.

574 Wunderer, N., Dürr, V., Frahm, A. et al. (2021). VSR demo cold string: recent developments and manufacturing status. *20th International Conference on RF Superconductivity (SRF 2021)*, Virtual (MSU), p. WEPTEV008. https://doi.org/10.18429/JACoW-SRF2021-WEPTEV008.

575 Conway, Z.A., Anderson, R., Holmes, D. et al. (2021). Upgrade of the RHIC 56 MHz superconducting quarter-wave resonator cryomoudle. *20th International Conference on RF Superconductivity (SRF 2021)*, East Lansing, MI, USA. Virtual (MSU), p. TUPTEV016. https://doi.org/10.18429/JACoW-SRF2021-TUPTEV016.

576 Krafft, G.A. and Old, D. (2008). Energy recovered linacs. *24th Linear Accelerator Conference (LINAC 2008)*, Victoria, BC, Canada. WE101. pp. 688–692. https://doi.org/10.18429/JACoW-SRF2021-TUPTEV016.

577 Hoffstaetter, G. (2020). The Cornell BNL ERL test accelerator (CBETA) 4 June 2020. *Virtual PERLE Collaboration Meeting*. https://indico.cern.ch/

event/923021/contributions/3878056/attachments/2051063/3437909/CBETA_Hoffstaetter_small.pdf (accessed 2 August 2022).

578 Deitrick, K., Hoffstaetter, G. H., Franck, D. et al. (2021). Intense monochromatic photons above 100 keV from an inverse Compton source. *Phys. Rev. Accel. Beams* 24 (5): 050701. https://doi.org/10.1103/PhysRevAccelBeams.24.050701.

579 Umemori, K., Egi, M., Enami, K. et al. (2017). Long-term operation experience with beams in compact ERL cryomodules. doi: 10.18429/JACOW-SRF2017-THYA06.

580 Kawata, H. (2015). Operational progress in compact-ERL and development of ERL-FEL for EUV light source at KEK. *17th International Conference on RF Superconductivity (SRF 2015)*, Whistler, BC, Canada, p. Talk TUAA01. https://accelconf.web.cern.ch/SRF2015/talks/tuaa01_talk.pdf.

581 Sakai, H., Adachi, M., Arakaws, D. et al. (2021). Stable beam operation in cERL for medical and industrial application at KEK. presented at the *20th International Conference on RF Superconductivity (SRF 2021)*, Virtual (MSU). https://doi.org/10.18429/JACoW-SRF2021-THOTEV02.

582 Sakai, H., Kako, E., Konomi, T. et al. (2018). Superconducting accelerator for ERL based FEL EUV light source at KEK. *18th International Conference on RF Superconductivity (SRF 2017)*, Lanzhou, China (17–21 July 2017), pp. 13–18. https://doi.org/10.18429/JACoW-SRF2017-MOXA04.

583 IRE and ASML. (2022). The SMART project: accelerator based production of Mo-99. https://ec.europa.eu/energy/sites/ener/files/documents/s1-3_ekollegger_ire.pdf (accessed 2 August 2022).

584 Knobloch, J. (2015). A demonstration energy recovery LINAC. *56th ICFA Advanced Beam Dynamics Workshopon Energy Recovery LINACs (ERL 2015)*, Stony Brook, NY, USA. paper MOPBTH006. https://accelconf.web.cern.ch/ERL2015/talks/mopbth006_talk.pdf.

585 Neumann, A., Anders, W., Burrill, A. et al. (2015). Update on SRF cavity design, production and testing for bERLinPRO. *17th International Conference on RF Superconductity (SRF 2015)*, Whistler, BC, Canada. Poster THPB026. pp. 1127–1131. https://accelconf.web.cern.ch/SRF2015/papers/thpb026.pdf.

586 Hug, F. (2021). High-gradient 'turn-key' SRF systems for the MESA ERL project. presented at the *20th International Conference on RF Superconductivity (SRF 2021)*, Virtual (MSU). https://indico.frib.msu.edu/event/38/attachments/158/1129/WEPTEV005.pdf.

587 Belomestnykh, S. (2015). Overview of recent SRF developments for ERLs, Whistler, BC, Canada, p. Paper MOAAO4, 24. https://accelconf.web.cern.ch/SRF2015/talks/moaa04_talk.pdf.

588 Arnold, A. (2021). RF experience from 6 Years of ELBE SRF-Gun II operation. *20th International Conference on RF Superconductivity (SRF 2021)*, Virtual (MSU), p. TUPTEV001. https://doi.org/10.18429/JACoW-SRF2021-TUPTEV001.

589 Leemann, C.W., Douglas, D.R., and Krafft, G.A. (2001). The continuous electron beam accelerator facility: CEBAF at the Jefferson Laboratory. *Annu. Rev. Nucl. Part. Sci.* 51 (1): 413–450. https://doi.org/10.1146/annurev.nucl.51.101701.132327.

590 Sundelin, R. (1985). High gradient superconducting cavities for storage rings. *IEEE Trans. Nucl. Sci.* NS-32 (5): 3570–3573.

591 Hogan, J., Burrill, A., Davis, G.K. et al. (2012). 12 GeV upgrade project cryomodule production. *International Particle Accelerator Conference (IPAC 2012).* New Orleans, LA. USA. https://accelconf.web.cern.ch/ipac2012/papers/weppc092.pdf.

592 Hovater, C., Allison, T., Bachimanchi, R. et al. (2012). Commissioning and operation of the CEBAF 100 MV cryomodules. *International Particle Accelerator Conference (IPAC 2012)*, New Orleans, LA, USA. *WEPPC093*. https://accelconf.web.cern.ch/IPAC2012/papers/weppc093.pdf.

593 Reece, C.E. and Ciovati, G. (2012). Superconducting radio-frequency technology R&D for future accelerator applications. In: *Reviews of Accelerator Science and Technology: Applications of Superconducting Technology to Accelerators*, vol. 5, (ed. A. Chao and W. Chou) 285–312. World Scientific. https://doi.org/10.1142/S1793626812300113.

594 Sekutowicz, J., Ciovati, G., Kneisel, P. et al. (2003). Cavities for JLab's 12 GeV upgrade.*2003 Particle Accelerator Conference (PAC 2003)*. Portland, OR. USA, Volume 2, pp. 1395–1397. https://accelconf.web.cern.ch/p03/PAPERS/TPAB085.PDF.

595 Sekutowicz, J., Ciovati, G., Kneisel, P. et al. (2002). Low loss cavity for the 12 GeV CEBAF upgrade, JLAB TN-02-023, June 2002. 2003 Particle Accelerator Conference PAC 2003. https://accelconf.web.cern.ch/p03/PAPERS/TPAB085.PDF.

596 Jefferson Lab. (2022). JLAB 12 GeV upgrade LINAC installation, *Jefferson Lab News Release*. https://www.jlab.org/news/releases/jefferson-lab-completes-12-gev-upgrade (accessed 2 August 2022).

597 Laxdal, R.E. et al. (2017). The 30MeV stage of the ARIEL e-Linac. *18th International Conference on RF Superconductivity (SRF 2017)*, pp. 6–12. https://doi.org/10.18429/JACoW-SRF2017-MOXA03.

598 Ma, Y., Ang, Z., Fong, K. et al. (2018). First RF test results of two-cavities accelerating cryomodule for Ariel Elinac at TRIUMF. *9th International Particle Accelerator Confernce (IPAC 2018)*, pp. 4512–4515. https://doi.org/10.18429/JACoW-IPAC2018-THPMK090.

599 Ma, Y. (2021). Status of ARIEL E-LINAC cryomodules. presented at the *TESLA Technology Collaboration Meeting (TTC 2021)*, DESY, Hamburg (Virtual). https://indico.desy.de/event/27572/contributions/94183/.

600 Laxdal, R.E., Ames, F., Baartmann, R. et al. (2014). Status of superconducting electron linac driver for rare ion beam production at TRIUMF. *Proceedings of LINAC2014*, Geneva, Switzerland, p. 31. https://accelconf.web.cern.ch/LINAC2014/papers/moioc01.pdf.

601 Jensen, E. (2015). ERL facility at CERN for applications. *56th ICFA Advanced Beam Dynamics Workshop on Energy Recovery Linacs, BNL ERL 2015.* https://accelconf.web.cern.ch/ERL2015/talks/weialh1043_talk.pdf.

602 Angal-Kalinin, D., Arduini, G., Bernauer, J. et al. (2018). PERLE. Powerful energy recovery linac for experiments. Conceptual design report. *J. Phys. G: Nucl. Part. Phys.* 45 (6): 065003. https://doi.org/10.1088/1361-6471/aaa171.

603 Aryshev, A., Behnke, T., Berggren, M. et al. (2022). The International Linear Collider: report to Snowmass 2021. *ArXiv Prepr. ArXiv220307622.* https://doi.org/10.48550/arXiv.2203.07622.

604 ICFA (2017). ICFA Statements (November 2017). http://icfa.fnal.gov/wp-content/uploads/ICFA-Statement-Nov2017.pdf (accessed 2 August 2022).

605 Behnke, T., Brau, J.E., Foster, B. et al. (2013). The international linear collider technical design report-volume 1: Executive summary. *ArXiv Prepr. ArXiv13066327*. https://doi.org/10.48550/arXiv.1306.6327.

606 Broemmelsiek, D., Chase, B., Edstrom, D. et al. (2018). Record high-gradient SRF beam acceleration at Fermilab. *New J. Phys.* 20 (11): 113018. https://doi.org/10.1088/1367-2630/aaec57.

607 Komamiya, S. (2016). International linear collider, latest status towards realization. *7th International Particle Accelerator Conference (IPAC 2016)*, Busan, Korea (8–13 May 2016), pp. 1–5. https://accelconf.web.cern.ch/ipac2016/papers/moxaa01.pdf.

608 Yamamoto, Y., Dohmae, T. Egi, M. et al. (2017). Achievement of stable pulsed operation at 36 MV/m in STF-2 cryomodule at KEK. *18th International Conference on RF Superconductivity (SRF 2017)*, Lanzhou, China, pp. 722–728. https://doi.org/10.18429/JACoW-SRF2017-THYA02.

609 Behnke, T., Buesser, K., and Mussgiller, A. (2020). Future developments of detectors. In: *Particle Physics Reference Library* (ed. C.W. Fabjan and H. Schopper), 1035–1078. Cham: Springer International Publishing. https://doi.org/10.1007/978-3-030-35318-6_22.

610 Benedikt, M., Mertens, V., Zimmermann, F. et al. (2018). FCC-ee: the Lepton collider: future circular collider conceptual design report volume 2. *Eur. Phys. J. Spec. Top.* 228 (CERN-ACC-2018-0057): 261–623. https://doi.org/10.1007/978-3-030-35318-6_22.

611 Robson, A., Burrows, P.N., Lasheras, N.C. et al. (2018). The compact linear collider (CLIC): accelerator and detector. *ArXiv Prepr. ArXiv181207987.* https://doi.org/10.48550/arXiv.1812.07987.

612 C. S. Group (2018). CEPC conceptual design report: Volume 1-accelerator. *ArXiv Prepr. ArXiv180900285.* https://doi.org/10.48550/arXiv.1809.00285.

613 Sosoho-Abasi, U., Calaga, R., Zadeh, G. et al. (2016). Optimization of Two-Cell Cavities for the W and H Working Points of the FCC-ee Considering Higher-Order Mode Effects. https://doi.org/10.18429/JACoW-IPAC2021-TUPAB341.

614 Delsolaro, W.V. (2018). Thin film research: CERN experience and possible future applications. *TESLA Technology Collaboration (TTC 2018)*, Milan, Italy. https://agenda.infn.it/event/13791/contributions/20933/attachments/14869/16809/Nb-Cu_CERN_experience_and_possible_future_applications.pptx.

615 Zhai, J., Chi, Y., Dai, J. et al. (2018). CEPC SRF system design and challenges. *18th International Conference on RF Superconductivity (SRF 2017)*, Lanzhou, China. https://doi.org/10.18429/JACOW-SRF2017-TUXAA01.

616 Amundson, J. and Sexton-Kennedy, E. (2019). Quantum computing. *23rd Internationa Conference on Computing in High Energy and Nuclear Physics (CHEP 2018) EPJ Web of Conferences*, Volume 214, p. 09010. https://doi.org/10.1051/epjconf/201921409010.

617 Lykken, J. (2019). Quantum science for HEP. presented at the *Aspen Winter Conference*, Aspen, Colorado (30 March 2019). https://indico.cern.ch/event/748043/contributions/3326052/attachments/1820777/2977976/lykken-Aspen-3-30-19-short.pdf.

618 Schneider, A. (2020). Quantum Sensing Experiments with Superconducting Qubits. https://doi.org/10.5445/KSP/1000118743.

619 Kjaergaard, M., Schwartz, M.E., Braumüller, J. et al. (2020). Superconducting qubits: current state of play. *Annu. Rev. Condens. Matter Phys.* 11: 369–395. https://doi.org/10.1146/annurev-conmatphys-031119-050605.

620 Romanenko, A. (2019). Employing SRF to boost coherence of 3D quantum systems. *19th International Conference on RF Superconductivity (SRF 2019)*, Dresden, Germany. https://accelconf.web.cern.ch/srf2019/talks/thfub5_talk.pdf.

621 Kutsaev, S.V., Agustsson, R., Carriere, P. et al. (2019). Quarter-wave resonator with the optimized shape for quantum information systems. *19th International Conference on RF Superconductivity (SRF 2019)*, Dresden, Germany (30 June-05 July 2019), pp. 430–433. https://doi.org/10.18429/JACoW-SRF2019-TUP016.

622 Kutsaev, S.V., Taletski, K., Agustsson, R. et al. (2020). Niobium quarter-wave resonator with the optimized shape for quantum information systems. *EPJ Quantum Technol.* 7 (1): 1–17. https://doi.org/10.1140/epjqt/s40507-020-00082-8.

623 Duckering, C., Baker, J.M., Schuster, D.I., and Chong, F.T. (2020). Virtualized logical qubits: a 2.5D architecture for error-corrected quantum computing. *2020 53rd Annual IEEE/ACM International Symposium on Microarchitecture (MICRO)*, pp. 173–185. https://doi.org/10.48550/arXiv.2009.01982.

624 Rigetti Computing. 775 Heinz Ave, Berkeley, CA 94710. rigetti.com.

625 Romanenko, A. (2021). Superconducting high coherence platforms for sensing and computing. presented at the *ECFA Task Force 5: Quantum and Emerging Technologies*. https://indico.cern.ch/event/999818/contributions/4253080/attachments/2224441/3767340/ECFA_DetectorBRN_Romanenko_2021.pdf.

626 Naik, R.K. (2018). *Multimode Circuit Quantum Electrodynamics*. The University of Chicago. http://schusterlab.uchicago.edu/static/pdfs/Naik_thesis.pdf

627 Holland, E.T. (2015). *Cavity State Reservoir Engineering in Circuit Quantum Electrodynamics*. Yale University. https://rsl.yale.edu/sites/default/files/files/RSL_Theses/Eric Holland_dissertation.pdf.

628 Grassellino, A. (2021). Building a quantum computer using SRF cavities as qubits. presented at the *International Conference on RF Superconductivity*, Virtual (MSU).

List of Symbols

C	definition speed of light
e	elementary charge electron mass
m	peak electric field
ε_0	vacuum dielectric constant
μ_0	vacuum magnetic permeability
E	electric field
E_p	peak electric field
H	magnetic field
B	magnetic field induction
$\mathbf{A}$	vector potential
B_p	peak magnetic induction
V	particle speed
β	ratio of the particle speed divided to c
$\hat{n}$	unit vector perpendicular to the surface
ω_0	angular resonance frequency
$\hbar$	reduced Planck constant (Planck constant over 2π)
k_b	Boltzmann constant
R	universal gas constant
v_F	fermi velocity
u	stored energy
P_c	cavity dissipated power
Q_0	unloaded quality factor
G	geometry constant for cavity
E_{PK}, B_{PK}	electric and magnetic field peak
Γ	dynes smearing parameter
D	diffusivity
f_s	free energy density of the superconductor
f_n	free energy density of the normal-conductor
τ	drude electron relaxation time
R_a	shunt impedance
R_s	surface resistance
n	density of charge carriers
n_n	density of normal-electrons

Superconducting Radiofrequency Technology for Accelerators: State of the Art and Emerging Trends, First Edition. Hasan Padamsee.

n_S	density of super-electrons
λ_L	London penetration depth
J	Current density
J_n	normal-electrons current density
J_s	super-electrons current density
H_c	thermodynamic critical field
H_{c1}	lower critical field
H_{c2}	upper critical field
H_{sh}	superheating critical field
T_c	critical temperature
Δ	energy gap
ξ	coherence length
$\lambda,$	penetration depth
ψ	order parameter
κ	Ginzburg–Landau parameter
Φ_0	magnetic flux quantum
$N(O)$	density of state at the Fermi level
σ_1	real part of the conductivity
σ_2	imaginary part of the conductivity
Z_s	surface impedance
Z_n	normal-conducting surface impedance
R_{BCS}	BCS surface resistance
R_{res}	residual resistance
R_n	normal-conducting surface resistance
R_{fl}	trapped flux surface resistance
η	flux trapping efficiency or Carnot efficiency
S	trapped flux sensitivity
Q_L	loaded *Q*-factors
ρ	resistivity
E_{acc}	accelerating field
N_{trap}	number of trapped vortices
g	Gibbs free energy density
f_p	pinning force
∇T_c	critical thermal gradient
ΔT_c	critical temperature difference along the cavity
B_{SC}/B_{NC}	flux expulsion ratio
ℓ	mean free path
f_L	Lorentz force
f_v	viscous force
U_p	pinning potential
p	pinning constant

f_v	viscous force
ℓ_p	mean distance between pinning sites
η	Viscosity
γ	coefficient of the normal state electronic specific heat
S	Positron annihilation spectroscopy parameter
W	Positron annihilation spectroscopy
I_c	critical tunnel current of a Josephson junction

List of Acronyms

ADS	accelerator-driven system
ANL	Argonne National Laboratory
ALD	atomic layer deposition
ALPI	acceleratore lineare per ioni
APS	advanced photon source
ARIEL	Advance Rare Isotope Laboratory
ASML	advanced semiconductor materials lithography
ATLAS	Argonne Tandem Linear Accelerator System
ATW	accelerator-based transmutation of waste
BARC	Bhabha Atomic Research Centre
BCP	buffered chemical polishing
BCS	Bardeen, Cooper, Schrieffer
BESSY	Berlin Electron Storage Ring Society for Synchrotron Radiation
BNL	Brookhaven National Laboratory
CAD	computer-aided design
CADS	China Accelerator Driven Systems
CAFÉ	China Accelerator Facility for new elements
CAT	Center for Accelerator Technology (Indore)
CBETA	Cornell Brookhaven ERL Test Accelerator
CBP	centrifugal barrel polishing
CDR	Conceptual Design Report
CEA	Commissariat å l'énergie atomique (Saclay, France)
CEBAF	Continuous Electron Beam Accelerator Facility
CepC	Circular Electron Positron Collider
cERL	Compact ERL
CERN	Conseil Européen pour la Recherche Nucléaire
CESR	Cornell Electron Storage Ring
CiADS	China Initiative Accelerator Driven System
CLIC	Compact Linear Collider
CM	cryomodule
CMM	coordinate measuring machine
cQED	Circuit Quantum Electrodynamics
DC	direct current

Superconducting Radiofrequency Technology for Accelerators: State of the Art and Emerging Trends,
First Edition. Hasan Padamsee.

dcMS	Direct Current Magnetron Sputtering
DESY	Deutsches Elektronen-Synchrotron
DOS	Density of States
DQW	Double Quarter Wave
DUNE	Deep Underground Neutrino Experiment
EBSD	electron backscatter diffraction
EBW	electron beam welding
ECR	Electron Cyclotron Resonance
EDM	electro discharge machining
EDX	energy dispersive X-ray
ELBE	ELectron accelerator for beams with high Brilliance and low Emittance
ELETTRA	Sincrotrone Trieste
EP	electropolishing
ERL	Energy Recovery Linacs
ESRF	European Synchrotron Radiation Facility
ESS	European Spallation Source
FAIR	Facility for Antiproton and Ion Research
FCCee	Future Circular Collider, electron–positron
FE	field emission
FELs	free electron lasers
FNAL	Fermi National Accelerator Laboratory
FPC	Fundamental Power Coupler
FREIA	Facility for Research Instrumentation and Accelerator
FRIB	Facility for Rare Isotope Beams
GANIL	Grand Accélérateur National d'Ions Lourds
GL	Ginzburg Landau
GSI	Gesellschaft für Schwerionenforschung
HELIAC	Helmholtz Linear Accelerator
HEP	high energy physics
HEPS	High Energy Photon Source
HERA	Hadron Elektron Ring Anlage
HFQS	High Field *Q*-Slope
HFSS	high-frequency structures simulator
HIAF	High Intensity Heavy Ion Accelerator Facility
HIE-ISOLDE	High-Intensity and Energy upgrade of ISOLDE
HIP	hot isostatic pressing
HIPIMS	High-power impulse magnetron sputtering
HL-LHC	High Luminosity-Large Hadron Collider
HOM	higher order mode
HPCVD	Hybrid physical–chemical vapor deposition
HPPMS	high power pulsed magnetron sputtering
HPR	high-pressure rinsing
HWR	half-wave resonator
HZB	Helmholtz-Zentrum Berlin

HZDR	Helmholtz-Zentrum Dresden-Rossendorf
IASF	Institute of Advanced Science Facilities
IBS	Institute of Basic Sciences (S. Korea)
IFMIF	International Fusion Material Irradiation Facility
IHEP	Institute of High Energy Physics
IJC	Irène Joliot-Curie – IJCLab, Laboratoire de Physique des 2 infinis
ILC	International Linear Collider
IN2P3	Institut national de physique nucléaire et de physique des particules
INFN	Istituto Nazionale di Fisica Nucleare
IP	interaction point
ISAC	isotope separator and acceleration
ISO	International Organization for Standardization
ISOL	Isotope Separation-On-Line
ISOLDE	Isotope mass Separator On-Line
ITER	International Thermonuclear Experimental Reactor
IUAC	Inter-University Accelerator Centre (Delhi)
J Lab	Jefferson Laboratory
JACOW	Joint Accelerator Conference Website
JAEA	Japan Atomic Energy Agency
JJ	Josephson Junction
KEK	Koh Ene Ken (Japanese: National Laboratory for High-Energy Physics)
KEKB	KEK B-Factory
LARP	LHC Accelerator Research Program
LASA	Laboratorio Acceleratori e Superconduttività Applicata
LBNF	Long Baseline Neutrino Facility
LCLS	Linac Coherent Light Source (Stanford Linear Accelerator Center)
LEP	Large Electron-Positron
LE-μSR	low energy muon spin relaxation/resonance/research
LFQS	low field Q-slope
LHC	Large Hadron Collider
LHeC	Large Hadron Electron Collider
LINAC	linear accelerator
LNL-INFN	Laboratori Nazionali di Legnaro – Italian National Institute for Nuclear Physics
LOM	lower order modes
MESA	Mainz energy-recovering superconducting accelerator
mfp	mean free path
MFQS	medium field Q-slope
MLC	main linac cavity
MP	multiple impact electron amplification
MSU	Michigan State University
MW	megawatt
NSCL	National Superconducting Cyclotron Laboratory

PAC	Particle Accelerator Conference
PAPS	Platform of Advanced Photon Source
PCTS	Point Contact Tunneling Spectroscopy
PERLE	Powerful ERL for Experiments
PIP	Proton Improvement Plan
PIP-II	proton improvement plan-II
PKU	Peking University
POLFEL	Polish free-electron laser
PPU	proton power upgrade
QWR	quarter-wave resonator
RAON	Rare isotope accelerator complex for ON-line experiment
RCC	4 rod crab cavity
ReA	re accelerator
RFD	radio frequency dipole
RFQ	radiofrequency quadrupole
RHIC	relativistic heavy ion collider
RIB	rare isotope beams
RIKEN	Rikagaku Kenkyusho (Institute of Physical and Chemical Research, Japan)
RILAC	RIKEN heavy Ion Linac
RISP	Rare Isotope Science Project (S. Korea)
RRCAT	Raja Ramanna Centre for Advanced Technology
RRR	residual resistivity ratio
SARAF	Soreq Applied Research Accelerator Facility
SASE	self-amplified-spontaneous-emission
SCL	Superconducting Linac
SEM	secondary electron microscopy
SHINE	Shanghai HIgh repetitioN rate XFEL and Extreme light facility
SIMS	secondary ion mass spectrometry
SIS	superconducting-insulator-superconducting
SLAC	Stanford Linear Accelerator
SLS	Swiss Light Source (Paul Scherrer Institute, Zurich, Switzerland)
SNS	Spallation Neutron Source at Oak Ridge National Laboratory
SOLEIL	Source optimisée de lumiére d'énergie intermédiaire du LURE
SPES	Selective Production of Exotic Species
SPIRAL	Systeme de Production d'Ions Radioactifs en Ligne (at GANIL, France)
SPX	Short-Pulse X-Rays
SQUID	superconducting quantum interference device
SRF	superconducting rf
SRIMP	Surface Resistance and Impedance
SRRC	Synchrotron Radiation Research Center (Taiwan) stainless steel
SS'	Superconducting-Superconducting (Layers)
SSR	single spoke resonator
SSRF	Shanghai Synchrotron Radiation Facility

STFC	Science and Technology Facilities Council
SW	Standing Wave
TDR	Technical Design Report
TE	transverse electric
TEM	Transverse Electromagnetic
TESLA	TeV Energy Superconducting Linear Accelerator
TeV	tera electron volt
TLS	Two Level Systems
TM	transverse magnetic
TPS	Taiwan Photon Source
TRIUMF	TRI-University Meson Facility (Vancouver, Canada)
TTC	Tesla Technology Collaboration
TUL	Lodz University of Technology
TW	Travelling Wave
UHV	ultrahigh vacuum
VECC	Variable Energy Cyclotron Centre
VSR	Variable Pulse Storage Ring
VUV	vacuum ultraviolet
WUST	Wrocław University of Science and Technology
WUT	Warsaw University of Technology
XFEL	X-ray free electron laser
XPS	X-Ray Photoelectron Spectroscopy
XRD	X-ray Diffraction

Index

Superconducting Radiofrequency Technology for Accelerators: State of the Art and Emerging Trends, First Edition. Hasan Padamsee.

d

e

f

j

k

l

m

t

u

v

w

x

z